Electronic Materials and Devices

David H. Navon
University of Massachusetts, Amherst

Electronic Materials and Devices

Houghton Mifflin Company   Boston
Atlanta   Dallas   Geneva, Illinois
Hopewell, New Jersey   Palo Alto   London

Library of Congress Catalog Card Number: 74-11948
ISBN: 0-395-18917-9

Contents

4 INTRODUCTION TO THE QUANTUM THEORY OF SOLIDS

5 INTRODUCTION TO SEMICONDUCTOR PHYSICS

6 INTRODUCTION TO SEMICONDUCTOR DIODES

7 DIODE APPLICATIONS AND FREQUENCY PERFORMANCE

8 INTRODUCTION TO THE BIPOLAR JUNCTION TRANSISTOR

9 BIPOLAR TRANSISTOR APPLICATIONS AND FREQUENCY PERFORMANCE

Preface

The trend toward the design of electronic circuits using solid-state devices began shortly after the invention of the transistor in 1948. This necessitated the introduction of transistor physics into the engineering curriculum. A good understanding of transistor operation requires an awareness of the physics of the semiconductors silicon and germanium.

More recently new solid-state devices have been employed in significant numbers in electronic circuitry. Often these devices are fabricated of solid-state materials other than silicon and germanium. It is becoming increasingly important for engineering students to have a more profound knowledge of the physics of solid-state materials (quantum mechanics), to be able to understand the operation of the new devices of today and to invent those of tomorrow. For example, the semiconductor gallium arsenide is employed in modern devices such as the semiconductor laser and the high-frequency Gunn oscillator. The operation of the latter must be described quantum mechanically. No doubt, future solid-state devices will be based on even more unusual materials and sophisticated quantum electronic principles.

Since one important objective of a text in a rapidly developing field is to prepare the student for future developments in his area of interest, this book presents a basic introduction to the physics of solids via quantum mechanics. Then the physical principles of existing semiconductor devices, such as diodes and bipolar and field-effect transistors, are discussed. Circuit models for these devices, useful in analyzing integrated circuits, are given. Finally descriptions of the operation of the newer quantum electronic devices, such as the semiconductor laser and Gunn oscillator, are presented in the language of quantum mechanics.

This text is designed for students who have had a good introductory course in physics as well as a brief introduction to modern physics including the Bohr atom and the periodic table of the elements. In addition, the student should have the mathematical maturity typical of a junior- or senior-level college student. The material can be covered conveniently in one semester if Chapters 2, 7, 9, and 11 are omitted. The full text can be offered in two college quarters. The problems presented at the end of each chapter are considered an integral part of the text and often fill in gaps left in the material of the chapter.

The subject matter is divided roughly in two parts: the first half is devoted to the physics of solid-state electronic materials, and the second half to solid-state devices.

Chapter 1 provides an introduction to the subject of solid-state materials and devices, illustrating the importance of an understanding of materials and device physics to both device designers and electronic circuit designers. The transistor as an amplifier and solid-state computer switching element is explained. An introduction to integrated circuits is presented. Other modern solid-state devices are then briefly introduced.

Chapter 2 discusses the crystal structure of those solids which are of interest for the fabrication of electronic devices. The forces which hold the atoms of solids in specific arrangements are described. Methods of describing and identifying the different crystalline configurations are presented. Finally typical crystal imperfections that affect device behavior are illustrated.

Chapter 3 is an introduction to quantum mechanics, leading to a description of the electrical properties of solids, according to the energy band theory, in Chapter 4. After an introduction to the philosophical ideas which separate quantum mechanics from classical physics, the Schrödinger equation, which embodies the ideas of this new physics, is presented. Typical solutions of this equation are offered to illustrate the manner of solution as well as the physical consequences of the results. The motion of an electron in a periodic potential is calculated to analyze the electrical conductivity of crystals which are a periodic arrangement of atoms in a solid. The simple harmonic oscillator and central-field problem are briefly mentioned because of applications in light emission from lasers and in generating the periodic table of the elements.

Chapter 4 explains the differences in electrical conductivity among insulators, metals, and semiconductors in terms of the quantum mechanical band theory of solids. Chapter 5 provides a general introduction to the physics of semiconductors since it deals with the materials from which transistors, integrated circuits, and other solid-state devices are constructed.

Chapter 6 provides a physical description of the semiconductor junction diode and the flow of minority carriers in the device. The diode current-voltage characteristic is derived. The metal-semiconductor (Schottky) diode is discussed. Chapter 7 describes some typical diode applications and discusses diode performance as a function of frequency.

Chapter 8 describes the physics of the bipolar junction transistor and presents some useful circuit models for the device. Chapter 9 discusses some typical transistor circuit applications as well as the effect of frequency on transistor performance. Also included is a discussion of the bipolar transistor as a switch.

Chapter 10 contains a description of the semiconductor field-effect transistor and its application to integrated circuits.

Chapter 11 describes some more solid-state devices in the light of the previously discussed band theory of solids. The operation of integrated circuits, semiconductor lasers, Gunn oscillators, and charge-coupled devices is described in quantum mechanical terms.

It is hoped that the manner and order in which topics are discussed in this text will clearly illustrate the integration of material properties with device and circuit behavior; this should make clear that both the device designer and the user need to have a knowledge of device physics for successful electronic system design. This is clearly indicated in the field of integrated circuits, where the device designer and circuit designer are merged into one individual.

David H. Navon
Amherst, Massachusetts

1 Introduction to Solid-State Devices

1.1 INTRODUCTION

The most important electronic device of the first half of the twentieth century was the electronic valve or vacuum tube. Knowledge of the operation of this device required an understanding of the classical theory of the motion of electrons in vacuum under the action of an electric field. This theory was developed during the second half of the nineteenth century. Electron emission from vacuum-tube cathodes could be described classically, although an explanation of the effect of different cathodic materials on emission had to await the discovery of the new physics of *quantum mechanics*. This theory was developed during the first half of the twentieth century. Electronics in the second half of the twentieth century thus far has been dominated by solid-state devices whose operation requires an understanding of the quantum physics of solids. Classically, once the physical laws of motion are understood, the *exact* position of a particle in a force field can be determined at any time. Quantum mechanically, the particle can be anywhere, but its *most probable* location can be predicted.

The ideas which provide the foundation of this twentieth-century physics are quite different from those of the classical physics of Isaac Newton and James Clerk Maxwell. Whereas classical physics is *deterministic*, quantum physics is *probabilistic*. This new physics not only has revolutionized the manner in which scientists view the world,[1] but it has had a significant effect on the music, art, and philosophy of the middle third of the twentieth century. For example, the " action " art of Jackson Pollock[2] and the " chance " music

---

[1] See, for example, B. Hoffman, *The Strange Story of the Quantum*, Dover Publications, Inc., New York (1947).
[2] F. V. O'Connor, *Jackson Pollock*, Museum of Modern Art, New York (1967).

of John Cage[3] no doubt have their roots in this nondeterministic physics. Pollock created colorful designs by dripping paint on canvas in a somewhat arbitrary fashion. Cage has pioneered musical expression based on chance, both in the process of composition and during the performance of the work. Many of the important modern philosophers such as Alfred N. Whitehead, Arthur S. Eddington, and Bertrand Russell were quantum mechanicians; probabilistic thinking pervades their philosophies, which express a certain randomness in man's view of the world.[4]

For the beginner the new physics may seem difficult because it is necessary to accept ideas which do not seem to be consistent with everyday experience. However, once the basic tenets of this modern physics are accepted, the door is opened to a new microscopic world. A knowledge of this subatomic realm is essential to an understanding of the solid-state materials from which most modern-day electronic devices are fabricated, for, in our era, it is the description of electrons moving in a three-dimensional periodic arrangement of atoms or ions, called a *crystal*, that is basic to the understanding of the operation of solid-state devices such as transistors and integrated circuits. The problem here is more difficult than the classical problem of the electron moving in vacuum; however, techniques have been developed whereby pseudo-classical formulations are possible after certain definitions are made.

This book introduces the ideas of quantum physics primarily to permit a more complete understanding of solid-state materials and hence the present and future devices in which they are used. A secondary purpose is cultural — to introduce the student to a mysterious but mathematically elegant exposition of the micro-world, to make him aware of a revolution that "shook the world" not many years ago.[5] Finally it is hoped that this introduction will motivate the student to learn more about this ever-expanding branch of physics so that he can perhaps make future contributions to the field of quantum electronics.

A. Materials and Electronic Devices

It is possible to design electronic circuits knowing only the terminal characteristics of devices such as transistors. That is, a circuit model can be constructed for the device using only the electrical characteristics measured at the three electrical terminals or leads. However, this circuit model does not have general applicability, being good at only one frequency or one temperature. One must understand the basic physics of the device if one is to be able to extrapolate circuit behavior to another frequency or temperature. In addition, the circuit designer should have some insight into the physics of the device to be better qualified to take advantage of its multifold uses.

[3] John Cage, *A Year from Monday*, Wesleyan University Press, Middletown, Conn. (1967).
[4] A. S. Eddington, *The Nature of the Physical World*, University of Michigan Press, Ann Arbor (1958).
[5] L. de Broglie, *The Revolution in Physics*, Noonday Press, New York (1953).

The device designer should, of course, have an even more profound knowledge of material physics, not only for the design of the devices of today, but to anticipate and invent the devices of tomorrow. Many of today's solid-state devices such as transistors, semiconductor lasers and light-emitting diodes, tunnel diodes, and Gunn oscillators use semiconductors as basic starting materials. Hence an introduction to the physics of semiconductors is essential.

Another argument in favor of the requirement that all electrical engineers have a basic knowledge of the science of materials is the need for the young engineer to remain current in a rapidly changing technological world. Since the obsolescence time for engineering information today is less than 10 years, the engineering student's training should be as broad and basic as possible within the time available in a college degree program. He must have enough knowledge of the physics of the solid state to enable him to work later with future devices — still unknown today — which probably will exploit still another electrical property of solid materials.

1.2 TRANSISTOR DEVICES AND INTEGRATED CIRCUITS

The invention of the transistor in 1948 by Bardeen, Brattain, and Shockley caused a revolution in electronics which took place in less than 10 years. Although the inventors were all physicists, their unusual insight into the requirements for a good amplifying device was essential in exploiting this physical discovery. At the time of the discovery the point-contact (whisker) semiconductor diode had already been developed. Now an active device[6] using a surface field-effect principle was sought. In fact, it was the basic study of semiconductor surfaces which was being pursued by these physicists at the time of their epoch-making invention. This dramatic development is well documented[7] and again pinpoints how an understanding of semiconductor material behavior can lead to new device development. In fact, the later invention of the p-n junction transistor by Shockley which superseded the point-contact device was another triumph of keen physical insight into the properties of semiconductor materials. In order to explain more explicitly the close connection between material properties and solid-state devices, the modern transistor will now be discussed. This is followed by a description of the most important electronic circuit building block of today — the *integrated circuit* — which derives from transistor devices and performs a complete electronic circuit function.

A. The Junction Transistor Structure

The junction transistor is fabricated from semiconductor materials such as silicon and germanium. A single crystal of semiconductor material conducts electric current either predominantly via negatively charged electron carriers

---

[6] An active electronic device is one which provides signal amplification.
[7] For example, see "Special Report: The Transistor," *Electronics*, Vol. 41, p. 77, Feb. 19, 1968.

Drs. William Shockley, Walter H. Brattain, and John Bardeen, the inventors of the transistor. (Photo taken in July, 1948. Courtesy of Bell Laboratories.)

A variety of packaged transistors and diodes. (Courtesy of Unitrode Corp., 580 Pleasant Street, Watertown, Mass. 02172.)

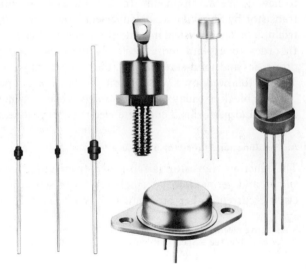

FIGURE 1.1

Schematic diagram of an
n-p-n junction transistor. The
actual device is constructed
of a single crystal of semiconductor material con-
taining the emitter, base, and
collector regions.

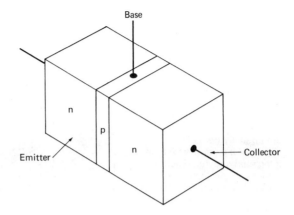

or via positively charged carriers called *holes*.[8] A semiconductor crystal which conducts predominantly via positively charged carriers is called *p-type*; a semiconductor which conducts predominantly via negatively charged carriers is called *n-type*. Either type of carrier can be provided by the introduction of the proper impurity atoms into the crystal. For example, if a single crystal of silicon has a few atoms of phosphorus, antimony, or arsenic introduced into its crystal lattice[9] for roughly each million silicon atoms, it becomes an n-type electron conductor. On the other hand, if a few atoms of aluminum, gallium, or boron are added to roughly each million atoms of single-crystal silicon, it will conduct via positively charged carriers or holes. The reason for this type of behavior will become more apparent when the band theory of solids is discussed in Chapter 4 and will be explained in detail in Chapter 5 on semiconductors.

The n-p-n junction transistor contains two n-type regions of semiconductor crystal coupled to a central p-region. The p-n-p junction transistor contains two p-type regions coupled by an n-region. Actually, all three regions are part of a single crystal of semiconductor material.

Figure 1.1 shows schematically a model of the n-p-n junction transistor invented by Shockley. The bar of silicon pictured has two p-n junctions in a single crystal of silicon, one at the left and one at the right. The left-hand junction will be called the *emitter* junction, that at the right the *collector* junction. Since the device shown is symmetric, either side could be called the emitter or collector. This, in fact, is indicative of the physical construction of early junction transistors, fabricated from the semiconductor element silicon. However, more recently the planar transistor[10] has been adopted as the industry

---

[8] The source of these positive carriers is an important result of semiconductor theory and will be discussed later.

[9] Crystal lattice is a term used for a three-dimensional periodic arrangement of points in crystal space.

[10] *Electronics*, Vol. 41, p. 84, Feb. 19, 1968. This important device structure was invented in 1959 by J. Hoerni; a patent was granted in 1962.

FIGURE 1.2

Schematic drawing and
cross section of an n-p-n
planar junction transistor.
The emitter, base, and collector contacts are normally
metallic (aluminum or gold)
and are alloyed to the semiconductor body. The emitter,
base, and collector regions
are normally germanium or
silicon of a p- or n-variety in
integrated circuits. Such
devices are protected from
contamination from the surrounding environment by a
coating of electric insulator
which in the case of integrated circuits is silicon
dioxide.

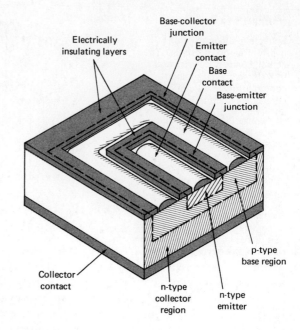

standard and is used in all integrated circuits. A sketch of an n-p-n planar
transistor is shown in Fig. 1.2. The reason why this type of structure is more
desirable will become apparent in a subsequent discussion on integrated circuits.

B. Transistor Amplifier Power Gain

The transistor structure shown is intrinsically capable of power gain. That is,
an electronic signal applied to the emitter (input) junction can be detected
somewhat later at the collector (output) junction, as an analog of the input
signal but at a higher power level. Signal transmission can be quite fast, of the
order of 10^{-9} sec. This rapid response can be used to achieve signal amplification even at frequencies as high as several gigahertz.

To understand why this simple structure yields power gain and why it is
constructed from semiconductor materials like silicon or germanium, assume
that the transistor device is *biased* or has electric potentials applied to it by
batteries as shown in Fig. 1.3. Here the emitter junction is biased in the
forward direction (n-type region, negative, with respect to the p-type region,
positive); the collector junction, on the other hand, is *reverse* biased (n-type
region, positive, with respect to the p-type region, negative). Hence the input
and output sections of the device consist of oppositely biased p-n junction
diodes, coupled by a thin p-type base region. The device is said to be biased
in the active or amplifying region. The typical current-voltage (I-V) characteristic of each of these p-n junction diodes is shown in Fig. 1.4. These diodes

FIGURE 1.3

Electric biasing of an n-p-n transistor. The emitter-base junction X_E is forward biased since the p-type base is positive relative to the negative n-type base. The collector-base junction X_C is reverse biased since the n-type collector is positive relative to the negative p-type base. The direction of current flow in the emitter, base, and collector leads is specified. The base width W must be very narrow for good power amplification. Also shown are circuit symbols for an n-p-n transistor and a p-n-p transistor.

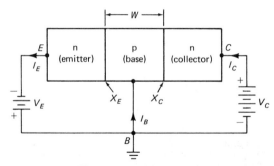

The n-p-n junction transistor

Symbol for
n-p-n transistor

Symbol for
p-n-p transistor

FIGURE 1.4

Current-voltage characteristic of a semiconductor p-n junction diode. The forward bias direction is specified when the p-region is positive relative to the n-region and the voltage and current are positive. Reverse bias is indicated where the voltage and current are negative, and the p-region is negative relative to the n-region. The forward bias condition gives low impedance and reverse bias yields high impedance.

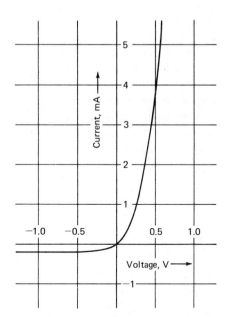

exhibit nonlinear I-V behavior compared to a linear resistor. That is, the I-V characteristic of the resistor graphs as a straight line of constant slope while the I-V characteristic of the diode has a slope which varies with current as well as voltage polarity. Note that the value of the slope (*dynamic conductance* dI/dV) of the diode characteristic is generally much less in the reverse $(-I, -V)$ direction than in the forward $(+I, +V)$ direction. (In fact, the slopes in these

FIGURE 1.5

Simple transistor circuit for demonstrating power gain. V_B is a battery used to establish proper emitter dc bias. The collector bias V_{CC} is such that the collector resistance is much greater than R_{out}. In this *common-base connection* the power gain is $(\Delta I_C/\Delta I_E)^2 (R_{\text{out}}/r_{\text{in}})$, where normally $\Delta I_C \approx I_E$ and $R_{\text{out}} \gg r_{\text{in}}$.

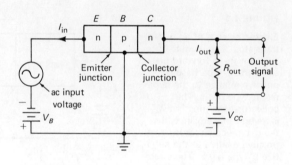

two regions can differ by as many as 12 orders of magnitude.) It is this difference in input and output diode conductance which provides a possibility for power gain. This nonlinear type of diode *I-V* behavior is typical of semiconductor p-n junctions.

Figure 1.5 shows a simple transistor amplifying circuit with a small ac input voltage signal superimposed on the dc voltage bias applied in the emitter circuit by battery V_B; the output voltage is detected by the potential-difference change, induced by the input signal, across the output resistor R_{out}. Consider an incremental increase in voltage, ΔV_{in}, supplied to the transistor emitter by the input signal. Because of the steep slope of the emitter diode *I-V* curve, this small change in applied voltage gives rise to a large change in emitter electron current, ΔI_E. Since the input and output circuits are coupled by transistor action (to be explained later), this emitter current of electrons will be transmitted nearly undiminished through the p-type base region to the collector region, which is positively charged to attract electrons. Hence a ΔI_E increase in emitter current will result in a ΔI_C rise in collector electron current.

The collector junction acts like a current source because of the high-resistance nature of the reverse-biased collector p-n junction. Hence the output resistance R_{out} may be within an order of magnitude of this high collector resistance without limiting the collector current rise ΔI_C. The voltage gain of this device is then

$$V_{\text{gain}} = \frac{\Delta V_{\text{out}}}{\Delta V_{\text{in}}} = \frac{(\Delta I_C) R_{\text{out}}}{(\Delta I_E) r_e}, \qquad (1.1a)$$

where r_e is the incremental or dynamic resistance (reciprocal of the dynamic conductance) of the emitter input junction.

In a well-designed transistor the coupling between the emitter and collector circuits is improved by making the p-type base region (W in Fig. 1.3) very narrow, of the order of a few micrometers (1 μm $= 10^{-6}$ meter). Under this condition ΔI_C is only a few percent less than ΔI_E. Hence Eq. (1.1a) indicates substantial voltage gain since $R_{\text{out}}/r_e \gg 1$. In a similar manner the power gain of this device can be written as

$$P_{\text{gain}} = \frac{\Delta P_{\text{out}}}{\Delta P_{\text{in}}} = \frac{(\Delta I_C)^2 R_{\text{out}}}{(\Delta I_E)^2 r_e}. \qquad (1.1b)$$

Again the power gain can be substantial because $R_{\text{out}} \gg r_e$. Hence the gain of transistor results from the following:

1) The dynamic resistance of the reverse-biased collector junction is much greater than that of the forward-biased emitter junction.
2) Electron current can be transmitted from the emitter to the collector, through the base region, practically undiminished in magnitude.

These two conditions are satisfied in a silicon or germanium junction transistor. Their physical basis will become apparent after a detailed study of semiconductor physics is undertaken in subsequent chapters.

Hence the transistor acts as an amplifier in that the output signal power exceeds the input signal power. Conservation of energy requires that the source of this gain must be the chemical energy supplied by the batteries.

C. The Transistor as a Switch

An equally important application of the transistor is its use as an electronic switch. The modern digital computer contains literally millions of transistor switches. The computational processes in a digital computer are carried out by circuit elements which can be placed in either of two states, ON or OFF. The ON state, in analogy with an ordinary household light switch, corresponds to the joining together of two points in an electronic circuit by means of a very low resistance, causing a substantial current to flow. In the OFF state the circuit is opened or a very high resistance is placed between two points in the circuit, practically cutting off current flow. In the transistor switch, these two points are normally the emitter and collector terminals. A small voltage signal applied to the third terminal or base lead can cause the normally nonconducting transistor to convert to the conducting state.

Figure 1.6 is a schematic diagram in which an n-p-n transistor is connected in the *common-emitter* circuit configuration. The high-resistance state of the transistor corresponds to the case in which practically no current flows between emitter and collector. The low-resistance state, on the other hand, corresponds to the case of substantial current flow between emitter and collector. With the transistor in the grounded-emitter connection, a small current I_{B1} introduced into the transistor base can cause a substantial increase in current flow from emitter to collector. This results from the current gain exhibited by a transistor in this circuit configuration.[11] Hence the transistor can be switched on or off by the application or removal of a small signal at the base connection. This could correspond to the application of a voltage pulse at the input indicated in Fig. 1.6.

---

[11] The intrinsic common-emitter current gain is defined as the ratio of the collector current increase to the base current which causes this increase, and can be 50 or more.

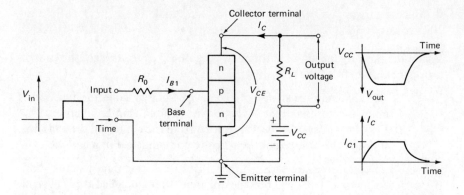

FIGURE 1.6

The common-emitter switching mode for an n-p-n transistor. A positive-going square input pulse drives the emitter junction into the forward direction and supplies emitter current which is transmitted to the collector. The collector current, after a time delay, rises to a steady-state value I_{C1}. When the input pulse shuts off, the collector current eventually decays back to zero. The inverted output voltage pulse is a natural consequence of transistor action.

FIGURE 1.7

Typical current-voltage characteristic of a common-emitter-connected n-p-n transistor for different values of base current I_B plotted as a function of collector-to-emitter voltage V_{CE}.

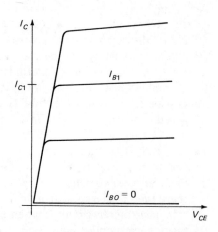

The influence of base current on collector current is best illustrated by a typical set of *I-V* characteristics for a transistor (Fig. 1.7). Note that for a particular value of collector-to-emitter voltage V_{CE}, as in the circuit of Fig. 1.6, increasing the base current from zero to I_{B1} can increase the collector current from nearly zero to a substantial value I_{C1}.

The fact that the transistor can easily be placed in either of two states, ON and OFF, quite distinct from each other, accounts for the device's significant use in digital computers. Digital computations are carried out in the binary system with only two numbers, **0** and **1**. The transistor in the ON or "high" state can represent the number **1**, while the OFF or "low" state corresponds to

0. Since the transistor can be switched from one state to the other in less than 10^{-9} sec, one billion computations can be carried out in a second. This accounts for the enormous computational speed of the modern digital calculator.

D. The Integrated Circuit

Although the transistor is indeed the basic building block of computers, other circuit elements such as diodes, resistors, and capacitors are also needed. In fact, a commonly used information-storage circuit, the *flip-flop*,[12] can contain perhaps 20 or 30 circuit elements such as transistors, diodes, resistors, and capacitors. The technology now exists to build all these elements into a single block of silicon, about 1 mm^2 in area. A device which performs a complete circuit function is known as an *integrated circuit*. The concept of the integrated circuit as one piece of solid material consisting of several components, passive as well as active, without external connection between these components, was proposed by both J. S. Kilby of the Texas Instruments Corporation and R. H. Noyce of the Fairchild Semiconductor Corporation in 1959.[13] Figure 1.8 shows a schematic diagram and a microphotograph of such a circuit. Some of these devices have as many as 40 terminals or leads, compared to only 3 for a transistor. Recently silicon chips less than 1 cm^2 in area, containing 10,000 and more devices, have been fabricated. These devices constitute the "computer on a chip" which performs most of the computational functions required by a hand-held calculator as well as the time counting for electronic wristwatches which contain no moving parts. Photographs of these instruments are shown in Fig. 1.9.

The introduction of integrated circuits about the year 1959 was a natural outgrowth of a development which had only begun 10 years earlier. These elements have now become the basic building blocks of electronic circuits. The circuit designer of the future must be able to choose appropriate integrated circuit blocks and interconnect them so as to create a complete electronic system such as a computer. Perhaps he will require the fabrication of integrated circuits which are not normally available. He must then communicate his circuit needs to the device fabricator. This communication will be much more efficient if the circuit designer has some detailed knowledge of the physics and technology of integrated circuits. Of course, the device designer now must have knowledge of electronic circuit design as well as the physics and technology of integrated circuits. The many aspects of this subject will be outlined in the later chapters of this book.

---

[12] This element is maintained in a particular state indefinitely, even after removal of input signal. Hence it exhibits a "memory."
[13] A recent court decision on a patent interference case between these corporations awarded a priority on four of the claims to Kilby (Texas Instruments) and two claims to Noyce (Fairchild Semiconductor). See footnote 7 and J. S. Kilby, "Semiconductor Solid Circuits," *Electronics*, Vol. 32, p. 110, Aug. 7, 1959.

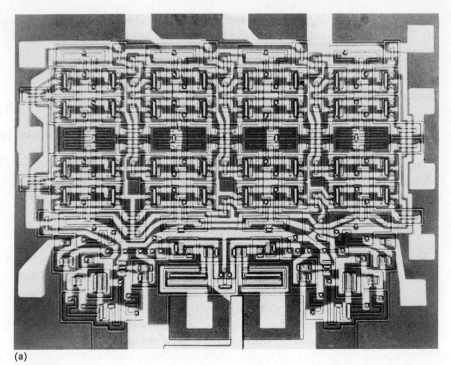

(a)

FIGURE 1.8

a) Actual photograph of a
silicon digital integrated circuit showing the various
planar transistors, diodes,
and resistors. The large
white islands on the periph-
ery are metalized contacts
for external connection.
b) A schematic diagram of
the circuit in (a).

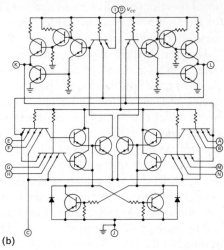

(b)

1.3 MODERN SOLID-STATE DEVICES

The transistor and integrated circuit are by far the most-used solid-state
devices at this time. However, there are numerous other important electronic
devices whose fundamental behavior also must be described quantum mechanically. In addition, one can expect that there are many electronic devices yet
to be invented whose operation will best be explained in the language of modern

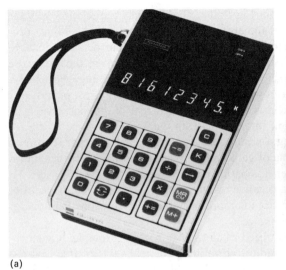

(a) (b)

FIGURE 1.9

a) Hand-held calculator. (Courtesy of Sharp Electronics Corporation.)
b) Electronic wristwatch. (Courtesy of Gruen Industries, Inc.)

physics. A brief description of three devices in use today will now be presented. The quantum mechanical concept that they are based on will be qualitatively discussed. In a later chapter, after the elements of modern solid-state physics are explained, a more complete description of some of these devices will be given.

A. The Tunnel Diode

It has already been mentioned that junction transistors operate at speeds of about 10^{-9} sec. In the frequency domain, operation is possible up to several gigahertz. The attempt has been made in recent years to extend the frequency spectrum to tens of gigahertz. One of the basic speed limitations on the junction transistor is the transit time of electrical carriers from emitter to collector. In 1956 Leo Esaki invented a device called the *tunnel diode* whose speed (and hence frequency) was not limited by carrier transit time.[14] The device physically appears to be just another variety of the p-n junction diode made by a metal-semiconductor micro-alloying process. A difference in fabrication technique, however, achieves this unique type of p-n junction; both the p- and n-regions must be made with high impurity concentrations. Figure 1.10a shows a cross section of a tunnel diode fabricated by soldering a metal sphere onto a block of germanium.

The dc current-voltage characteristic of this device as measured between the

[14] L. Esaki, "New Phenomenon in Narrow Germanium p-n Junction," *Phys. Rev.*, Vol. 109, p. 603 (1958).

p- and n-terminals is shown in Fig. 1.10b. To illustrate the relation between this characteristic and that of the ordinary p-n junction, the dashed line in the figure represents the ordinary p-n junction *I-V* forward curve. Note that the only difference between these two characteristics is a bump in the curve at a voltage value of about 100 mV. *Vive la différence!* For voltages just above 100 mV, the current decreases with increasing voltage. This is in contrast to a normal resistor or diode characteristic, where current always increases with increasing voltage. Hence in the region *BC* in the figure the slope of the *I-V* curve is negative, indicating a *negative resistance*. This means that the tunnel diode is intrinsically capable of gain and can be used as an electric generator. For if I^2R represents power dissipation or power *loss* in a resistor (where *I* is current and *R* resistance), a negative resistance indicates power *gain*. Electrons lose energy to the crystal lattice in an ordinary resistor; electrons gain energy in the case of a negative resistance.

The tunnel diode can be utilized as an oscillator — for example, in a superheterodyne radio circuit or radio transmitter. The real value of this device is that very high-frequency oscillations can be achieved. For this diode the high-frequency limitation is given by

$$f_{max} = \frac{1}{2\pi R_{neg} C_{jct}},$$

(1.2)

FIGURE 1.10

a) Schematic drawing and cross section of a germanium tunnel diode showing an alloyed p-n junction.
b) The dc electrical characteristic of a tunnel diode showing the peak current at *B* and the valley current at *C* as indicated by the solid line. Also shown, by the dashed line from *A* to *C*, is the normal p-n junction diode curve for forward bias.

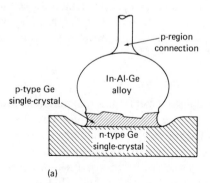

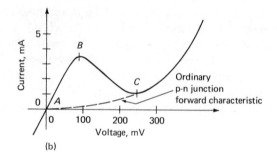

where R_{neg} is the negative dynamic resistance of the device, and C_{jct} the p-n junction capacitance. Small-area junctions mean low capacitance and hence high-frequency possibilities. Devices oscillating at frequencies exceeding 10 GHz in the microwave region are available. Since the device has gain, it can also be utilized as a high-frequency amplifier or high-speed switch.

The fundamental mode of operation of the tunnel diode was explained by Esaki in terms of quantum physics. This scientist was actually involved in investigating some unrelated properties of conventional p-n junction diodes when he noticed the now-famous bump in the forward I-V diode characteristic. As the name suggests, this diode characteristic depends on the phenomenon of quantum mechanical tunneling through a potential barrier. The concept is discussed in a subsequent chapter on quantum theory.

B. The Gunn Oscillator

Another two-terminal device capable of generating oscillations at frequencies exceeding 10 GHz was discovered by J. B. Gunn in 1963.[15] He was investigating the high-voltage bulk behavior of a semiconductor compound gallium arsenide. Gunn noticed that a negative resistance was indicated in the current-voltage characteristic of a small bar of this material which became unstable and broke into electronic oscillations above a certain voltage. He observed high-frequency radiation emanating from this structure. The I-V characteristic of such a device is shown in Fig. 1.11. Gunn soon realized that a high-frequency oscillator could be constructed from such a device because of the intrinsic negative resistance. Gigahertz frequency performance was made possible because the high electric field present in the material caused the electrons to move at high speed and hence have a short transit time. In contrast to the tunnel diode, which must have a small, low-capacitance p-n junction to achieve high-frequency oscillation, Gunn's device contained no junction and could have a relatively large area (1 mm^2) for cooling purposes. Hence both high power and high-frequency oscillations are possible using this device. Whereas powers of the order of a milliwatt are possible with tunnel diodes, Gunn oscillators can handle many watts of power.

The physical explanation of this high-frequency oscillation again depends on a strictly quantum mechanical description of electrical carrier transport in a semiconductor material like gallium arsenide. This phenomenon does not occur in semiconductors such as germanium and silicon. This fact is explained in terms of a subtle difference between the crystal structures of gallium arsenide and germanium.

[15] J. B. Gunn, "Microwave Oscillations of Current in III-V Semiconductors," *Solid-State Commun.*, Vol. 1, p. 88 (1963). See also J. A. Copeland, "Bulk Negative-Resistance Semiconductor Devices," *IEEE Spectrum*, Vol. 4, p. 71, May, 1967.

FIGURE 1.11

Current-voltage characteristic
of a typical Gunn oscillator.
The device breaks into oscillations at voltages of $+25$
or -25 V, where a negative
resistance occurs. Very high-
frequency (gigahertz) oscillations are normally observed.
Also shown is a sketch of a
typical design of a Gunn-
oscillator chip. The dimensions of the GaAs chip and
the impurity content of the
different device regions are
indicated. The purest section,
the central region, is the
active part of the device,
while the end sections provide low-resistance electrical
contacts.

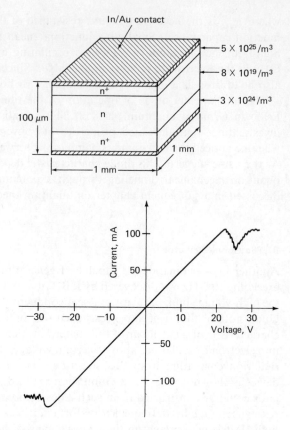

A simple physical description of the cause of electrical oscillations produced
in this device is as follows: The electrical conductivity σ of a bar of any
material is given by Ohm's law as $\sigma = J/\mathcal{E}$ or

$$\sigma = \frac{nqv}{\mathcal{E}}, \qquad (1.3)$$

where n represents the density of electrical carriers, q is the electric charge, v is
the carrier velocity, and $\mathcal{E}$ is the value of the applied electric field. Normally, as
the electric field is raised, the carrier velocity increases correspondingly. This is
consistent with Eq. (1.3) since the electrical conductivity usually is a constant,
characteristic of the material and independent of the electric field. In gallium
arsenide at the critical field of about 300 kV/m, the velocity does not increase in
proportion to the field.[16] Hence at the voltage corresponding to that field, the
I-V characteristic does not have constant slope. In fact, it has a negative slope
corresponding to a negative resistance, as may be seen in Fig. 1.11. Again as

[16] The field in a uniform sample of length l is given by the applied voltage V divided by the
length, or $\mathcal{E} = V/l$.

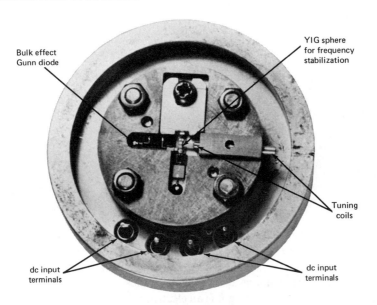

A complete Gunn oscillator circuit in a microwave package. (Courtesy of Hewlett Packard.)

in the case of the tunnel diode, this negative resistance offers the possibility of generating electrical oscillations.

The prediction of the tendency for a semiconductor crystal to exhibit this type of behavior requires detailed analysis of the electrical conductivity of the material, using quantum mechanics. A more extensive description of this effect will be given later, after the energy band theory of solids has been explained.

C. The Semiconductor Laser

It is well known that the laser is a device which can produce an enormous concentration of energy in the form of light. Gas lasers can produce heat locally exceeding the temperature of the sun. The semiconductor laser is an extremely small device which, for its size, also produces large amounts of light energy.[17] The light is emitted from a semiconductor p-n junction (often made in gallium arsenide) by the application of a voltage. Figure 1.12 is a schematic diagram of a lasing diode indicating the highly directional nature of the emitted light. This latter property explains the interest in the junction laser for communications. That is, an accurately aimed light signal can be emitted from a gallium arsenide p-n junction and received or detected at some remote point

[17] H. Kressel et al., "Semiconductor Lasers," *Electronics*, Vol. 43, p. 78 (1970).

FIGURE 1.12

Schematic drawing of a p-n
junction semiconductor laser.
The light emitted from the
p-n junction region is highly
directional, within an
angle of about 10 degrees.
The coherency of emitted
light is ensured by polishing
the emitting faces of the
crystal to mirrorlike
perfection.

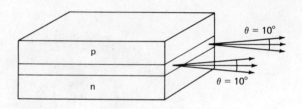

by a photodetector.[18] If the light signal transmitted is modulated or varied by a voice signal or a television picture signal, the possibility exists of "transmitting intelligence over a light beam." The modulation of the solid-state laser is extremely simple since it only involves varying the electric field applied to the p-n junction.

The tremendous advantage of laser communication over normal radio or television transmission is that light is an electromagnetic radiation much higher in frequency than radio or television waves. A theorem of communication theory states that the higher the frequency of transmission, the greater the amount of information that can be transmitted. For example, the wavelength λ of yellow light is about 6000 angstroms (Å) (6×10^{-7} m). Since the velocity c of light is 3×10^8 m/sec, the frequency v of light can be calculated from

$$v = \frac{c}{\lambda}. \qquad (1.4)$$

This turns out to be 5×10^{14} Hz. This is about a million times higher than the frequency of television carrier waves and hence accounts for the enormous excitement about laser communication systems.

Recently a perhaps less sophisticated but commercially more exciting light-emitting solid-state device has come into general use. This is the light-emitting diode (LED) which uses the semiconductor gallium arsenide phosphide (GaAsP) as starting material. When about 1 V and 20 mA are applied to a p-n junction of this material, it emits visible red light and hence constitutes a miniature solid-state lamp which has no fundamental burn-out mechanism. Because of the very high electric power-to-light conversion efficiency, only milliwatts of power are required to produce a significant amount of light. A 5×7 matrix of these devices provides a small numeric readout, visible in daylight, of the type used in the hand calculator of Fig. 1.9.[19] The principle of operation of this device is exactly that of the semiconductor laser except that no

[18] This is usually in the form of a silicon p-n junction diode.
[19] For a survey of numeric readout devices, including the GaAsP LED, see "Special Report: Numeric Readout Displays," *Electronics*, Vol. 44, p. 65, May 24, 1971.

Four-digit light-emitting diode (LED) display shown along with the electronic "move-ment" for a digital wristwatch. (Courtesy of Litronix, Inc.)

special polishing of the faces of the crystal is necessary to ensure light coherency, which is unnecessary for a simple light source.

The explanation of light emission from p-n junctions and the prediction of the most efficient light-emitting material require quantum mechanics. The fact that GaAs and GaAsP are excellent light emitters while Ge and Si are poor in this respect must be discussed in terms of energy band theory. A fuller account of these devices will be given in a later chapter.

PROBLEMS

1.1 Determine the number of atoms in one cubic meter of

a) silicon,
b) germanium.

Some facts about silicon and germanium are:

	Density, gm/cm^3	Atomic weight, gm/gm-atom
Silicon	2.33	28.1
Germanium	5.33	72.6

Note: A gram-atom of any material contains 6.02×10^{23} atoms (Avogadro's number). The impurity density in a crystal of silicon is estimated by electrical measurements to be 1.0×10^{21} arsenic atoms/m^3.

c) Determine the impurity level in atomic percent of arsenic in the silicon crystal.

1.2 Assume that the diode characteristic of Fig. 1.4 corresponds to the emitter junction of the transistor shown schematically in Fig. 1.5. Take the emitter junction forward bias voltage as 0.25 V and assume the load resistance R_{out} equals 10,000 Ω.

a) Calculate the dynamic emitter input resistance r_{in} $(= dV/dI)$ at 0.25 V forward bias.
b) If the output-to-input signal current ratio $\Delta I_C/\Delta I_E$ is 0.96, find the power gain of the transistor in this circuit.

1.3 The base width of the n-p-n transistor of Fig. 1.6 is 1.0×10^{-6} m, and the average time for signal transmission from the emitter to the collector junction is 1.0×10^{-9} sec.

a) Calculate the average speed of transmission of the signal from emitter to collector.
b) What is the absolute maximum number of switching operations that can be performed by this transistor in 1 sec if this transit time limits the switching speed?

1.4 The p-n junction capacitance of the tunnel diode shown in Fig. 1.10 is 1.0 picofarad (10^{-12} F).

a) Calculate the maximum operating frequency of this device using Eq. (1.2).
b) Estimate the range of bias voltages over which this device can operate as an oscillator.

1.5 Estimate the dc operating power level of

a) the tunnel diode whose *I-V* characteristic is shown in Fig. 1.10.
b) the Gunn oscillator whose *I-V* characteristic is given in Fig. 1.11.

1.6 Consider a chip of gallium arsenide whose dimensions and *I-V* characteristic are shown in Fig. 1.11.

a) Calculate the maximum thickness of the region in the crystal where the high electric field exists so that the critical field of 300 kV/m necessary for Gunn-type oscillation is achieved at 25 V.
b) Compute the electrical conductivity of the active region of this gallium arsenide crystal.

1.7 The primary wavelength of light emitted from a gallium arsenide laser diode is in the infrared at 9600 Å.

a) Calculate the frequency of this electromagnetic radiation. The threshold current necessary to achieve lasing action in this type of diode is typically 20 A at room temperature with a potential of 1 V applied to the junction.

b) Compute the heat in joules which must be removed from the diode per second in order to maintain the device at room temperature, given the mass of this laser is 1.0 gm and the specific heat of GaAs is 0.076 cal/gm-°C.

c) Calculate the temperature rise of this laser crystal in 1 sec if no heat is removed.

2 The Crystal Structure of Solids

2.1 CRYSTAL STRUCTURE AND BONDING

The vast majority of electronic devices in use today are fabricated from single-crystal solid-state materials. The device applications of single crystals of silicon, germanium, gallium arsenide, and gallium arsenide phosphide were discussed in Chapter 1. This chapter defines the form of the solid known as the single crystal, describes typical crystal structures, and explains the forces that hold the crystal together. Techniques for identifying the atomic arrangement in solids are given, as are methods for specifying the structures.

A. Crystallinity

Solids occur in the *crystalline* or *amorphous* state. In the crystalline solid about 10^{28} atoms/m³ are arranged in three dimensions in a regular manner. This structure may be obtained by repeating in three dimensions an elementary arrangement of some atoms or building blocks called *unit cells*. This is shown in two dimensions (for simplicity) in Fig. 2.1. Such a periodic arrangement occurring throughout the volume of a solid sample constitutes a single crystal. However, if the regular structure occurs only in portions of a solid and the different portions are aligned arbitrarily with respect to each other, the material is said to be *polycrystalline*; the individual regular portions are referred to as *crystallites* or *grains* and are separated from each other by *grain boundaries*. If the individual crystallites are reduced in extent to a point where they approach the size of a unit cell, periodicity is lost and the material is called *amorphous*.

B. The Single Crystal

A system of atoms in thermal equilibrium at a particular temperature tends toward a specific crystalline state if this is the lowest free-energy configuration

FIGURE 2.1

Two-dimensional represen-
tation of a single crystal. The
lattice structure is indicated
by the dashed lines. There
are two atoms (black dots)
indicated per lattice site.
Three possible unit cells are
indicated by the parallelo-
grams *ABCD, EFGH,* and
JKLM. Each cell contains
two atoms. Any of these
basic blocks placed side by
side will reconstruct the total
crystal. The basic vectors for
defining the lattice points are
denoted by **a** and **b.** Every
point in the lattice may be
determined by translating
from point to point by a
vector **a** or **b.**

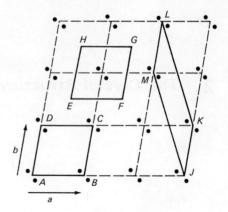

of the solid.[1] Of course, different types of arrangements represent different
energies, and hence a certain atomic configuration is favored more than
others by a given material at a particular temperature. The most stable
arrangement at any temperature depends in a complex manner on the forces
between atoms, the space between them, the size of the atoms, etc. In common
usage the words crystal and gem are sometimes used interchangeably. This has
led to the idea that all single crystals must exhibit the natural geometric shape
typical of gems. If all crystals were grown under truly equilibrium conditions
without the action of outside forces, this would tend to be true. However, good
single crystals do not always have a completely regular outward appearance
when grown under normal laboratory conditions. Figure 2.2a, b shows photographs of a few single crystals of a semiconductor material prepared for use in
device fabrication. Some tendency toward regular gemlike facets is observed,
but not perhaps like that expected of natural quartz or rochelle salt crystals.
A sketch of a typical crystal-pulling apparatus for "growing" these crystals is
shown in Fig. 2.2c.

Most solid-state devices employ single-crystalline substances as starting
materials. The reason for this is that it is easier to control the properties and
electrical behavior of crystals whose properties are everywhere the same; in
polycrystalline materials foreign substances tend to collect along grain boundaries. The electrical properties of single crystals are not only more uniform, but
in some respects better for devices. Also, the crystal form makes analysis and

---

[1] The fact that a system in equilibrium tends toward a minimum free energy is one of the
most important laws of physics. Symmetry arguments dictate that a regular array of atoms
will have a lower total energy than a corresponding group of randomly arranged atoms. The
free energy of a system represents the energy available to do work.

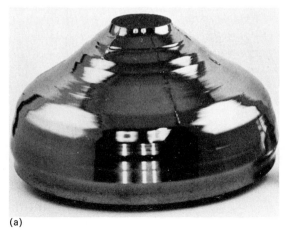

(a)   (b)

FIGURE 2.2

a) Large (6 in. diameter) single crystal of silicon.
b) Top view of a silicon crystal just withdrawn from a melt of liquid silicon. The single-crystal "seed" used to trigger single-crystal formation is shown still attached to the top of the grown crystal.
c) Typical crystal-pulling apparatus. A crystal is grown by dipping a seed (small piece of a single crystal) into the molten material and slowly withdrawing it so that the remainder freezes out on the seed. (From W. R. Runyan, *Silicon Semiconductor Technology*. Copyright 1965 by McGraw-Hill Book Company. Used with permission of McGraw-Hill Book Company.)

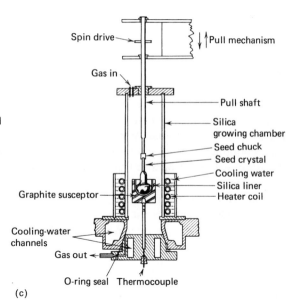

(c)

characterization of the material easier. For example, extraneous impurities are then easier to identify and detect using the electrical properties of crystals. Nevertheless, single-crystallinity is not essential for solid-state devices and, for example, amorphous semiconductor materials are presently being studied for future device applications.[2]

[2] S. R. Ovshinsky, "Reversible Electrical Switching Phenomena in Disordered Structures," *Phys. Rev. Letters*, Vol. 21, p. 1450 (1968).

FIGURE 2.3

Schematic drawing indicating
three types of liquid crystals
having different molecular
alignments. Also indicated is
the random molecular orientation in these liquid crystals
at high temperature when
they become isotropic
liquids. The nematic form is
used in constructing electronically activated numeric
readout display devices
which, however, become
inoperative at higher
temperatures when the
material becomes isotropic.

Isotropic liquid

Nematic

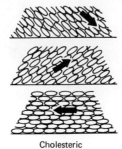

Cholesteric

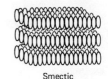

Smectic

C. Liquid Crystals

Although the topic of crystalline solids will be treated exclusively here, there is a
class of "liquid crystals" which has been known for some time but has only
recently become of interest for device applications. These comprise a class of
materials that have only two- or one-dimensional regularity. The term *liquid
crystal* refers to materials which pour like a liquid and assume the shape of
their container, yet possess a degree of regularity in their molecular ordering.
Their optical reflection and transmission properties indicate regularities unlike
an ordinary isotropic[3] liquid. They occur generally as organic molecules,
cigarlike in shape and organized or packed in "layered structures" as shown in
Fig. 2.3. The recent interest in *nematic* liquid crystals is due to their use as the
working material in the numeric readout displays such as are employed in the
electronic wristwatch of Fig. 1.9; such displays are important because they
consume very little battery energy.[4] The operation of the device is based on the
material being transparent or opaque, depending on whether or not an electric
field is applied to it.

2.2 SPACE LATTICES AND CRYSTAL STRUCTURE

A single crystal has already been defined as a regular repetition of a basic
building block of atoms throughout the volume of a solid. The geometric
shape of the building block is a three-dimensional parallelepiped. It contains
one atom in simple crystals like copper, silver, and sodium but may contain
many thousands of atoms in complex organic protein crystals. The enormous

---

[3] One having no particular directional properties.
[4] See footnote 19, Chapter 1.

FIGURE 2.4

Two-dimensional represen-
tation of a space lattice. The
lattice points are at the intersections of the parallel lines.
A cluster of two atoms is
shown at each lattice point.
The basic translation vectors
a and **b** are indicated. The
lattice looks identical when
viewed from the point de-
noted by the vector **r** and
when viewed from **r′**.

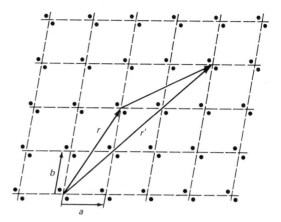

variety of crystal structures may be defined by arranging atoms systematically
about a regular or periodic arrangement of points in space called a *space
lattice.* The lattice is defined by three fundamental translation vectors **a**, **b**, and
c so that the arrangement of atoms in a crystal infinite in extent looks identically
the same when observed from any point, displaced **r** from an origin, as when
viewed from the point **r′** where

$$\mathbf{r'} = \mathbf{r} + n_1\mathbf{a} + n_2\mathbf{b} + n_3\mathbf{c}, \tag{2.1}$$

and n_1, n_2, and n_3 are integers. This is shown in a two-dimensional representation in Fig. 2.4. The set of points defined by the vector **r′** for all integer values
of n_1, n_2, and n_3 is periodic in space and constitutes the space lattice.

A. The Primitive and Unit Cell

The volume defined by the vectors **a**, **b**, and **c** constitutes the *primitive cell* of
the lattice if it is the *smallest* cell which, when periodically stacked, fills all of the
crystal space. Since the angles of this volume are oblique in general, it is
sometimes more convenient, but not always possible, to work with a unit cell
which is defined by three axes at right angles to each other (orthogonal axes).
Periodic stacking of the unit cell will also map out the whole crystal lattice,
but it is not always the smallest volume with which this task can be accomplished. The length of the edge of the unit cell is called the *lattice constant.*

In 1848 Bravais defined the concept of the space lattice in order to classify the
structure of all crystals.[5] He proved that it is possible to define no more than
14 space lattices in three dimensions. The 14 Bravais lattices are shown in
Fig. 2.5. The unit cells defined by these lattices are not always primitive.[6]
Simple crystal structures are obtained by placing atoms only at the corners of

---

[5] A. Bravais, "Etudes cristallographiques," Gautier-Villars, Paris (1866).
[6] Nevertheless they represent configurations having different degrees of symmetry.

these cells. Other structures are obtained by placing atoms at the center of the cell faces or at the center of the volume or body. In general, the lattice points do not represent the positions of single atoms, but groups of two or even more atoms (see Fig. 2.4).

B. Five Simple Crystal Structures

Simple cubic (sc) If single atoms and groups of atoms are placed at the corners of the simple cubic Bravais lattice, this constitutes a *simple cubic* crystal as shown in Fig. 2.5, part 12. This unit cell also happens to be a primitive cell because it contains only one lattice point and hence is the smallest volume which, when repeated, describes the whole crystal. One possible simple cubic structure has a single atom per corner site and hence contains one atom per primitive cell. The one atom per cell derives from the fact that each of the eight corner atoms contributes only one-eighth of an atom to each of the eight cells surrounding each lattice point. CsCl, TlBr, and NH_4Cl form crystals of the simple cubic type. The only element to crystallize in this form is polonium.

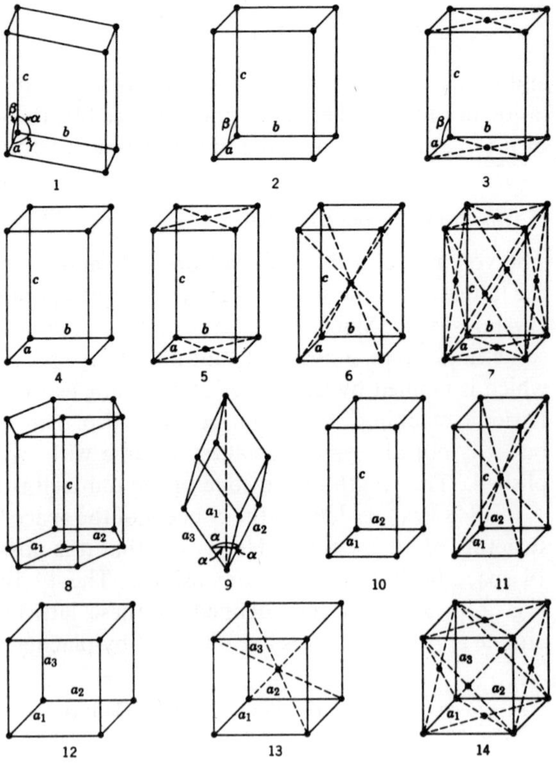

FIGURE 2.5

Possible space lattices of the seven crystal systems.
(1) Triclinic, simple.
(2) Monoclinic, simple.
(3) Monoclinic, base-centered. (4) Orthorhombic, simple. (5) Orthorhombic, base-centered. (6) Ortho-rhombic, body-centered.
(7) Orthorhombic, face-centered. (8) Hexagonal.
(9) Rhombohedral.
(10) Tetragonal, simple.
(11) Tetragonal, body-centered. (12) Cubic, simple.
(13) Cubic, body-centered.
(14) Cubic, face-centered.
(From T. S. Hutchison and D. C. Baird, *The Physics of Engineering Solids*, 2nd edition, John Wiley & Sons, New York, 1968.)

FIGURE 2.6

The primitive (solid lines) and conventional (dashed lines) unit cells of the bcc lattice.

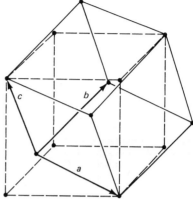

FIGURE 2.7

The primitive (solid lines) and conventional (dashed lines) unit cells of the fcc lattice.

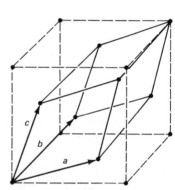

Body-centered cubic (bcc) The cubic cell of the body-centered crystal lattice not only contains a site in each corner of the cube but also one at the center of the volume, as shown in Fig. 2.5, part 13. Hence the bcc unit cell contains two lattice points and is not primitive. The primitive cell for this crystal lattice, shown in Fig. 2.6, contains only one lattice site. Note that the angles of this structure are oblique and not identical. This accounts for the conventional usage of the cubic unit cell for describing bcc crystal structures. Typical materials which crystallize in bcc form are Na, Rb, Cr, Ta, and W.

Face-centered cubic (fcc) The unit cell of the face-centered Bravais lattice contains eight corner positions plus one site in each of the six cube faces, as shown in Fig. 2.5, part 14. Since the face-centered sites are shared between two adjacent cells and the corner sites are shared by eight cells, there are four lattice points in the conventional face-centered cubic unit cell. Also, as in the body-centered case, one can define a primitive fcc cell and this is shown in Fig. 2.7. Again this primitive cell contains one site but has oblique angles and is more difficult to visualize than the conventional unit cell. Some materials which crystallize in the fcc form are Al, Ca, Ni, Cu, Pb, Ag, NaCl, LiH, PbS, and MgO.

FIGURE 2.8

The hexagonal close-packed structure. The atom positions in this structure do not constitute a Bravais space lattice owing to the three extra body atoms. The space lattice is simple hexagonal with a cluster of two atoms associated with each lattice point. These form a cluster of the type shown in Fig. 2.4. One atom of the cluster is at the origin (0, 0, 0) ; the other atom is at ($\frac{2}{3}$ **a**, $\frac{1}{3}$ **b**, $\frac{1}{2}$ **c**), where **a**, **b**, and **c** are the translation vectors.

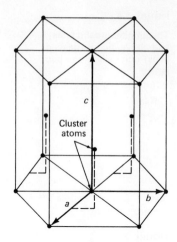

Hexagonal close-packed (*hcp*)   The hexagonal unit cell of the Bravais lattice contains 12 corner positions plus one site at each of the upper and lower cell faces (see Fig. 2.5, part 8). Since each of the 12 corner sites is shared by six surrounding cells, they contribute a total of two sites per unit cell. The two face positions each are shared by two adjoining cells and hence contribute one site to the unit cell. The most common hexagonal crystal structure (close-packed), however, contains three additional body sites in this unit cell so that there are a total of six sites per cell. This is shown in Fig. 2.8. Typical materials which crystallize with this structure are Be, Cd, Mg, Co, Ti, Tl, and Zn.

The close-packed hexagonal structure is the most dense arrangement of spherical atoms possible. This may be seen as follows: Consider the atoms as spheres of equal diameter arranged in a single layer by placing each sphere in contact with six others. Now a second similar layer is placed on top of this by allowing each sphere to touch three neighboring spheres in the bottom layer. The positions of the atoms in the lower layer are called *A* sites and are indicated in Fig. 2.9. The positions of the second-layer atoms are called *B* sites and are also shown in the figure. Now if a third layer is added above the second so that these spheres lie directly over the *A* sites of the first layer, a close-packed hexagonal structure is obtained. These third-layer sites are hence again *A* sites and the complete crystal structure is built up by the *ABABAB* · · · packing. Note, however, that there is another possibility for stacking the third layer. The sites marked *C* in Fig. 2.9 constitute such possible sites, and this crystal structure is referred to as *ABCABC* · · · . Careful examination of this latter structure will show it to be face-centered cubic with the normal to these layers being the cube diagonal.

Any atom in the fcc and hcp structures can be shown to be surrounded by 12 adjoining atoms or *nearest neighbors*. This is the maximum possible in an arrangement of solid spheres. In contrast, the bcc structure has only eight nearest neighbors and hence is more loosely packed. This is so because the bcc atoms touch each other along the cube diagonal rather than along a cube edge.

Diamond structure The conventional unit cell of the diamond structure is basically cubic as pictured in Fig. 2.10a. This configuration is even more loosely packed than the structures previously discussed. Each atom has only four nearest neighbors forming a *tetrahedral* bond. This basic configuration may be constructed by placing an atom at the body center of a cube, two atoms at opposite corners of the top face of this cube, and two atoms at opposite corners of the bottom cube face, but twisted 90° with respect to the top face atoms. This configuration is shown in Fig. 2.10b but is not the unit cell. The tetrahedral bonds of the four corner atoms to the central atom are very strong and highly directional, occurring at angles of about 109.5 degrees. The total diamond structure can be visualized as two interpenetrating fcc lattices, one displaced from the other by one-fourth the length and along a cube diagonal.

This type of crystal structure is of particular interest in the study of electronic materials since many important properties of the elemental semiconductors silicon and germanium derive from the basic tetrahedral structure. A closely related structure is the so-called *zinc-blende* crystalline configuration of the intermetallic semiconductors gallium arsenide, indium arsenide, and cadmium sulfide. Here the center atom in the tetrahedral configuration is of atomic species *A* while that of the four corner atoms is of species *B* (for example, *A* is Ga, *B* is As). Silicon and germanium form the tetrahedral covalent bond by joining atoms, each with valence four; one atom shares an electron with each of the four corner atoms. Gallium arsenide consists of a compound containing the atom gallium with three outer electrons, and arsenic with five outer electrons. Here again, they join together to form, on the average, a stable shell of eight electrons. This also applies to cadmium with two electrons and sulfur with six electrons to yield cadmium sulfide.

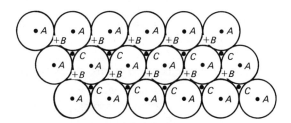

FIGURE 2.9

A two-dimensional represention of a close-packed layer of spheres with centers at points labeled *A*. Two different arrangements of similar spheres can be placed on top of this, one with centers over the points marked *B*, the other with centers over the points marked *C*. If the second layer goes over *B*, there are then two choices for a third layer. It can go over *A* or over *C*. If it goes over *A*, the sequence is *ABABAB* ··· and the structure is defined as hexagonal close-packed. If the third layer is placed over *C* sites, the sequence is *ABCABCABC* ··· and the structure is face-centered cubic with the normal to these layers being the cube diagonal.

FIGURE 2.10

a) Diamond crystal structure. For C, Si, and Ge, $a = 3.56$, 5.43, and 5.66 Å, respectively. Nearest-neighbor spacing is $a\sqrt{3/4}$, and occurs with tetragonal symmetry (dashed cube). The large cube (volume a^3) constitutes a unit cell and effectively contains 8 atoms. (After W. Shockley, *Electrons and Holes in Semiconductors*, Van Nostrand Rheinhold, New York, 1950. Copyright 1950 by Litton Educational Publishing, Inc. Used by permission of Van Nostrand Rheinhold Company.)
b) The tetrahedral bond joins the center atom to four nearest-neighbor atoms along the lines indicated by the arrows.

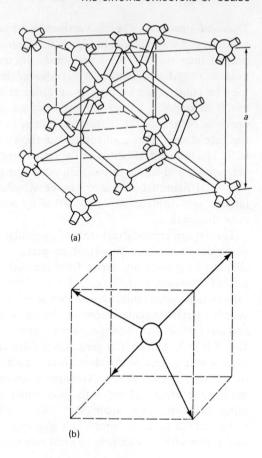

(a)

(b)

EXAMPLE 2.1

Silicon crystallizes in the diamond structure shown in Fig. 2.10a. The dimension of the unit cell of this basically cubic structure is 5.43 Å (1 Å $= 10^{-10}$ m). The atomic weight of silicon is 28.1. Find:

a) The nearest-neighbor distance between atoms (the bond length).

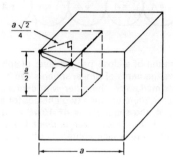

b) The atomic radius of a silicon atom in this structure.
c) The density of silicon, using the above data.

SOLUTION a) From Fig. 2.10a:

$$r^2 = \left(\frac{a}{4}\right)^2 + \left(\frac{a\sqrt{2}}{4}\right)^2, \qquad r = \frac{a\sqrt{3}}{4} = \frac{5.43(1.73)}{4} = 2.35 \text{ Å}.$$

This is the closest approach of any two atoms in the lattice to each other.
b) For the calculation of atomic radius, it is assumed that the atoms are stacked as rigid spheres. The atoms in closest proximity to each other are treated as spheres in contact. Hence the atomic radius $R = r/2 = 1.17$ Å.
c) Per unit cell, there are

6/2 atoms on the 6 cube faces (each atom is shared by 2 adjoining unit cells),
8/8 atoms on the 8 corners (each atom is shared by 8 adjoining unit cells),
4 atoms contained entirely within the body of the unit cell.

Hence the total number of atoms per unit cell is $6/2 + 8/8 + 4 = 8$.

$$\text{Mass/unit cell} = \frac{\text{no. atoms/unit cell}}{\text{no. atoms/gm-atom}} \times \text{gm/gm-atom}$$

$$= \frac{8}{6.02 \times 10^{23}} \times 28.1 = 3.74 \text{ gm/unit cell}.$$

Therefore,

$$\text{Density} = \frac{\text{mass/unit cell}}{\text{volume/unit cell}} = \frac{3.74 \text{ gm/unit cell}}{(5.43 \times 10^{-10})^3 \text{ m}^3/\text{unit cell}}$$

$$= 2.34 \text{ gm/m}^3 \qquad (\text{experimental value} = 2.33 \text{ gm/m}^3).$$

C. Miller Indices and Crystal Plane Identification

Parallel sets of planes passing through specific lattice points in a space lattice can be identified by a set of numbers called *Miller indices*. These numbers can be obtained by determining the intercepts of one of these planes on the three coordinate axes. For example, choose the origin of the coordinate system to coincide with a lattice site in one of these planes. Assume that the next parallel plane in this set has intercepts on the x-, y-, and z-axes equal to x_1, y_1, and z_1, respectively. Now take the *reciprocals* of these numbers and multiply each by a common integer so as to obtain three new numbers which are integers. Commonly, the three smallest such integers are referred to as the Miller indices h, k, and l. This is normally written in parenthesis as $(h\ k\ l)$.

An example of this procedure of plane identification is shown in Fig. 2.11. The problem is to determine the Miller indices of a set of parallel planes, one of

FIGURE 2.11

The plane shown is identified
by the Miller indices (1 2 3).
This plane intersects the
x-axis at 6 unit distances, the
y-axis at 3 units, and the
z-axis at 2 units. The axes are
orthogonal to each other and
hence the lattice indicated is
basically cubic. However, the
Miller scheme is more
general and may be applied
to oblique axes as well.

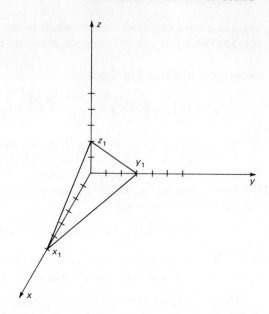

which passes through the origin of the coordinate system shown; the next
parallel plane intersects the x-, y-, and z-axes at 6, 3, and 2 unit distances,
respectively. The reciprocals of these numbers are $\frac{1}{6}$, $\frac{1}{3}$, and $\frac{1}{2}$; multiplying each
of these by 6 yields the Miller indices (1 2 3). If the x-intercept were negative
then the three reciprocal intercept numbers would be $-1/6$, $1/3$, and $1/2$ and the
Miller notation would be ($\bar{1}$ 2 3). Here the bar above the number 1 denotes a
negative intercept. It can be seen that the family of planes (1 2 3), (1 2 $\bar{3}$),
(1 $\bar{2}$ 3), ($\bar{1}$ 2 3), (1 $\bar{2}$ $\bar{3}$), ($\bar{1}$ 2 $\bar{3}$), ($\bar{1}$ $\bar{2}$ 3), and ($\bar{1}$ $\bar{2}$ $\bar{3}$) are all equivalent in a cubic
lattice in the sense that the number of atoms per square meter and the inter-
planar spacing are the same. Furthermore, ($\bar{1}$ 2 3) and (1 $\bar{2}$ $\bar{3}$) denote the
same set of parallel planes; likewise, ($\bar{1}$ $\bar{2}$ $\bar{3}$) and (1 2 3), ($\bar{1}$ $\bar{2}$ 3) and (1 2 $\bar{3}$), etc.
Similarly, for a cubic lattice, permuting the numbers 1, 2, and 3 in the Miller
notation also results in planes equivalent in the above sense. The equivalent
set of crystallographic planes is then written as {1 2 3}. Examples of several
sets of cubic crystal planes are given in Fig. 2.12.

A plane perpendicular to the x-axis intersecting that axis at 2 unit distances
from the origin, for example, intersects neither the y- nor z-axis. For purposes
of identification, this plane is assumed to intersect these axes at infinity. The
coordinate-axis intersections for this plane are at 2, ∞, and ∞ units; the reciprocals are then $\frac{1}{2}$, 0, and 0 and the Miller notation is (2 0 0).

D. Plane Spacing and Orientation

A crystallographic direction in a solid is denoted by the bracket notation
[h k l]. Here the numbers h, k, and l refer to the x-, y-, and z-components of a
vector defining a particular direction. Normally the three smallest numbers

whose ratio corresponds to the three intercepts are used. For example, the direction of the negative x-axis corresponds to the notation $[\bar{1}\ 0\ 0]$, while the positive y-axis is denoted by $[0\ 1\ 0]$. In cubic lattices it can be shown that the $[h\ k\ l]$ direction is normal to the $(h\ k\ l)$ plane. That is, the $[1\ 2\ 3]$ direction is perpendicular to the $(1\ 2\ 3)$ plane. In terms of Miller indices, a set of planes in a cubic lattice denoted by $(h\ k\ l)$ are spaced a distance d apart, where d is given by

$$d = a/(h^2 + k^2 + l^2)^{1/2}. \tag{2.2}$$

Here a is the unit distance (lattice constant) measured along the coordinate axis. An example of this for $(1\ 2\ 3)$ planes is shown in Fig. 2.13. Also the

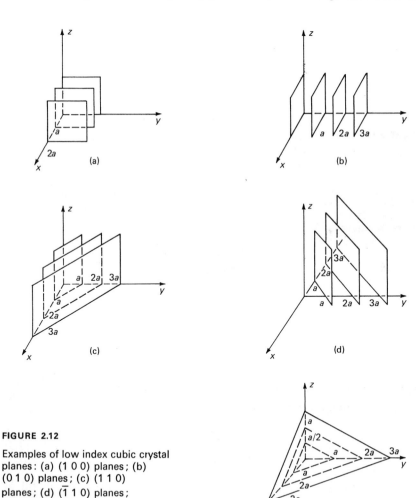

FIGURE 2.12

Examples of low index cubic crystal planes: (a) $(1\ 0\ 0)$ planes; (b) $(0\ 1\ 0)$ planes; (c) $(1\ 1\ 0)$ planes; (d) $(\bar{1}\ 1\ 0)$ planes; (e) $(1\ 1\ 2)$ planes. The lattice constant in each case is a.

FIGURE 2.13

The distance d between two adjacent (1 2 3) planes is given by Eq. (2.2) as $a/\sqrt{14}$. The two planes are indicated by the different shadings.

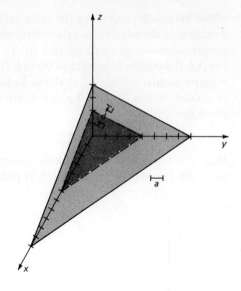

FIGURE 2.14

The crystal directions [1 1 2], [1 1 1], [1 1 0], and [0 $\bar{1}$ 0] are indicated by vectors. The angle θ is the angle between the [1 1 1] and [1 1 0] directions. Equation (2.3) gives $\cos\theta = \sqrt{6/3}$.

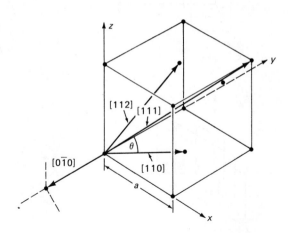

angle θ between two crystallographic directions denoted by $[h_1\ k_1\ l_1]$ and $[h_2\ k_2\ l_2]$ is given for a cubic lattice by

$$\cos\theta = \frac{h_1 h_2 + k_1 k_2 + l_1 l_2}{(h_1^2 + k_1^2 + l_1^2)^{1/2}(h_2^2 + k_2^2 + l_2^2)^{1/2}}. \tag{2.3}$$

This is illustrated in Fig. 2.14. The Miller notation is extremely useful in describing the crystalline structure of a solid as determined by analytic methods, which will be discussed next.

EXAMPLE 2.2

Prove that for a cubic lattice with lattice constant $a = 5.43$ Å the $[h\ k\ l]$ direction is normal to the $(h\ k\ l)$ plane for the $(0\ \bar{1}\ 0)$ and $(1\ 1\ 0)$ planes. Then determine the angle between the $[0\ \bar{1}\ 0]$ and $[1\ 1\ 0]$ directions and the distance between two adjacent $(1\ 1\ 0)$ planes.

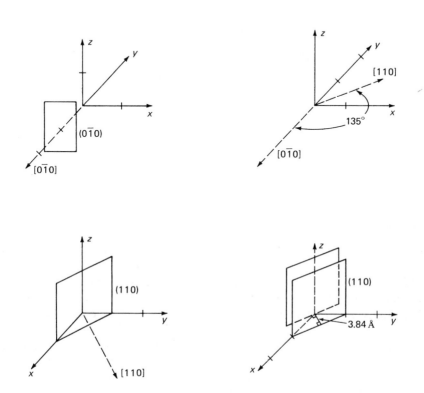

SOLUTION The direction of the negative y-axis may be represented by a unit vector whose y-component is -1 and whose x- and z-components are 0. This direction in Miller notation is indicated as $[0\ \bar{1}\ 0]$. In Miller notation a $(0\ \bar{1}\ 0)$ plane intercepts the negative y-axis at 1 lattice distance and the x- and z-axes at ∞, and hence is perpendicular to the y-axis. Therefore the $[0\ \bar{1}\ 0]$ direction is normal to the $(0\ \bar{1}\ 0)$ plane.

The $[1\ 1\ 0]$ direction is that of a vector whose x- and y-components are 1 and whose z-component is 0. It is a vector in the xy-plane at an angle of 45 degrees with the x- and y-axes. In Miller notation a $(1\ 1\ 0)$ plane intercepts the x- and y-axes at 1 lattice distance and intercepts the z-axis at ∞, and hence is a plane which intersects the xy-plane in a line which makes an angle of 45 degrees with both the x- and y-axes. Therefore the $[1\ 1\ 0]$ direction is normal to the $(1\ 1\ 0)$ plane.

The $[0\ \bar{1}\ 0]$ and $[1\ 1\ 0]$ directions are specified by

$$h_1 = 0, \qquad h_2 = 1,$$
$$k_1 = -1, \qquad k_2 = 1,$$
$$l_1 = 0, \qquad l_2 = 0.$$

The angle θ between these two directions is obtained from Eq. (2.3):

$$\cos \theta = \frac{(0)(1) + (-1)(1) + (0)(0)}{[0^2 + (-1)^2 + 0^2]^{1/2}(1^2 + 1^2 + 0^2)^{1/2}} = -\frac{1}{\sqrt{2}};$$

$$\theta = 135 \text{ degrees.}$$

This can be verified by inspection of the figure.

The distance d between two adjacent $(1\ 1\ 0)$ planes as given by Eq. (2.2) is

$$d = \frac{5.43 \text{ Å}}{(1^2 + 1^2 + 0^2)^{1/2}} = \frac{5.43 \text{ Å}}{\sqrt{2}} = 3.84 \text{ Å.}$$

This can be verified by inspection of the figure.

2.3 CRYSTAL STRUCTURE ANALYSIS

To understand the basis of some of the physical, electrical, and optical properties of crystals it is necessary to know the atomic arrangement in the solid. Since the atomic spacings are of the order of a few angstrom units (10^{-10} m), normal optical-microscope techniques do not provide sufficient resolution. This is the case since the resolution of a microscope is of the order of the wavelength of the light used for viewing. Since the optical microscope uses light of a few thousand angstroms, atoms cannot be seen directly with visible light. Instead they must be "viewed" with the aid of shorter-wavelength electromagnetic radiation. X-rays are a form of electromagnetic radiation with wavelength of the order of a few angstrom units and hence are useful for crystal structure determination.

A. Bragg Reflection

Von Laue discovered in 1912 that x-rays could be diffracted by single-crystal materials. He explained this phenomenon in terms of the interaction of these electromagnetic waves with each of the atoms in the crystal taken as a dipole, setting each of them into vibration, yielding a new set of wave fronts, etc. In the same year W. L. Bragg[7] offered a simple explanation of the angular distribution of a parallel beam of x-rays diffracted from a crystal. He postulated that the atomic planes in the crystal tend to reflect the x-rays specularly,

[7] W. L. Bragg, "The Structures of KCl, NaCl, KBr and KI," *Proc. Roy. Soc.* (*London*), Vol. A89, p. 248 (1913).

FIGURE 2.15

a) Bragg reflection from crystal planes spaced a distance d apart. The figure can be used to derive Bragg's law.

b) A number of sets of reflecting planes in a simple cubic lattice, labeled by their Miller indices. The number of such sets of planes participating in Bragg reflection is infinite. These planes are normal to this page.

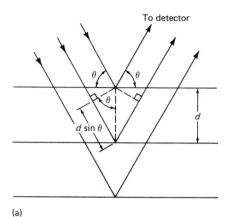

(a)

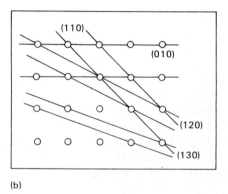

(b)

similar to a mirror, so that the angles of incidence and reflection are equal. Each plane only reflects a portion of the radiation like a semi-silvered mirror. This is shown in Fig. 2.15a. After deflection, the beams are collected in some detecting instrument sensitive to x-rays, such as a Geiger counter or photosensitive film. The rays arrive at the detector after traversing different path lengths. If these path lengths differ by an integral number of wavelengths, they will *interfere constructively* and a maximum signal will be detected by the counter. Otherwise, there will be some destructive interference among rays, resulting in a reduced signal. Mathematically, the condition for constructive interference is given by the Bragg law,

$$2d \sin \theta = n\lambda. \tag{2.4}$$

Here d is the spacing between planes, θ is the angle of incidence of the x-rays with the crystal planes, λ is the x-ray wavelength, and n is an integer representing the order of the interference. A number of sets of these reflecting planes in a simple cubic lattice are illustrated in Fig. 2.15b.

The experimental verification of the Bragg law is a demonstration of the periodicity of the crystal lattice. Note that the formula predicts that constructive interference cannot occur for $\lambda > 2d$. This indicates why visible light (6000 Å) cannot be used, since d is only of the order of a few angstroms. Although mainly x-rays are used for structure determination, other types of radiation are sometimes employed. For example, an electron or neutron beam will result in diffraction patterns similar to x-ray patterns. This results from the fact that these particles have associated wave properties given by $E = h\nu$ and $p = h/\lambda$ (see Chapter 3).

B. Three Methods of Structure Determination

There exist three different important methods for x-ray analysis of materials: (1) the Laue technique, (2) the rotating-crystal method, and (3) the powder method. The *Laue* technique involves holding the crystal under investigation fixed in a beam of x-rays containing a large range of wavelengths (white radiation). Here the crystal picks out the specific values of λ corresponding to planes of spacing d, for an x-ray incidence angle θ. For three-dimensional crystals the x-ray diffraction pattern consists of a series of bright interference spots on a sheet of film exposed to the diffracted beam. This technique is useful mainly for the determination of crystal orientation as sometimes required in electronic device fabrication. This is accomplished by studying the symmetry of the Laue pattern. Information about crystal structure perfection can be obtained from the spot sharpness. The apparatus for creating this pattern is shown in Fig. 2.16a.

The *rotating-crystal* method involves rotating the crystal about a fixed axis in a beam of monochromatic (single-wavelength) x-rays. The angle θ is varied until a particular plane is in position for constructive interference, and then a strong signal is detected upon reflection. At that position θ and λ are known and the specific atomic plane spacing can be determined from Bragg's law. A schematic diagram of the apparatus for this method is shown in Fig. 2.16b.

In the *powder* method a monochromatic x-ray beam is directed onto a specimen ground into fine particles. These little crystals will be randomly oriented with respect to the beam direction. Since λ is fixed, there will be constructive interference observed at definite angles θ to the incident beams, owing to reflection from crystal planes of spacing d which happen to be oriented to satisfy the Bragg law. This results in circular patterns produced on film, corresponding to the intersection of a cone of diffracted x-rays and the plane of the film. This technique is particularly useful for polycrystalline samples which contain crystal grains oriented randomly.

By determining its plane spacing and crystal symmetry, one can match the crystal structure of a sample with one of the Bravais lattices. When this is identified and the intensity of the Laue spots studied, the chemical composition of the sample often can be determined. This is true since the intensities of the

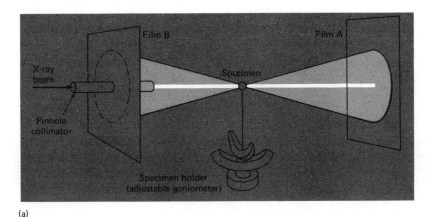

(a)

FIGURE 2.16

a) An apparatus for obtaining Laue patterns using a continuous spectrum of x-rays and a single-crystal sample. The adjustable holder is useful in analysis methods where the crystal must be oriented.
b) Apparatus for structure determination by the rotating-crystal method. Here a monochromatic beam of x-rays is used. (From C. S. Barrett, *Structure of Metals*, 3rd edition. Copyright 1966 by McGraw-Hill Book Company. Used with permission of McGraw-Hill Book Company.)

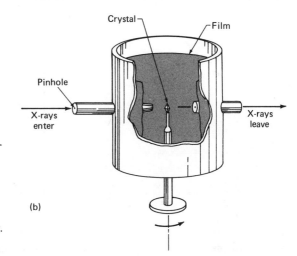

(b)

Laue spots depend on the distribution of electrons in their shells in the atom. Since this is characteristic of the different elements, they can be identified by x-rays.

2.4 BONDING FORCES IN SOLIDS

The term *solid* refers generally to rigid materials which maintain their shape unless stressed externally. The material can be deformed elastically when subjected to small hydrostatic, tensile, or shear forces. As early as 1784, R. J. Haüy[8] published a treatise postulating that crystal solids are formed by a regular repetition of building blocks. However, the serious study of the physical properties of solids started at the beginning of this century as a branch of

---

[8] R. J. Haüy, "Essai d'une théorie sur la structure des cristaux," Paris (1784).

FIGURE 2.17

Schematic representation of
(a) the force and (b) the
energy of interaction between
two atoms, plotted versus
their separation r. The
dashed lines denote the total
force and energy, obtained
by adding the attractive and
repulsive components. Note
the equilibrium separation of
the atoms, denoted by r_0,
representing the minimum-
energy condition.

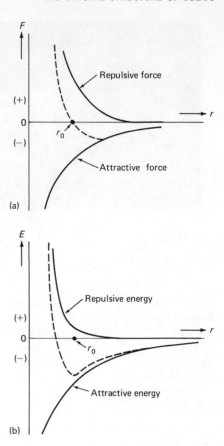

atomic physics. This study was given a tremendous boost about 1912 when
x-ray diffraction was discovered as a tool for determining atomic arrangements
in crystals. The internal forces holding crystals together were calculated in
detail beginning about 1918 and continuing into the 1920's. The great advance
toward understanding the mechanical and electrical properties of solids was
begun with the introduction of the Schrödinger equation of quantum mechanics
in 1926. Felix Bloch applied this equation to the crystalline solid in 1928.
Many quantum mechanical calculations for determining the mechanical and
electrical properties of solids are still being pursued today.

A. The Form of the Forces between Atoms in a Solid

Attractive forces act between the atoms in a solid to hold it rigidly together.
It must be true that there are repulsive forces acting, too, because of the manner
in which solids resist forces of compression. The attractive force between
atoms decreases as the distance r between them increases. The repulsive force is

generally of a shorter range than the attractive force; that is, these forces can be represented approximately as

$$F(r) = A/r^m, \tag{2.5}$$

where A is a constant and can be negative or positive, depending on whether the force is attractive or repulsive, and m is greater for the repulsive than the attractive force (m can be as high as 12 in the former case). The form of these types of forces is shown graphically in Fig. 2.17a. Here the net force, obtained by summing the two forces, is seen to be attractive as expected for atom spacings greater than the equilibrium spacing, for otherwise the crystal solid would fly apart. Since the bonding energy is given by $E = \int F\, dr$, the energy of the solid is obtained by integrating the force curve; the result is shown in Fig. 2.17b. Note that at the atom separation r_0 the crystal energy is at a minimum. This then represents the equilibrium spacing of the atoms. If the atoms are pushed closer together than r_0, the repulsive force between atoms rises rapidly. Any attempt to pull the crystal apart is resisted by an attractive force which falls off somewhat more slowly with distance.

B. The Nature of the Bonding Force

The force that holds a crystal together is primarily electrical. The repulsive force is primarily of a quantum mechanical nature and becomes effective at very close atomic spacing. Magnetic forces are small, and gravitational forces are completely negligible. The cohesive or attractive force which keeps the crystal atoms together results from electrostatic interaction between the negatively charged electrons and the positively charged atomic nuclei. The *cohesive energy* of the crystal is defined as the difference between the energy of the atoms when free and their total energy when bound in a crystal. This quantity must always be positive since the crystal energy must be less than that of the free atoms if the crystal is to be stable. This energy is usually expressed in kilocalories per mole.

The principal types of crystal bonding, listed in order of increasing bond strength, are

1) van der Waals,
2) metallic,
3) covalent, and
4) ionic.

These bonds differ in strength and directionality, as will be described next. Some typical values for crystal materials whose bonding can be placed primarily in one of these four broad categories are shown in Table 2.1. It should be pointed out that no crystal can be placed completely into one of these categories. Generally each of these types of forces contributes at least a small part of the crystal bonding.

TABLE 2.1

Approximate bonding
energies of solids (in
electron-volts per atom)
at 0°K (from C. Kittel,
*Introduction to Solid-
State Physics,* 4th edition,
John Wiley & Sons,
1971)

Bond				
Van der Waals	He 0.002	Ne 0.02	A 0.08	Kr 0.116
Metallic	Na 1.13	Cu 3.50	Fe 4.29	Al 3.34
Covalent	C (graphite or diamond) 7.36	Si 4.64	Ge 3.87	
Ionic	LiF 10.7	NaCl 8.05	KBr 6.92	Rbl 6.29

C. Van der Waals Forces

Crystals of the inert gases such as neon, argon, and krypton exist only at very
low temperatures and are held together by van der Waals forces. The unique
thing about these crystals is that they are formed of very stable atoms having
very high electron ionization energies. The electronic shells, as per the atomic
theory of Bohr, are completely filled, and the charge distribution is hence
quite symmetric. Since the electrons are pictured as circulating around the
nucleus, this may be viewed *at any instant* as an electric dipole[9] consisting of
the negative electrons and a positively charged nucleus ion. The weak interaction of this dipole with the induced electric dipole of the nearest atoms provides the attractive force between these atoms. The van der Waals bonding
energy is typically about 0.1 eV/atom.

D. The Metallic Bond

Metals like copper and sodium characteristically have an atom with one
loosely bound valence electron. This electron is essentially free to wander, like
the atoms in a gas, through the crystal consisting of metallic nuclei. It is the
electrical attraction between a swarm of such electrons and the ionic nuclei
that accounts for the cohesive energy of metals. The metallic bond is weaker
than the ionic or covalent bonds but stronger than the van der Waals type
of interaction. It is the relative weakness of the bond that accounts for the
mechanical malleability of many metals.

E. The Covalent Bond

The covalent bond is typified by the chemical bond, the result of the mutual
sharing of electrons by neighboring atoms. For example, the hydrogen mole-

---

[9] The strength of the electric dipole is defined as the product of the positive and negative
charges multiplied by the distance between them.

cule is held together by a covalent bond. Each hydrogen atom has only one electron but requires two electrons to fill its electronic shell and form a highly stable structure. This extra electron is obtained by sharing the electron of a neighboring hydrogen atom. The carbon bond in a crystal of diamond is a covalent bond. Here a central carbon atom with four outer-shell electrons forms a stable shell (eight electrons) by sharing an electron with each of four neighboring carbon atoms. This type of bond is highly directional, with the four neighboring atoms forming a symmetric tetrahedral structure. It is a strong bond, comparable to the ionic bond, even though it acts between electrically neutral atoms. Its importance in the field of electronics stems from the fact that this tetrahedral covalent bond accounts for many of the interesting properties of semiconductors such as silicon and germanium.[10] At moderate temperatures, thermal energy tends to break some of these bonds, causing the electrons thus freed to contribute to the electrical conductivity of the material.

F. The Ionic Bond

Ionic crystals are composed of alternate arrangements of positive and negative ions. The most common example of this type of material is sodium chloride. Here the binding or cohesive energy is mainly determined by Coulomb (electrostatic) interaction between the positive sodium ions and the negative chlorine ions. A simple order-of-magnitude calculation of this energy is given by $q^2/4\pi\epsilon_0 r_0$, where q is the ionic charge, ϵ_0 the permittivity of free space, and r_0 the ionic spacing. In the case of NaCl, the result is a value of several electron-volts, typical of these crystals. Since an atom of sodium has one loosely bound electron and a chlorine atom lacks one electron to complete its outer electron shell, an electronic transfer occurs when these atoms are brought together. The result is two stable interacting ions. The repulsive force is quantum mechanical and varies rapidly with ionic spacing.

At high temperatures these crystal ions become mobile, and such materials characteristically exhibit ionic conductivity. Ionic crystals absorb light strongly in the infrared portion of the light spectrum, owing to the significant interaction of these electromagnetic waves and the charged ions.

2.5 IMPERFECTIONS IN SOLIDS

This entire chapter thus far has been devoted to the description of crystal structures, with every atom in a precisely defined periodic position. This is, of course, a mathematical expedient. Real crystals have several types of imperfections or defects. In fact, the mechanical, electrical, and magnetic properties of solids often depend on these crystal defects in a significant way.

---

[10] The covalent bond was previously discussed in connection with the diamond type of crystal structure (see Fig. 2.10).

A. Crystal Lattice Vibrations

At any finite temperature the atoms in a crystal lattice have thermal energy imparted to them and may be pictured as vibrating in a random fashion about their normal lattice sites. This departure from strict periodicity of atom arrangement is a form of crystal imperfection that accounts in part for the electrical resistivity of a material, for these *lattice vibrations* tend to scatter electrons as they move through a crystal under the influence of an electric field. These oscillations also account for much of the heat capacity of a solid.

B. Point Defects

Although thermal lattice vibrations cause the atoms to deviate temporarily from their normal crystal sites, on the average they maintain their lattice positions. Atoms that are more permanently misplaced from their normal lattice sites at isolated locations in the crystal are referred to as *point defects*. Lattice sites where atoms are missing are called *vacancies*. Extra atoms which are located in the crystal between normally occupied lattice sites are called *interstitials*. Foreign atoms which occur in lattice sites normally occupied by the host crystal atoms are called *substitutional impurities*. Foreign atoms which are present in a host crystal between occupied lattice sites are known as *interstitial impurities*. These imperfections have dimensions of the order of the lattice spacing and yet can be detected in semiconductor lattices by careful measurements of the electrical properties of the crystal. Figure 2.18 illustrates these different types of point defects.

Atoms in a crystal can be knocked out of place in a radiation environment consisting of high-speed electrons, neutrons, or protons. When an atom is displaced from its normal lattice site to a nearby interstitial position, forming a *vacancy-interstitial pair*, this is known as a *Frenkel* defect. Should a displaced atom find its way to the surface, this is known as a *Schottky* defect. These defects account for the sensitivity of the electrical properties of semiconductor materials and devices to atomic radiation.

The energy necessary to create a vacancy is of the order of an electron-volt. Hence atoms can be removed from their normal lattice positions by thermal energy if the crystal is raised to an elevated temperature. The number of vacancies N_v in a crystal containing N atomic sites in thermal equilibrium at an absolute temperature T is given essentially by

$$N_v = \gamma N e^{-E_v/kT}, \qquad (2.6)$$

FIGURE 2.18

Three common point defects:
(a) vacancy; (b) interstitial;
(c) substitutional impurity.

(a)   (b)   (c)

FIGURE 2.19

Schematic representation of
an edge dislocation. An
extra half-plane of atoms is
wedged between two adja-
cent planes of atoms. The
dislocation line is indicated
and is normal to the plane of
the page. (From C. A. Wert
and R. M. Thomson, *Physics
of Solids*, 2nd edition. Copyright 1970 by McGraw-Hill
Book Company. Used with
permission of McGraw-Hill
Book Company.)

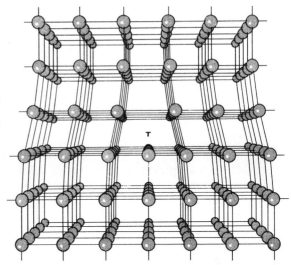

where E_v is the energy required to produce a vacancy, k is the Boltzmann gas
constant, and γ is a structure constant. Hence a crystal at any temperature will
contain a certain number of defects in thermal equilibrium with the crystal
lattice.

C. Line Defects: Dislocations

A *line defect* is a displacement of whole rows of atoms from their regular
lattice positions. This more complicated type of gross crystal imperfection is
known as a *dislocation*. An *edge dislocation* is the insertion of an extra half-
plane of atoms into a regular crystal. This is pictured in Fig. 2.19; it causes the
crystal to distort. Most of the stress is concentrated at the furthest edge of
the extra half-plane penetrating the crystal. This line of maximum stress is the
dislocation line.

A *screw dislocation* is formed by the motion of one part of a crystal with
respect to another. To illustrate its formation, consider making a fine cut
partway into a regular crystal. Now displace the crystal on one side of the cut
relative to the other side by one atomic distance. This is illustrated in Fig. 2.20.
Maximum stress is present along the edge of the cut, and this constitutes the
dislocation line. A plane of atoms in the crystal perpendicular to the screw
dislocation line tends to form a surface which spirals around the dislocation line.
Hence the designation "screw dislocation."

Line defects in solids tend, in general, to be made up of combinations of
edge and screw dislocations. Note that in the forming of the defect, the crystal
displacement is parallel to the screw dislocation line, whereas the crystal
displacement is perpendicular to the edge dislocation line. Since the energy
stored in a dislocation is generally several electron-volts per atom, compared to

FIGURE 2.20

Structure of a crystal containing a screw dislocation. The height of the step on the top surface is usually one lattice spacing. The atom rows perpendicular to the dislocation are on a spiral ramp.

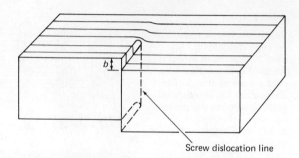

Screw dislocation line

about 1 eV in point defects, the dislocated solid is in a highly stressed state. Whereas point defects are in thermal equilibrium with the crystal lattice, dislocations are not. Hence dislocations cannot be removed from a solid by thermal *annealing*, although this possibility exists for point defects.

D. Effect of Crystal Defects on the Electrical Properties of Crystals with Diamond Structure

A vacancy in a diamond-type crystal yields four unsatisfied bonds (see Fig. 2.10b). Hence there is a tendency for the immobilization of electrons on these sites, which normally contribute to the electrical conductivity of the crystal. This is the case since the resistivity of n-type silicon under atomic or nuclear radiation is observed to increase. Radiation of this type is known to produce point defects in crystals.

Line defects are observed to provide a focus for the precipitation of impurity atoms on the half-plane edge comprising the dislocation. Dislocation lines piercing p-n junctions tend to short-circuit these junctions since metallic impurity atoms often collect on such structural defects, providing a highly conducting path. These impurities can also scatter electrons in a uniform n-type crystal, impeding their motion and reducing their mobility.

Both point and line defects have been found to cause highly localized distortion of the crystal lattice leading to the formation of "trapping" sites where the recombination of positive (holes) and negative (electrons) carriers is enhanced. This may cause the electrons from the n-p-n transistor emitter to recombine with holes in the p-type base region before they can be collected at the n-type collector region, reducing the transistor current gain. The electron "lifetime" may be significantly reduced when as few as one out of 10^{11} germanium atoms are removed from their normal lattice sites by radiation.

PROBLEMS

2.1 Distinguish between the terms:

a) unit cell *and* primitive cell,
b) space lattice *and* crystal structure,
c) ionic bonding *and* covalent bonding.

2.2 a) Perform a simple order-of-magnitude calculation of the binding energy in eV of NaCl, assuming that this cohesive energy is mainly electrostatic and determined by the Coulomb interaction between the positive sodium ions and negative chlorine ions. The separation of the ions is 2.82 Å.

b) If $m = 12$ in Eq. (2.5), by how much must the separation between the atoms be reduced in order for the repulsive force to increase 10 times?

2.3 Aluminum has a density of 2700 kg/m³ and an atomic weight of 26.98.

a) How many gram-atoms are contained in 1.0 m³ of the solid? The dimension of the unit cube (lattice constant) for this fcc metal as determined by x-ray techniques is 4.04 Å.
b) Calculate the atomic radius of aluminum.
c) How many atoms are contained in 1.0 m³ of an aluminum crystal?
d) Determine Avogadro's number from these data.

2.4 Repeat Problem 2.3 for the metal sodium whose crystal structure is bcc, unit cell dimension is 4.28 Å, density is 970 kg/m³, and atomic weight is 22.99.

2.5 Assuming that the atoms are treated as rigid spheres, show that the ratio of the volume occupied by the atoms to the volume available in various crystal structures is

a) $\pi/6$ (52 percent) for the simple cubic lattice,
b) $\pi\sqrt{2}/6$ (74 percent) for the hexagonal close-packed structure,
c) $\pi\sqrt{3}/16$ (34 percent) for the diamond cubic structure.

2.6 Show that the ratio of c to a (see Fig. 2.8) for the ideal hexagonal close-packed structure is 1.633.

2.7 Calculate the number of atoms per square meter for the following planes in a crystal of aluminum (see Problem 2.3):

a) (1 0 0),   b) (1 1 0),   c) (1 1 1),
d) ($\bar{1}\,\bar{1}\,\bar{1}$),   e) (1 2 0).

2.8 Repeat Problem 2.7 for sodium (see Problem 2.4).

2.9 Compute the distance between the following set of planes in aluminum where the lattice constant a is 4.04 Å:

a) (1 0 0),   b) (1 1 1),   c) (1 2 0).

2.10 Determine the angle between the two crystallographic directions

a) [1 0 0] and [1 1 0],   b) [1 1 0] and [1 1 1],
c) [1 1 1] and [1 2 0].

2.11 In connection with structure determination using x-rays or elementary particles for radiation, determine:

a) the wavelength associated with an electron of kinetic energy 10 kiloelectron-volts (keV),
b) the wavelength associated with a neutron of 300°K temperature (kinetic energy $= \frac{1}{2}kT$).
c) the minimum wavelength, in variable wavelength (white) x-ray radiation, if the applied voltage to the x-ray tube anode is 30 kilovolts (kV).

2.12   What is the maximum wavelength of radiation that can be used to achieve Bragg reflection from (1 1 1) crystal planes in aluminum? (See Problem 2.9.)

2.13   Monochromatic x-rays whose wavelength is 1.537 Å are used to irradiate an aluminum crystal. These x-rays are diffracted off the (1 1 1) planes of the crystal, and the first-order interference is observed at a Bragg angle of 19.2 degrees, producing constructive interference. From these experimental data calculate a value of Avogadro's number (see Problem 2.3 for the density and atomic weight of aluminum).

2.14   Distinguish between:

a) vacancies *and* interstitials,
b) Schottky defect *and* Frenkel defect,
c) point defect *and* dislocation.

2.15   A vacancy requires an expenditure of energy of about 0.75 eV for formation in aluminum. Determine the ratio of the number of vacancies that exist in thermal equilibrium at 600°C to the number present at 400°C.

3 Introduction to the Quantum Mechanics of Solids

3.1 THE NEW PHYSICS

To effectively explain the operation of solid-state devices it is necessary to develop the physics of the solid state which, as presently understood, is based on quantum physics. First, though, the historical background of this new physics will be discussed.

By the end of the nineteenth century, fabulous success had been achieved in explaining macroscopic phenomena using the classical Newtonian laws of mechanics, including problems ranging from the motions of the stars and planets to those involving interacting atoms in a gas. The elegance and simplicity of Newton's laws led some to believe that all physical phenomena could ultimately be explained within this framework. However, microscopic phenomena on a subatomic level were often incapable of being described with the classical laws of physics.

At the beginning of the twentieth century, Max Planck found these classical laws to be inadequate in accurately describing the dependence of blackbody radiation on wavelength. Also, soon after the turn of the century, Niels Bohr postulated a theory of the structure of the hydrogen atom which abandoned some of the fundamental ideas of classical physics to explain available spectroscopic data on hydrogen gas. In addition, at about that time it was realized that electric current was conducted in some metal and semiconductor solids by charge carriers similar to electrons but apparently positively charged. This surprising postulate of the existence of both positively and negatively charged current carriers in semiconducting materials is basic today to our understanding of modern junction transistor behavior.

It soon became clear that a general physical theory was needed which could adequately account for the mounting number of phenomena that were unexplainable in classical terms. A giant step in this direction was provided in 1926 by Erwin Schrödinger[1] when he published his mathematical theory for predicting the spectral lines of the hydrogen atom based on postulates which were previously stated by Planck and de Broglie. This new physics took the form of a second-order partial differential equation known today as Schrödinger's equation. This formulation still serves as the basis of our understanding of many microscopic phenomena, including electric conduction processes in metals and semiconductors. For example, the fact that the electric conductivity of a class of materials known as semiconductors increases sharply as the temperature increases above a certain "intrinsic" temperature is not explained by the classical theory of electric conduction in solids, but is explained with Schrödinger's equation. The proposing and/or understanding of many future solid-state devices will be facilitated if the engineer has some knowledge of this modern branch of physics, called *wave mechanics* or *quantum physics*.

This chapter begins with an explanation of the basic postulates of the theory, leading to a statement of the Schrödinger equation. This formulation bears the same relationship to wave mechanics as Newton's laws do to classical physics. Several problems whose solutions can be obtained by solving the Schrödinger equation in closed form are presented to demonstrate methods of obtaining these solutions and of investigating the properties of the solutions. Finally, it is shown that the one-dimensional wave mechanical treatment of an electron in a periodic potential characteristic of a crystal lattice results in the prediction of allowed energy bands for the electron. This concept is essential for understanding most modern solid-state materials and devices such as transistors and integrated circuits, tunnel diodes, and semiconductor lasers.

A. Energy Quanta

At the end of the nineteenth century, there were two well-known formulations of the problem of the spectral distribution of radiation from a blackbody[2] — one the Wien law and the other the Rayleigh-Jeans law. The former accurately described the variation with wavelength of the radiated energy of a blackbody for wavelengths shorter than the wavelength of maximum radiation, while the latter worked for longer wavelengths.[3] In 1901 Max Planck[4] found a formula which was correct for all wavelengths and which reduced to the Wien and Rayleigh-Jeans expressions in the limits of short wavelength and long wavelength, respectively. Planck's empirical formulation fitted the precisely determined

---

[1] E. Schrödinger, "Quantization as an Eigenvalue Problem," *Ann. Phys.*, Vol. 79, p. 489 (1926).
[2] A blackbody may be defined as an ideally perfect emitter and absorber of radiation.
[3] For a description of the dilemma presented by the purely classical explanation of blackbody radiation, see F. Richtmyer, E. Kennard, and T. Lauritsen, *Introduction to Modern Physics*, 5th edition, McGraw-Hill Book Company, New York (1955).
[4] M. Planck, "Distribution of Energy in the Spectrum," *Ann. Phys.*, Vol. 4, p. 553 (1901).

spectrographic data then available with amazing accuracy. The Wien and Rayleigh-Jeans formulas were both based on a straightforward application of classical mechanics and thermodynamics. Planck was able to arrive at his formula only by postulating that the blackbody consisted of oscillators which radiated energies that were integral multiples of a certain amount of energy. This basic amount or *quantum* of energy was given by

$$E = hv, \tag{3.1}$$

where v is the frequency of the radiated energy, and h is a constant now associated with Planck's name. These steps in energy values were in sharp contrast with classical radiation theory in which all energy values are permitted.

The idea of discrete quanta of energy was successfully applied by Einstein in 1905[5] to explain the photoelectric effect. He related the energy threshold for electron emission from metals to the minimum frequency of the incident light according to the Planck formula. Light of a frequency below this minimum would then be totally ineffective in producing photoelectrons, regardless of its intensity. The experimental evidence supporting this quantum theory of light radiation could not be explained by the classical radiation theory derived from Maxwell.

About 1913 Niels Bohr[6] was able to predict accurately the major spectral lines emitted by excited hydrogen atoms by applying a pseudo-classical theory with the Planck quantum idea as a basic postulate. Although his physics was a strange mixture of the old and new physics, the prediction of the major sharp spectral lines was indisputably in excellent agreement with accurate spectroscopic observations made at the time. Hence, in spite of some inherent contradictions in the theory, it appeared to have some elements of truth.

B. Wave-Particle Duality

Working in the totally unrelated area of relativity considerations, Louis de Broglie in 1924 proposed[7] that particles should be assigned a wavelength

$$\lambda = \frac{h}{p}, \tag{3.2}$$

where p is the particle momentum, and h is the same Planck constant. De Broglie's reasoning was that in classical physics electromagnetic radiation was always considered a wave phenomenon; yet the work of Planck and Einstein dictated the consideration of light radiation in terms of discrete particles of energy or quanta. He suspected then that the inverse might also be

[5] A. Einstein, "Generation and Transformation of Light," *Ann. Phys.*, Vol. 17, p. 132 (1905).
[6] N. Bohr, "On the Constitution of Atoms and Molecules," *Phil. Mag.*, Vol. 26, p. 1 (1913).
[7] L. de Broglie, "A Tentative Theory of Light Quanta," *Phil. Mag.*, Vol. 47, p. 446 (1924).

Diffraction patterns due to reflection of electrons from a crystalline nickel surface.

The sequence of electron diffraction patterns is produced by the absorption of oxygen on the crystalline nickel surface. Each pattern can be explained by the surface atom arrangement shown below it, in which the open circles represent nickel, the closed circles oxygen. The upper left pattern represents a clean nickel surface. In the upper right pattern, the oxygen atom concentration has increased, the streaks implying random oxygen positions. The lower left pattern represents one-half oxygen coverage; the surface is ordered and the streaks have coalesced into sharp spots halfway between the original spots. At two-thirds oxygen coverage (lower right), a triple periodic pattern and atomic arrangement occur. (Courtesy of Bell Laboratories.)

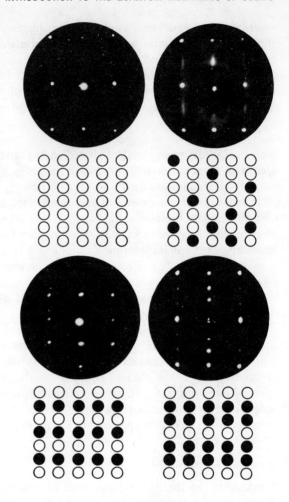

true — that the classical particle might also be represented by a characteristic wavelength, according to his formula. The fact that particles such as electrons do, indeed, exhibit wave properties and undergo diffraction by a grating was demonstrated by Davisson and Germer in 1927.[8] The diffraction grating used was not the ruled grating that was commonly used for light, but a periodic arrangement of atoms in a nickel crystal provided by nature. This choice was necessary because of the small wavelength of the electron predicted by de Broglie and the requirement that the periodicity be of the order of this

[8] C. Davisson and L. Germer, "Diffraction of Electrons by a Crystal of Nickel," *Phys. Rev.*, Vol. 30, p. 705 (1927).

wavelength to obtain a strong diffraction pattern. This observation finally resolved the nineteenth-century debate between Isaac Newton, who favored a corpuscular theory of light, and Christian Huygens, who emphasized its wave nature. There exists, in fact, a *wave-particle duality*. Light, or an electron, can be considered to have some of the properties of both particles and waves. Hence, an electron ejected from the hot filament of an electron gun has a definite charge and mass, is deflected according to the usual laws of ballistics, and yet shows some diffraction properties when interacting with a periodic crystal structure. Matter and light can be considered as particles. However, the behavior of these corpuscles when interacting with atoms may not be described by Newton's mechanics but by wave or quantum mechanics. Also, it is impossible to observe *simultaneously* the wave and particle properties of matter or light.

Actually, even more drastic changes in our mode of thinking about sub-atomic processes are necessary if we are to understand this strange new microscopic world. We shall now present arguments to show that the physical laws of quantum mechanics which govern electrons and *photons* (light quanta) interacting with atoms, are not deterministic as are the classical Newtonian laws, but are *probabilistic* and *statistical*.

3.2 HEISENBERG'S UNCERTAINTY PRINCIPLE

In classical mechanics the one-dimensional motion of a particle is described by a position function of time $x(t)$ and a momentum function of time $p(t)$. According to Newton, we can fully determine the motion of a particle such as an electron given its initial momentum and position and the mechanical laws which describe its motion. This is sometimes referred to as a deterministic or *causal* point of view. Quantum mechanics, on the other hand, takes a probabilistic or statistical point of view. That is, it is not possible to predict a definite trajectory and motion of a particle but only the *probability* that it will behave in a specific way.

This may be expressed in terms of the *uncertainty* or *indeterminacy* principle of Werner Heisenberg (1927), which states[9] that it is not possible to measure simultaneously the position and momentum of a particle with arbitrary accuracy for the purpose of predicting future behavior. If the position is precisely identified, then the momentum will be somewhat uncertain, and vice versa. Mathematically, this is expressed as: given that the position of a particle can be measured within Δx, then its momentum uncertainty in the x-direction, Δp_x, cannot be ascertained to any greater accuracy than

$$\Delta p_x = \frac{h}{\Delta x}, \qquad (3.3)$$

[9] W. Heisenberg, "The Actual Content of Quantum Theoretical Kinematics and Mechanics," *Zeit. Phys.*, Vol. 43, p. 172 (1927).

where h is the famous Planck constant. Similar expressions apply to the y- and z-directions. This indeterminacy demonstrates the noncausal nature of quantum mechanics.

The indeterminacy concept can be demonstrated by noting the disturbance (and uncertainty) introduced into the motion of a particle by the instrument used to study its trajectory. Consider shining light on a particle in an attempt to locate it accurately. The momentum of the light photons disturbs the particle, changing its position and momentum. Nor can we reduce this influence by lowering the light intensity in a continuous manner to zero, since the smallest quantity of light is still one photon with discrete energy $h\nu$. Hence the quantization hypothesis and the uncertainty principle follow from each other. Obviously, this idea has serious consequences when we deal with small particles, such as electrons, with small energy.

A. A Demonstration of the Uncertainty Principle

Bohr proposed a "gedanken" (or thought) experiment to demonstrate the Heisenberg principle exactly. (We shall see later that this concept derives

FIGURE 3.1

Apparatus for the gedanken experiment to determine the position of an electron using a microscope with illumination by light of wavelength λ.

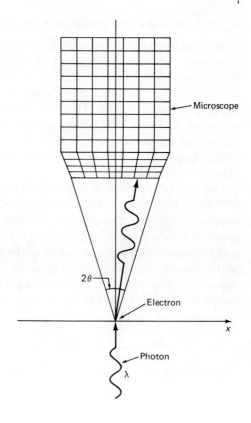

automatically from the Schrödinger formulation of problems involving the wave mechanical treatment of the electron.) Imagine that we have a powerful microscope that is used to determine the x-position of an electron (see Fig. 3.1). Suppose the objective lens of the microscope subtends an angle of 2θ from the point where the electron is located. The classical theory of optics dictates that the minimum separation Δx between two adjacent positions of the electron that can be resolved by observing with light of wavelength λ is[10]

$$\Delta x \sim \frac{\lambda}{\sin \theta}. \tag{3.4}$$

That is, the uncertainty in locating the electron position in this experiment is $\lambda/\sin \theta$. Since the light used in viewing the electron may be considered to be composed of photons, these particles interact with the electron, cause it to recoil (Compton effect), and are correspondingly deflected into the microscope objective lens and we "see" the electron.

By conservation of momentum, the electron recoil momentum in the x-direction must just equal in magnitude the maximum interacting photon x-component of momentum. If the photons which are seen enter the objective anywhere within an angle 2θ, the range of the x-component of their momenta (de Broglie) is between $(+h/\lambda) \sin \theta$ and $(-h/\lambda) \sin \theta$. Hence the electron recoil momentum total range is given by

$$\Delta p_x \sim \frac{2h}{\lambda} \sin \theta. \tag{3.5}$$

By combining Eqs. (3.4) and (3.5), we get

$$\Delta p_x \, \Delta x \sim 2h > h, \tag{3.6}$$

and the uncertainty principle is demonstrated.

The physical quantities momentum and position comprise one set of variables governed by this principle. It turns out that the uncertainty principle also refers to a number of other sets of variables. One other such set is energy and time, and it can be shown also that

$$\Delta E \, \Delta t \geq h. \tag{3.7a}$$

Here ΔE can be the uncertainty in the energy of a photon, and Δt the accuracy limits of time estimation of its emission. Combined with the Planck formula this becomes

$$\Delta v \, \Delta t \geq 1. \tag{3.7b}$$

This relation is useful in estimating the sharpness of a laser spectral line, where Δv is the line width, and Δt is the electron transition time.

---

[10] See, for example, *Fundamentals of Optics* by F. A. Jenkins and H. E. White, 2nd edition, McGraw-Hill Book Company, New York (1950).

EXAMPLE 3.1

The average duration of an excited state of an atom in a gas laser is about 10^{-8} sec. This laser emits a spectral line whose central wavelength is 6328 Å.

a) Compute the minimum frequency variation of the emitted red light, according to the uncertainty principle.

b) Express the possible sharpness of the spectral line by the ratio of the frequency variation to the fundamental radiated frequency in parts per million (ppm).

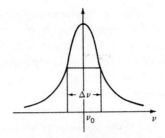

SOLUTION a) From Eq. (3.7b) the minimum frequency variation is given by

$$\Delta v \geq \frac{1}{\Delta t} = \frac{1}{10^{-8} \text{ sec}} = 10^8 \text{ Hz}.$$

The longer the time that the atom spends in an excited state, the narrower the spectral line.

b) The central frequency is given by

$$v_0 = \frac{c}{\lambda_0} = \frac{3.0 \times 10^8 \text{ m/sec}}{6328 \times 10^{-10} \text{ m}}$$

$$= 4.7 \times 10^{14} \text{ Hz}.$$

Hence the sharpness is expressed by

$$\frac{10^8}{4.7 \times 10^{14}} = 0.21 \times 10^{-6} \quad \text{or} \quad 0.21 \text{ ppm}.$$

3.3 THE SCHRÖDINGER EQUATION

What has been discussed thus far is a series of ideas and concepts which are peculiar to the modern physics of quantum mechanics. What is needed now is a mathematical formulation that will permit us to predict and explain the many subatomic processes which govern the behavior of the electronic materials of interest to us. As previously indicated, the Schrödinger wave equation expresses the essence of this modern physics. Obtaining solutions to this equation for a variety of physically interesting cases will permit us to analyze problems which may be subject to experimental verification. (The ability to stand up to experimental verification is, of course, the final test for a physical theory.)

The Schrödinger equation refers to the motion of an electron in various force fields. It replaces for microscopic phenomena the Newtonian expression $F = ma$ for the motion of particles on a larger scale and permits a description of effects not previously explainable in classical terms. Since we have already indicated that the electron can be pictured as having a wavelength, it should not be surprising that the Schrödinger equation takes the form of a wave equation.

The *time-dependent* Schrödinger wave equation for an electron having a potential energy V which[11] in general can be a function of position and time is

$$\frac{h^2}{8\pi^2 m}\left(\frac{\partial^2 \Psi}{\partial x^2} + \frac{\partial^2 \Psi}{\partial y^2} + \frac{\partial^2 \Psi}{\partial z^2}\right) - V\Psi = \frac{h}{2\pi j}\frac{\partial \Psi}{\partial t}. \qquad (3.8)$$

(time-dependent Schrödinger equation)

Here h is the Planck constant, m the electron mass, $j = \sqrt{-1}$, and Ψ is the wave function which depends in general on position and time; that is,

$$V = V(x, y, z, t)$$

and

$$\Psi = \Psi(x, y, z, t), \qquad (3.9)$$

where Ψ is in general a complex quantity. This is a complicated second-order differential equation which can be solved in closed form in only a limited number of special cases. However, the equation may be greatly simplified in a large number of problems[12] by dealing separately with position and time. This is possible when the potential energy depends only on position, and not on time. Let us introduce this condition into the one-dimensional Schrödinger equation, which we consider for additional simplicity, although the three-dimensional formulation may be handled similarly. The one-dimensional equation is

$$\frac{h^2}{8\pi^2 m}\frac{\partial^2 \Psi(x, t)}{\partial x^2} - V(x)\Psi(x, t) = \frac{h}{2\pi j}\frac{\partial \Psi(x, t)}{\partial t}. \qquad (3.10)$$

A. Separation of Variables

Let us postulate a solution of the form

$$\Psi(x, t) = \psi(x)\phi(t), \qquad (3.11)$$

---

[11] Note that V here refers to potential energy, and in other chapters V is used for electric potential. The electric potential (in volts) is potential energy (in joules) per unit charge (in coulombs).
[12] We will deal only with such problems.

assuming that the space and time parts of the wave function Ψ can be separated. Substituting Eq. (3.11) into (3.10) gives

$$\frac{h^2}{8\pi^2 m}\left(\frac{d^2\psi(x)}{dx^2}\right)\phi(t) - V(x)\psi(x)\phi(t) = \frac{h}{2\pi j}\frac{d\phi(t)}{dt}\psi(x). \qquad (3.12)$$

Dividing both sides of the equation by $\psi(x)\phi(t)$, we get

$$\frac{h^2}{8\pi^2 m}\frac{1}{\psi(x)}\frac{d^2\psi(x)}{dx^2} - V(x) = \frac{h}{2\pi j}\frac{1}{\phi(t)}\frac{d\phi(t)}{dt}. \qquad (3.13)$$

Notice that the left-hand side of the equation is a function of x only, while the right-hand side is a function of t only. A useful theorem of mathematical physics states that in such a case, where x and t are independent variables, each side of this equation must be separately equal to a constant. This is so because, if we arbitrarily choose a value of x, the left-hand side of Eq. (3.13) takes on a value which is not necessarily equal to the right-hand side, since t can be chosen arbitrarily as any value. Hence for the equality to hold, both sides must be equal to a constant, which we will take[13] as $-E$.

This then reduces Eq. (3.13) to two ordinary differential equations, one in position and one in time, whose solutions are more easily obtainable. That is,

$$\frac{h^2}{8\pi^2 m}\frac{d^2\psi}{dx^2} + (E - V)\psi = 0 \qquad (3.14a)$$

(time-independent Schrödinger equation)

and

$$\frac{d\phi}{dt} = -\frac{2\pi j}{h}E\phi. \qquad (3.14b)$$

After some algebraic manipulation the second equation integrates to

$$\ln\phi = -2\pi j\frac{Et}{h} + A. \qquad (3.15)$$

We can take the integration constant $A = 0$ without loss of generality and introduce a constant later in reconstructing the product $\Psi = \psi\phi$. Then Eq. (3.15) becomes

$$\phi = e^{-(2\pi jE/h)t}, \qquad (3.16)$$

which is oscillatory in time; $E\ (=h\nu)$ is the total electron energy.

The problem of finding the solution for the motion of an electron having a potential energy V due to some force field now reduces to solving the second-order linear ordinary differential equation (3.14a), and the time-dependent solution can always be written by tacking on the results of Eq. (3.16).

---

[13] The reason for this peculiar choice of arbitrary constant will become clear later. E then will turn out to be the total electron energy.

B. Boundary Conditions on Ψ

We must now specify some boundary conditions on Ψ so that we can determine the two arbitrary constants necessary for the solution of this second-order differential equation. Although we have not specified the physical significance of Ψ, ultimately some connection must be found between this quantity and the real world, and hence Ψ must have intuitively acceptable behavior.[14] The usual conditions imposed are

$$\Psi(x, t) \quad \text{and} \quad \frac{\partial \Psi}{\partial x} \quad \begin{array}{l} \text{must be continuous, finite,} \\ \text{and single-valued for all } x \end{array} \qquad (3.17a)$$

and

$$\int_{-\infty}^{+\infty} |\Psi|^2 \, dx \quad \text{must be finite.} \qquad (3.17b)$$

Since these conditions are defined for $\Psi(x, t)$, they will apply to $\psi(x)$ as well. The physical meaning of these mathematical requirements will become clear when we discuss a physical interpretation of Ψ.

C. Physical Interpretation of Ψ

Max Born[15] in 1926 proposed an important physical interpretation of the wave function. Since the Schrödinger equation refers to matter waves à la de Broglie, Ψ must represent the amplitude of these matter waves. Born reasoned, by analogy with the theory of electromagnetic waves, that the absolute value of Ψ squared gives the probability of finding the electron between x and $x + \Delta x$ at a given time.[16] This interpretation provides us with a means of comparing the predictions of the Schrödinger equation with experiment. Contained in the wave function Ψ is a great deal of information about the motion of the electron. For example, motion parameters such as momentum and energy can be calculated once this function is known.

Another test of the validity of this new physics was proposed in 1923 by Bohr in pronouncing his *correspondence principle*. This requires that the laws of quantum mechanics, in the classical limit where many quanta are involved, lead to the equations of classical physics, on the average. For example, in macroscopic problems the Schrödinger equation gives the same result as Newton's laws. This is demonstrated by a problem at the end of this chapter which shows that the uncertainty in macroscopic particle motion is negligible.

---

[14] Indeed Schrödinger himself did not realize its full significance in his first paper.
[15] M. Born, "Quantum Mechanics of Collision," *Zeit. Phys.* Vol. 37, p. 863 (1926).
[16] Since mathematically Ψ in general is a complex quantity and it must have physical significance, $|\Psi|^2$, which is always real and positive, is most useful. According to the definition, $\int |\Psi| \, d\Omega = 1$, when taken over all space Ω.

3.4 THE FREE ELECTRON PROBLEM

Many of the quantum and wave mechanical ideas that have already been discussed are contained intrinsically in the Schrödinger equation. The consideration of a mathematically simple case such as the motion of an electron in field-free space will be used to illustrate this point.

The spatial (time-independent) part of the Schrödinger equation for an electron moving in one dimension with no forces acting on it [$V(x) = 0$] reduces to

$$\frac{d^2\psi}{dx^2} + \left(\frac{8\pi^2 m}{h^2} E\right)\psi = 0. \tag{3.18}$$

This may be recognized as the form of the differential equation describing the motion of a harmonic oscillator or the current oscillations in a lossless circuit containing inductance and capacitance. The solution can be written in terms of sine and cosine functions. However, since differentiating and integrating and multiplying and dividing solutions of this form are often necessary, the more convenient exponential form is chosen. That is,

$$\psi(x) = A \exp\left(\frac{jx\sqrt{8\pi^2 mE}}{h}\right) \tag{3.19}$$

is a particular solution of Eq. (3.18) and can be demonstrated to be so by substituting it back into that equation. Here A is a constant to be determined. A time-dependent particular solution then is

$$\Psi(x, t) = A \exp\left[\frac{-2\pi j}{h}(Et - x\sqrt{2mE})\right], \tag{3.20}$$

derived by tacking on the time-dependent part of Ψ. This is the equation of a *traveling wave*, and a general solution may be constructed by combining many solutions of this form.[17] To put Eq. (3.20) into a more useful form, let us make use of the Planck and de Broglie laws and take the electron energy as $E = p^2/2m$, if in fact E represents the total electron energy. Then

$$p = \frac{h}{\lambda} = \frac{h}{2\pi}\frac{2\pi}{\lambda} = \frac{h}{2\pi}k = \hbar k \tag{3.21a}$$

and

$$E = hv = \frac{h\omega}{2\pi} = \hbar\omega, \tag{3.21b}$$

where $\hbar = h/2\pi$ and k (wave vector) $= 2\pi/\lambda$. Then Eq. (3.20) becomes

$$\Psi(x, t) = A \exp\left[j(kx - \omega t)\right]. \tag{3.22}$$

The manner in which A is chosen to normalize the solution is explained later.

---

[17] The general solution of a linear partial differential equation such as Eq. (3.18) is constructed by summing all the particular solutions.

FIGURE 3.2

The propagation of a monochromatic wave during the
time t_1 as the wave as a
whole moves a distance x_1.
The phase velocity of the
wave is x_1/t_1.

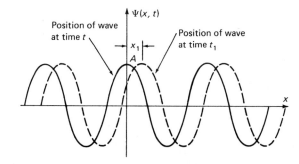

A. The Free Electron as a Traveling Wave

To show that this latter expression indeed represents a traveling wave, and to
determine the velocity with which it moves, refer to Fig. 3.2. Here the solid line
represents the real part of Eq. (3.22) at time $t = 0$. At $x = 0$ the amplitude
$\Psi(0,\ 0) = A$. Suppose at some later time t_1 the wave represented by the
exponential function has moved to the right a distance x_1. The new position of
the wave is represented by the broken curve. The amplitude $\Psi(x_1,\ t_1)$ must
again equal A. In light of Eq. (3.22), this means that

$$kx_1 = \omega t_1 \quad \text{or} \quad \frac{\omega}{k} = \frac{x_1}{t_1}. \tag{3.23}$$

But x_1/t_1 is just the speed with which the wave has moved to the right. Hence
Eq. (3.22) is said to represent a plane wave traveling in the positive x-direction
with a so-called *phase velocity* $v_p = \omega/k$.

B. The Electron as a Wave Packet: Group Velocity

As previously mentioned in footnote 17, the general solution of the Schrödinger
equation is obtained by summing particular solutions such as the one just
discussed. The solution is reminiscent of the analysis of a complicated wave
form expressed in terms of its Fourier series. To illustrate simply (although not
rigorously) the properties of this type of solution, let us consider summing only
two of the waves with slightly different wave vectors k and angular frequencies
ω. These component waves are indicated in Fig. 3.3. The rigorous solution
would require the summing of infinitely many such waves having slightly
different wave vectors and frequencies. However, it turns out that the essence
of the result is indicated effectively by our simplistic approach. What is
indicated is that this summing of many wavelets gives rise to the concept of a
wave packet which moves with a group velocity which is different from the
velocity of each of the component wavelets. The motion of this wave packet is
interpreted as representing the motion of the electron and provides a means of
comparison with experimental data. The group velocity of such a wave packet
expresses the speed with which energy is transmitted by this wave phenomenon.

FIGURE 3.3

Schematic representation of
two wavelets with slightly
different wave vectors k and
angular frequency ω is
shown above. A wave packet
formed by summing many
such wavelets (in fact, an
infinite number) is shown
below.

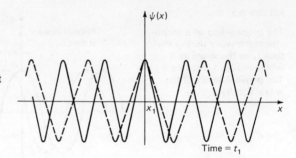

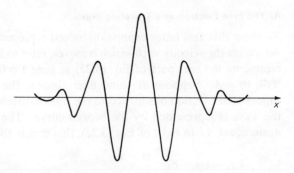

Referring now to Fig. 3.3, let us assume that the wavelets indicated by the
solid and broken lines agree in amplitude at position x_1 at the time t_1 indicated
in the diagram. Let us consider the progress of the wave packet in terms of the
motion of this one point of coincidence which has the maximum amplitude. If
the two wavelets differ in frequency by $d\omega$ and in wave vector by dk, then due to
amplitude agreement at x_1, t_1,

$$j(k_0 x_1 - \omega_0 t_1) = j[(k_0 + dk)x_1 - (\omega_0 + d\omega)t_1]. \tag{3.24a}$$

Here ω_0 and k_0 correspond to the frequency and wave vectors, respectively, of
one of the waves. Now at a later time t_2, coincidence will occur at some other
point, say x_2, and then

$$j(k_0 x_2 - \omega_0 t_2) = j[(k_0 + dk)x_2 - (\omega_0 + d\omega)t_2]. \tag{3.24b}$$

Subtracting Eq. (3.24a) from (3.24b) yields

$$\frac{x_2 - x_1}{t_2 - t_1} = \frac{d\omega}{dk}. \tag{3.25}$$

But since the left-hand side of this equation represents the speed with which this
group of two waves travels, this is referred to as the *group velocity* v_g of the wave
packet. Also $v_g = d\omega/dk$ turns out to be valid even when an infinite group of
wavelets is considered and hence represents physically the velocity of the

electron and the speed with which its energy is propagated. The result of summing an infinite number of wavelets is shown at the bottom of Fig. 3.3.

Now Planck's and de Broglie's law applied to the electron gives

$$v = \frac{\omega}{2\pi} = \frac{E}{h} = \frac{p^2}{2mh} = \frac{h}{2m\lambda^2} \tag{3.26}$$

or

$$\omega = \frac{hk^2}{4\pi m}. \tag{3.27}$$

Then from Eq. (3.25) the wave-packet velocity becomes

$$v_g = \frac{d\omega}{dk} = \frac{hk}{2\pi m} = \frac{p}{m}. \tag{3.28}$$

This indicates a relationship between the rate at which the wave packet propagates and the ratio of particle momentum to particle mass which is the classical particle velocity.[18] This shows that the group velocity of the de Broglie matter waves represents the electron velocity.

3.5 THE ELECTRON IN A BOX WITH IMPENETRABLE WALLS

This section considers another potential-energy function for the electron which leads to a somewhat simple solution of the one-dimensional Schrödinger equation. Because the problem is one-dimensional and the potential function is somewhat ideal, the solution has limited usefulness in real situations. However, the calculation does indicate how energy quantization and the uncertainty principle follow directly from, and are inherent in, the Schrödinger equation and illustrates the normalization procedure. Surprisingly, this simple solution will also supply an estimate of the binding energy of an electron in an atom as well as the nuclear binding energy. (This will be demonstrated by a problem solved later in this chapter.)

A. Boundary Conditions

Consider an electron bound in a square potential well with infinite sides as shown in Fig. 3.4. Here the electron potential energy $V(x)$ is plotted as a function of x. Note that the potential energy is taken as zero for $|x| < a/2$, that is, when the absolute value of x is less than $a/2$. At $x = \pm a/2$, the potential energy takes on an infinite value. This will confine the electron within the "walls" of the potential well because of the infinitely high barriers there. (An

[18] Note that if the electron were considered to be represented by a single wavelet, this phase velocity would be $v_p = \omega/k = p/2m$, which is in disagreement by a factor of two with the classical result. This verifies that indeed a wave packet must be formed to represent the electron, rather than a single wavelet.

FIGURE 3.4

Graph of potential energy versus position in a square potential well with infinite sides.

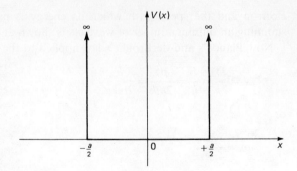

analogy would be a ball moving with finite energy attempting to climb an infinitely steep hill or a ball bouncing in a box with impenetrable walls.)

Hence the probability of an electron, once in the well, being found outside the walls is zero. In terms of the Born hypothesis, this means that $|\psi|^2 = 0$ there, and so $\psi = 0$ at $|x| \geq a/2$.[19]

B. Solution of the Schrödinger Equation

The one-dimensional time-independent Schrödinger equation for this case for $|x| < a/2$ is then

$$\frac{d^2\psi}{dx^2} + \frac{2mE}{\hbar^2}\psi = 0, \qquad (3.29)$$

where $\hbar$ is defined as $h/2\pi$.

Again this is similar to the equation of harmonic motion, and it is convenient here to write the solution as

$$\psi = A \sin\left(\frac{\sqrt{2mE}}{\hbar}\right) x + B \cos\left(\frac{\sqrt{2mE}}{\hbar}\right) x. \qquad (3.30)$$

To solve for the two constants of integration A and B, the boundary condition $\psi = 0$ at $x = \pm a/2$ is applied. This gives, at $x = +a/2$,

$$\psi = A \sin\left(\frac{\sqrt{2mE}}{\hbar}\right)\frac{a}{2} + B \cos\left(\frac{\sqrt{2mE}}{\hbar}\right)\frac{a}{2} = 0, \qquad (3.31)$$

and, at $x = -a/2$,

$$\psi = -A \sin\left(\frac{\sqrt{2mE}}{\hbar}\right)\frac{a}{2} + B \cos\left(\frac{\sqrt{2mE}}{\hbar}\right)\frac{a}{2} = 0. \qquad (3.32)$$

---

[19] This result should be established more rigorously by considering a finite barrier at $x = \pm a/2$, applying the continuity of ψ and $d\psi/dx$ there, and then letting $V \to \infty$. It will then become clear that $\psi \to 0$ for $V \to \infty$. This takes the place of the requirement of the continuity of ψ and $d\psi/dx$ at $x = \pm a/2$ in this case.

Hence from Eq. (3.31) or (3.32) either $A = 0$ or $\sin(\sqrt{2mE}/\hbar)(a/2) = 0$, and either $B = 0$ or $\cos(\sqrt{2mE}/\hbar)(a/2) = 0$. To avoid the trivial solution $\psi = 0$ for all x, it must be true that either

$$A = 0, \quad \cos\left(\frac{\sqrt{2mE}}{\hbar}\right)\frac{a}{2} = 0 \qquad (3.33a)$$

or

$$B = 0, \quad \sin\left(\frac{\sqrt{2mE}}{\hbar}\right)\frac{a}{2} = 0. \qquad (3.33b)$$

From Eq. (3.33a), $(\sqrt{2mE}/\hbar)(a/2) = n\pi/2$, where n is an odd integer. Similarly, from Eq. (3.33b), $(\sqrt{2mE}/\hbar)(a/2) = n\pi/2$, where n is an even integer. Hence the general solution can be obtained by summing solutions of the form

$$\psi = B\cos\left(\frac{\sqrt{2mE}}{\hbar}\right)x, \qquad (3.34a)$$

where

$$\frac{\sqrt{2mE}}{\hbar} = \frac{n\pi}{a} \quad \text{for } n = 1, 3, 5, 7, \ldots \qquad (3.35a)$$

or

$$\psi = A\sin\left(\frac{\sqrt{2mE}}{\hbar}\right)x, \qquad (3.34b)$$

where

$$\frac{\sqrt{2mE}}{\hbar} = \frac{n\pi}{a} \quad \text{for } n = 2, 4, 6, 8, \ldots. \qquad (3.35b)$$

Finally we can solve for the constants A and B by normalizing these solutions. This results from the Born probability interpretation of ψ. One of the conditions on ψ previously stated (see Eq. 3.17b) was that the integral of $|\psi|^2$ over all space was required to be finite. Since the probability is normally a fractional quantity less than one,[20] the probability of finding the electron somewhere in space must be identically one, and so in one dimension

$$\int_{-\infty}^{+\infty} |\psi|^2 \, dx = 1. \qquad (3.36)$$

[20] For example, the probability of rolling a one when throwing a single die is "one in six" or $\frac{1}{6}$ or 0.167.

Applying this normalization condition to Eq. (3.34a) and (3.34b) and noting that $\psi = 0$ for $-a/2 > x > a/2$ gives

$$A = B = \sqrt{\frac{2}{a}}. \tag{3.37}$$

Solving for E from Eqs. (3.35a) and (3.35b) yields

$$E_n = \frac{n^2 h^2}{8ma^2}, \tag{3.38}$$

where n is now any integer, and E_n refers to an infinite number of discrete energy values or levels corresponding to each integral value of n. These energy values are often referred to as *eigenvalues*[21] and occur in all quantum mechanical problems concerning spatially bound or constrained particles; the number n is called a *quantum number*. The electron is then said to be in any one of these energy states or *eigenstates* denoted by ψ_n at a given time. The lowest eigenstate is called the *ground state*, and the energy in this case is often referred to as the *zero-point energy*;[22] here, $E_1 = h^2/8ma^2$. Hence at absolute zero all motion does not cease.

Note that as the particle is less constrained as the width of the well is increased, the energy levels come closer together, indicating that quantum mechanics predicts a continuum of energy values for nearly free electrons. One thus would also expect that the higher energy levels would become closer together as the electron is excited to these nearly free states. The one-dimensional model presented here doesn't predict this. The three-dimensional model corrects this dilemma and is the subject of a problem at the end of this chapter. Note too that as the particle mass m becomes very large, a continuum of energy values is again predicted, in agreement with the correspondence principle.

C. Zero-Point Energy and Uncertainty

Note that in the quantum mechanics of the particle in a box, as a result of the particle energies being discrete or quantized and not continuous as in classical mechanics, the particle will never have zero energy. The fact is that the smallest particle energy is the ground-state energy. The difference between this lowest energy and zero is an expression of the uncertainty principle. To illustrate the relationship of this zero-point energy and the Heisenberg uncertainty principle, consider the latter as applied to the problem of the electron in the infinite potential well. It is clear that the bound electron is located at some moment somewhere between $x = +a/2$ and $x = -a/2$; that is, $\Delta x = a$. Now since in the potential well the electron energy is totally kinetic ($V = 0$), $E = p^2/2m$,

[21] *Eigen* is the German word for unique.
[22] This concept gives rise to a prediction of very peculiar behavior at near absolute zero temperature which is actually observed experimentally in liquid helium.

FIGURE 3.5

Graph of the lowest and next highest mode of the ψ function for an electron in an infinite potential well.

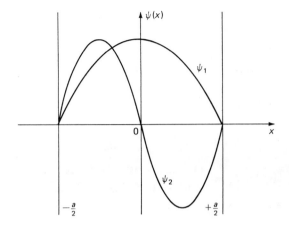

where p is the electron momentum and m is the electron mass. Its ground-state momentum is given by $p_1 = \sqrt{2mE_1} = h/2a$. Since momentum is a vector quantity, the electron can possibly be moving in the $+x$ or $-x$ direction, and hence its momentum is said to be uncertain to the extent $\Delta p_x = h/2a - (-h/2a) = h/a$. The product $\Delta p_x \, \Delta x$ must then be equal to h, which is the lower limit prescribed by Heisenberg.

D. The Electron in a Box as a Standing Wave

Physically, one might have surmised the eigenvalues in this problem by applying the de Broglie wave concept of the electron. Since the wave amplitude ψ at the boundaries $x = \pm a/2$ has to be zero, this corresponds to standing waves as in a vibrating string of length a, constrained at $x = \pm a/2$. The lowest frequency vibrational mode, ψ_1, corresponds to that shown in Fig. 3.5, with a wavelength of $2a$. Also shown is the next highest mode, ψ_2, with wavelength a. The modes can hence be expressed as $2a/n$, where $n = 1, 2, 3, \ldots$.

The de Broglie electron wave is pictured as reflected back and forth off the walls of the potential well, setting up a standing electron wave. Since the electron wavelength is $\lambda = h/p$ and in this problem $E = p^2/2m$, then $E = h^2/2m\lambda^2$. The possible standing-wave modes require that $\lambda = 2a/n$, where n is any integer, and so $E = n^2h^2/8ma^2$, which agrees with Eq. (3.38).

EXAMPLE 3.2

An electron is moving freely in a one-dimensional box of width a, having impenetrable walls. (a) Write an expression for the force on the walls of the box due to the trapped electron, when in its lowest state.

A particle of nucleonic mass (1.66×10^{-27} kg) is trapped in an infinitely deep one-dimensional well. The width of the well is the diameter of a small nucleus

$(10^{-15}$ m). (b) Calculate the force that this particle must be exerting on the wall when the particle is in its lowest state, in order to achieve an appreciation of the magnitude of a typical nuclear force.

SOLUTION   a) Equation (3.38) indicates that as the well narrows, the energy of the system increases. Therefore, the electron must exert a force F_x on the well walls to oppose this energy increase. Hence by Eq. (3.38),

$$F_x = -\frac{dE_n}{dx} = -\frac{d(n^2h^2/8ma^2)}{dx} = \frac{n^2h^2}{4ma^3}.$$

b) $$F_x = \frac{(1)^2(6.62 \times 10^{-34})^2 \, J^2\text{-sec}^2}{4(1.66 \times 10^{-27})kg(10^{-15})^3 \, m^3}$$

$$= 6.60 \times 10^4 \text{ newtons} \quad \text{or a force of 6740 kg.}$$

3.6   THE STEP POTENTIAL

The problem of an electron wave impinging on a potential barrier of finite height as indicated in Fig. 3.6 is of interest from a number of viewpoints. First, it serves as an introduction to the problem of the interaction of an electron with a potential barrier, which is the problem of the motion of electron waves through a crystal lattice. Here the scattering of an electron by the atoms in a crystal gives rise to the concept of the band theory of solids, which will be developed later. Also, this problem illustrates simply how to apply the condition of the continuity of ψ and $d\psi/dx$ to solve for the constants of integration of the Schrödinger equation. Finally, the solution serves as an illustration of a case where the quantum mechanical approach yields a surprising result compared to Newtonian physical reasoning. That is, the electron will be shown to have a finite (though normally small) probability of penetrating a potential barrier although its total energy is less than the barrier height. (This would be equivalent macroscopically to a ball with 10 J of energy rolling over a tall hill whose peak corresponds with a ball potential energy of 20 J.)

FIGURE 3.6

The step potential energy assumed for studying an electron wave impinging on a barrier of finite height.

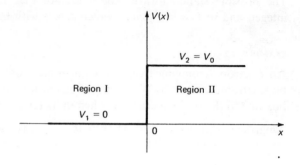

A. Solution of the Schrödinger Equation

In Fig. 3.6, an electron moves freely in the positive x-direction for $-\infty < x < 0$ (Region I). Its potential energy then is $V_1(x) = 0$ for $x < 0$; for values of $x > 0$ (Region II), the electron is assumed to travel through a potential field where it has a constant potential energy $V_2(x) = V_0$. Hence, the electron potential energy is assumed to change discontinuously at $x = 0$ from zero to V_0 (see Fig. 3.6).

Let us take the case where the total electron energy (kinetic plus potential) $E < V_0$, the barrier height. Classically the electron is reflected from the barrier and has zero possibility of penetrating it and appearing in Region II. It will be seen that quantum mechanics predicts some probability of barrier penetration or "tunneling" in this case.

The time-independent Schrödinger equation in Region I where $V = 0$ is

$$\frac{d^2\psi}{dx^2} + \alpha^2\psi = 0, \tag{3.39}$$

where $\alpha^2 = 2mE/\hbar^2$ is a constant quantity[23] independent of x. As before, it is convenient to write the solution of this type of equation in terms of exponentials as

$$\psi = Ae^{j\alpha x} + Be^{-j\alpha x}. \tag{3.40}$$

In Region II the Schrödinger equation is

$$\frac{d^2\psi}{dx^2} - \beta^2\psi = 0, \tag{3.41}$$

where $\beta^2 = 2m(V_0 - E)/\hbar^2$ is a real constant for $x > 0$. Now the solution of this equation may be written as

$$\psi = Ce^{\beta x} + De^{-\beta x}. \tag{3.42}$$

The constants of integration, A, B, C, and D, must be chosen to satisfy the boundary conditions, as follows:

$$\psi \text{ must be finite and continuous for all } x; \tag{3.43a}$$

$$\frac{d\psi}{dx} \text{ must be finite and continuous for all } x. \tag{3.43b}$$

Note that in Eq. (3.42) with β positive the first term on the right-hand side grows infinitely large as $x \to \infty$, which is in disagreement with condition (3.43a). This term can be eliminated by taking C as zero. Now to ensure the continuity of ψ at $x = 0$ from Eqs. (3.40) and (3.42),

$$[Ae^{j\alpha x}]_{x=0} + [Be^{-j\alpha x}]_{x=0} = [De^{-\beta x}]_{x=0} \tag{3.44}$$

---

[23] Conservation of energy dictates the constancy of E.

which gives

$$A + B = D. \tag{3.45}$$

Similarly, for $d\psi/dx$ to be continuous at $x = 0$, Eqs. (3.40) and (3.42) must be differentiated, yielding

$$[j\alpha A e^{j\alpha x}]_{x=0} - [j\alpha B e^{-j\alpha x}]_{x=0} = -[\beta D e^{x-\beta x}]_{x=0}. \tag{3.46}$$

This gives

$$\frac{j\beta D}{\alpha} = A - B. \tag{3.47}$$

Equations (3.45) and (3.47) are two equations in three unknowns; hence we can solve for A and B in terms of D as[24]

$$A = \frac{D}{2}\left(1 + \frac{j\beta}{\alpha}\right),$$

$$B = \frac{D}{2}\left(1 - \frac{j\beta}{\alpha}\right). \tag{3.48}$$

Now that the integration constants are known, the complete wave function can be written by tacking on the time-dependent part as

$$\Psi(x, t) = \frac{D}{2}\left(1 + \frac{j\beta}{\alpha}\right) e^{j(\alpha x - Et/\hbar)} + \frac{D}{2}\left(1 - \frac{j\beta}{\alpha}\right) e^{-j(\alpha x + Et/\hbar)}, \qquad \text{for } x \leq 0 \tag{3.49}$$

and

$$\Psi(x, t) = D e^{-\beta x} e^{-jEt/\hbar}, \qquad \text{for } x \geq 0. \tag{3.50}$$

Physically the solution in Region I ($x < 0$) consists of an "incident" wave traveling in the positive x-direction, plus a "reflected" wave traveling in the negative x-direction. In Region II ($x > 0$) the solution indicates a standing wave, i.e., at any position x, the wave oscillates harmonically with time.

Combined with the mathematical identity $e^{j\theta} = \cos\theta + j\sin\theta$, the complete wave functions (Eqs. 3.49 and 3.50) become

$$\Psi(x, t) = D\left(\cos\alpha x - \frac{\beta}{\alpha}\sin\alpha x\right) e^{-jEt/\hbar}, \qquad \text{for } x \leq 0$$

and

$$\Psi(x, t) = D(e^{-\beta x}) e^{-jEt/\hbar}, \qquad \text{for } x \geq 0. \tag{3.51}$$

---

[24] A third condition which would supply a third equation and permit evaluation of D is the normalization condition (see Eq. 3.36). This will not be necessary for our purposes here.

FIGURE 3.7

A plot of the Ψ function versus distance at $t = 0$ for an electron wave traveling to the right toward a finite potential barrier located at $x = 0$, where the total electron energy is less than the barrier height. Note the finite probability of finding the electron beyond the barrier ($x > 0$).

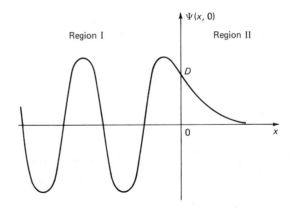

Hence it is seen that the wave functions are standing waves for all values of x. For $x < 0$ the incident and reflected traveling waves interact to form a standing wave in a manner identical to the interference of equal-amplitude light waves. A plot of these eigenfunctions versus x at $t = 0$ is shown in Fig. 3.7.

B. Tunneling

Note that, in interpreting $|\Psi|^2$ as representing the probability of finding the electron at some place between x and $x + dx$ at time t, there is a finite chance of finding the electron beyond the barrier ($x > 0$). This probability decreases exponentially with distance past the barrier. Nevertheless, this still contradicts the classical theory, which would make it impossible for the particle to penetrate the barrier since its total energy E is specified as less than the barrier height V_0. Note that the probability of penetration decreases greatly as the particle energy decreases relative to the barrier potential energy.

This surprising phenomenon is known as *tunneling* and forms the basis of a number of electronic devices. One of the most familiar of these is the tunnel diode (discussed in Chapter 1), which is used as a high-frequency (gigahertz) oscillator. Other devices whose operation can be explained by this effect are the Zener diode and the Josephson junction.

EXAMPLE 3.3

Equation (3.51) expresses the finite probability that an electron impinging on a barrier whose height is V_0, with an energy $E < V_0$, will penetrate the barrier. What is the probability of finding the electron at a distance of 3 Å beyond the barrier relative to its being 10 Å beyond the barrier? Take the electron energy as 0.6 eV and the barrier height as 1.0 eV.

SOLUTION From Eq. (3.51) the probability of finding the electron between x and $x + dx$ is

$$|\Psi|^2 = D^2 \left| e^{-\beta x} \right|^2 \left| e^{-jEt/\hbar} \right|^2 = D^2 e^{-2\beta x} = D^2 e^{-(2/\hbar)\sqrt{2m(V_0 - E)}\, x},$$

since the absolute value of the term of the form $e^{-j\theta}$ is 1 as is known from phasor theory. The relative probability of finding the electron at 3 Å compared with 10 Å beyond the barrier is

$$\frac{|\Psi|^2_{3\text{Å}}}{|\Psi|^2_{10\text{Å}}} = e^{-(2/\hbar)\sqrt{2m(V_0-E)}(3 \times 10^{-10} - 10 \times 10^{-10}\,\text{m})}.$$

Now

$$-\frac{2}{\hbar}\sqrt{2m(V_0-E)} = -\frac{2}{6.62 \times 10^{-34}\,\text{J-sec}}$$
$$\times \sqrt{2(9.1 \times 10^{-31})\,\text{kg}(1.0 - 0.6)(1.6 \times 10^{-19})\,\text{J}}$$
$$= -6.65 \times 10^9\,\text{m}^{-1}.$$

Hence

$$\frac{|\Psi|^2_{3\text{Å}}}{|\Psi|^2_{10\text{Å}}} = e^{(-6.65 \times 10^9\,\text{m}^{-1})(-7 \times 10^{-10}\,\text{m})} = 105.$$

This result demonstrates the extremely rapid falloff of tunneling probability beyond the barrier. Since 3 Å is the order of the spacing between atoms, the calculation may be used as an aid in understanding the motion of an electron through a series of periodically spaced barriers to be discussed next.

3.7 MOTION OF AN ELECTRON IN A PERIODIC POTENTIAL: THE KRÖNIG-PENNEY MODEL

The flow of electrons in a crystal will now be considered from a quantum mechanical viewpoint. This is necessary since countless physical phenomena observed in solid materials have no classical explanation. The crystal considered consists of a periodic arrangement of atoms, some stripped of an electron (ions), with these electrons free to move through the crystal lattice. Later we will see that the properties of solid crystals such as metals, semiconductors, and insulators can be derived from this analysis.

The model which we will assume is drastically simplified. We will reduce the problem to that of a single electron traveling in one dimension, impinging on a series, infinite in extent, of periodically spaced potential wells. The potential-energy function representing the force field created by the periodically spaced ions (nuclei plus immobile core electrons) in a sea of free electrons is shown in Fig. 3.8a. The simplified mathematical representation of this periodic potential function assumed by Krönig and Penney in 1930[25] is given in Fig. 3.8b. Note that the x-space is broken down into regions of widths a (Region I) and b

[25] R. de L. Krönig and W. G. Penney, "Quantum Mechanics of Electrons in Crystal Lattices," *Proc. Roy. Soc. (London)*, Vol. A130, p. 499 (1930).

FIGURE 3.8

a) Graph of the potential-energy function representing the force field created by periodically spaced ions.
b) Simplified mathematical representation of the periodic function of (a) assumed by Krönig and Penney.

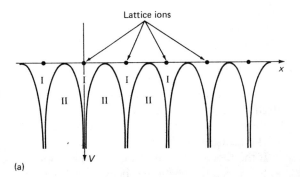

(a)

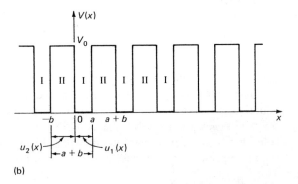

(b)

(Region II), with the distance between nuclei being $a + b$. In spite of the extreme simplicity of this formulation, the essential ideas of the quantum mechanical theory of the three-dimensional solid become clear when the Schrödinger equation is solved for this problem. Although the model is simple, the mathematics required to arrive at a solution in closed form is nevertheless somewhat complicated.

A. The Schrödinger Equation

The time-independent Schrödinger equation for the motion of one electron in a periodic potential field is, in one dimension,

$$\frac{d^2\psi}{dx^2} + \frac{2m}{\hbar^2}[E - V(x)]\psi = 0, \tag{3.52}$$

where $V(x)$ is given in Fig. 3.8b as

$$V = V_0 \quad \text{for } -b < x < 0, \quad a < x < a + b, \quad \text{etc.}$$
$$V = 0 \quad \text{for } 0 < x < a, \quad a + b < x < 2a + b, \quad \text{etc.} \tag{3.53}$$

A theorem proved by Felix Bloch in 1928[26] states that all one-electron wave functions for problems involving periodically varying potential-energy functions must be of the form

$$\psi(x) = u(x)e^{jkx}, \qquad (3.54)$$

where k is a constant of motion and $u(x)$ is a periodic function (which repeats itself in the present case after every distance $a + b$).[27] For example, $u(x)$ may be a function of the form $\sin(2\pi x/d)$, d being the distance after which the function $u(x)$ repeats itself.

If the Bloch function of Eq. (3.54) is a solution of the Schrödinger equation (3.52), it can be substituted back into the latter. This substitution leads to the requirement that $u(x)$ satisfies

$$\frac{d^2 u}{dx^2} + 2jk \frac{du}{dx} - \left[k^2 - \alpha^2 + \frac{2mV(x)}{\hbar^2} \right] u = 0, \qquad (3.55)$$

where $\alpha^2 = 2mE/\hbar^2$; that is, the Bloch function is a solution of the Schrödinger equation if $u(x)$ satisfies Eq. (3.55).

It is now necessary to introduce the proper potential-energy values into this equation. Since the potential function repeats itself with period $a + b$, it will only be necessary to solve the equation in the interval $-b < x < a$. The solution in any other similar interval will be identical with this one. Substituting the appropriate potential-energy values from Eq. (3.53), we get

$$\frac{d^2 u_1}{dx^2} + 2jk \frac{du_1}{dx} - (k^2 - \alpha^2)u_1(x) = 0 \qquad \text{(Region I, } 0 < x < a) \quad (3.56a)$$

and

$$\frac{d^2 u_2}{dx^2} + 2jk \frac{du_2}{dx} - (k^2 - \beta^2)u_2(x) = 0 \qquad \text{(Region II, } -b < x < 0).$$

$$(3.56b)$$

Here $u_1(x)$ and $u_2(x)$ are the values of $u(x)$ in Regions I and II respectively, and $\beta^2 = 2m(E - V_0)/\hbar^2$. Equations (3.56a) and (3.56b) are recognizable as being of the form of the differential equation for an electric circuit containing resistance, inductance, and capacitance; they can be solved by the method of the Laplace transform. One can show that the solutions of these two equations are of the form

$$u_1(x) = Ae^{j(\alpha - k)x} + Be^{-j(\alpha + k)x}, \qquad (0 < x < a) \qquad (3.57a)$$
$$u_2(x) = Ce^{j(\beta - k)x} + De^{-j(\beta + k)x}. \qquad (-b < x < 0) \qquad (3.57b)$$

---

[26] F. Bloch, "On the Quantum Mechanics of Electrons in Crystal Lattices," *Zeit. Phys.* Vol. 52, p. 555 (1928).
[27] When the time-dependent part of the wave function is tacked onto this function, it is recognized as that of a traveling wave, whose amplitude is periodically modulated in space.

B. Boundary Conditions

To determine the integration constants A, B, C, and D, it is convenient to apply the conditions of finiteness and continuity of ψ and $d\psi/dx$ at $x = a$, $x = -b$, and $x = 0$. It therefore follows that the functions $u(x)$ obey these requirements since e^{jkx} is known to be a bounded and continuous function for all x. Also because of the periodicity of $u(x)$ in the distance $a + b$ (see Eq. 3.54), it follows that $u_1(a) = u_2(-b)$ and $u_1'(a) = u_2'(-b)$. Introducing these boundary conditions in a straightforward but somewhat laborious manner gives

$$A + B = C + D,$$

since $u_1(0) = u_2(0)$;

$$j(\alpha - k)A - j(\alpha + k)B = j(\beta - k)C - j(\beta + k)D,$$

since $u_1'(0) = u_2'(0)$;

$$e^{j(\alpha-k)a}A + e^{-j(\alpha+k)a}B = e^{-j(\beta-k)b}C + e^{j(\beta+k)b}D,$$

since $u_1(a) = u_2(-b)$;

$$j(\alpha - k)e^{j(\alpha-k)a}A - j(\alpha + k)e^{-j(\alpha+k)a}B$$
$$= j(\beta - k)e^{-j(\beta-k)b}C - j(\beta + k)e^{j(\beta+k)b}D, \quad (3.58)$$

since $u_1'(a) = u_2'(-b)$.

The simultaneous solution of these equations is very complicated and, in fact, unnecessary. It is presently more interesting physically to obtain the energy values for the electron in a periodic crystal lattice for which valid solutions of Eqs. (3.58) are possible. A theorem of algebra states that in a set of simultaneous linear homogeneous equations (like 3.58), there is only a non-trivial solution if the determinant of the coefficients of A, B, C, and D equals zero. The evaluation of this determinant again is straightforward but laborious and leads to

$$-\frac{\alpha^2 + \beta^2}{2\alpha\beta} \sin \alpha a \sin \beta b + \cos \alpha a \cos \beta b = \cos k(a + b). \quad (3.59)$$

Now the electron energy E can, in general, be less than or greater than V_0. If $E > V_0$ then β is previously defined as a real quantity. In cases where $E < V_0$, it is convenient to introduce $\beta = j\gamma$, and then Eq. (3.59) becomes, with the use of $\cos jx = \cosh x$ and $\sin jx = j \sinh x$,

$$\frac{\gamma^2 - \alpha^2}{2\alpha\gamma} \sinh \gamma b \sin \alpha a + \cosh \gamma b \cos \alpha a = \cos k(a + b). \quad (3.60)$$

Here γ^2 is real and positive for electron energies such that $0 < E < V_0$. β^2 is real and positive for energies such that $V_0 < E < \infty$, where Eq. (3.59) is most conveniently applied.

C. Allowed Energy Bands

Equations (3.59) and (3.60) are complicated transcendental equations involving the electron energy E through the quantity α. To solve for α, the method of graphical solution is suitable. However, to obtain an equation more susceptible to solution, Krönig and Penney suggest a special case. Let the potential-barrier width b shrink to zero, and the barrier height V_0 at the same time grow to infinity, with the product of the two, $V_0 b$, remaining finite. Then Eq. (3.60) is most appropriate and, as $\gamma \to \infty$, Eq. (3.60) reduces to

$$\frac{P \sin \alpha a}{\alpha a} + \cos \alpha a = \cos ka, \qquad (3.61)$$

where $P = mV_0 ba/\hbar^2$ is sometimes referred to as the *scattering power* of the potential barrier. It is a measure of the strength with which electrons in a crystal are attracted to the ions on the crystal lattice sites.

There are only two variables in this equation, namely α and k. The right-hand side of Eq. (3.61) is bounded since it can only assume values between $+1$ and -1. If we plot the left-hand side of this equation against αa, it will be possible to determine those values of α (and hence energy) which are permissible; that is, permit $P \sin \alpha a/\alpha a + \cos \alpha a$ to take values between $+1$ and -1. When each of these same values is set equal to $\cos ka$, k is determined. Then α can be found from Eq. (3.61).

Figure 3.9 shows this plot for a value of P assumed arbitrarily as $3\pi/2$. Indicated in the figure are the permitted values of this function, shown as a solid line. This then gives rise to the concept of ranges of permitted values of α for a given ion lattice spacing a, and since $E = \alpha^2\hbar^2/2m$, permitted bands of

FIGURE 3.9

Graph of $(P/\alpha a)(\sin \alpha a + \cos \alpha a)$ versus αa with $P = 3\,\pi/2$, for the purpose of solving transcendental equation (3.61). The allowed values of αa are indicated by the crosshatched regions. Since α is related to electron energy, the allowed energy values are determined.

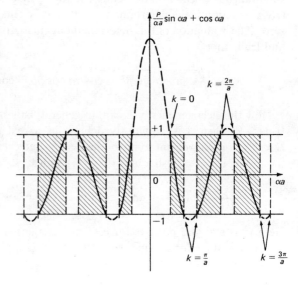

FIGURE 3.10

The permitted values of energy for an electron traveling in a one-dimensional periodic potential field are indicated by the solid lines. This indicates the results of a Krönig-Penney solution. The E-versus-k relationship for a free electron is indicated by the dashed parabola. The deviations from the free electron motion caused by the periodic field give rise to the allowed energy bands shown.

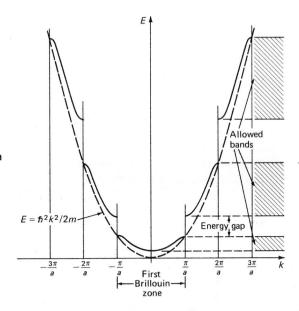

energy are predicted. Note that the width of each allowed band decreases as P increases; i.e., tighter binding results in narrower energy bands. In fact, as $P \to \infty$, since the left-hand side of Eq. (3.61) must remain bounded and hence finite, $\sin \alpha a = 0$. The solutions of this latter equation are $\alpha a = \pm n\pi$, where n is any integer; then the permitted energies reduce to unique values, $E_n = n^2 h^2 / 8ma^2$. These are recognized as the eigenvalues for the electron in a box (see Eq. 3.38) with impenetrable walls. This seems reasonable since in this case the electron is prohibited from leaving its ion attachment. Finite values of P lead to a smearing out of these discrete levels into energy bands.

At the other extreme, that is, $P \to 0$, we have $\cos \alpha a = \cos ka$ or $k = \alpha$. The latter value, α, leads to $E = \hbar^2 k^2 / 2m$ and corresponds physically to the "free" electron with all energies permitted. Since $E = p^2/2m$ for the free electron, k can be identified with the electron momentum as $p = \hbar k$. The general relationship (for any P) between the electron energy E and the wave vector k is of course contained in Eq. (3.61).

D. Brillouin Zones

A typical plot of E versus k for a value of $0 < P < \infty$ is given in Fig. 3.10. [The dashed curve refers to the case $P = 0$, is parabolic, and corresponds to the $E(k)$ relationship for a free electron.] Indicated are energy discontinuities at values of

$$k = \frac{n\pi}{a},$$ (3.62)

where $n = \pm 1, \pm 2, \pm 3, \ldots$. These are referred to as the *Brillouin-zone* boundaries and correspond to values of electron momenta where discontinuities in the relation of energy to wave vector occur. Equation (3.62) physically indicates the condition for a standing wave due to an electron wave impinging and being reflected from the potential barrier represented by P. Since the electron momentum is given by $p = \hbar k$ and $p = h/\lambda$ by the de Broglie formula, k is identified with the electron wavelength as $k = 2\pi/\lambda$. Then the Brillouin-zone boundaries defined by Eq. (3.62) correspond to $\lambda = 2a/n$. That is, maximum interaction with the lattice ions occurs when the electron wavelength is twice the lattice spacing, equal to the lattice spacing, etc. This is exactly reminiscent of the electron-in-the-box problem discussed in Section 3.5. The concept of the Brillouin zone is extremely useful in describing the electrical and magnetic properties of solids and will be used in the next chapter.

The problem is also related to the case of x-rays (electromagnetic waves) diffracted by a crystal lattice. The Bragg condition (see Section 2.3A) is directly related to Eq. (3.62), which refers to electron wave diffraction in a crystal. This will now be shown. The one-dimensional Bragg condition ($\theta = 90$ degrees) for reflection from crystal planes where the planes are spaced a distance a apart can be written as (see Eq. 2.4)

$$2a \sin \theta = 2a = n\lambda. \tag{3.63}$$

Dividing by 2π gives

$$\frac{a}{\pi} = \frac{n\lambda}{2\pi} = \frac{n}{k} \quad \text{or} \quad k = \frac{n\pi}{a}. \tag{3.64}$$

Hence by the wave-particle duality, electrons in a periodic lattice and x-rays or photons incident on a periodic structure *cannot* be transmitted through the lattice when the electron or photon wave vector (or momentum) bears a specific relationship to the lattice spacing. This corresponds to standing electron waves and a maximum reflection (minimum transmission) of x-rays.

3.8 THE SIMPLE HARMONIC OSCILLATOR

The solution of the Schrödinger equation for a particle moving with a restoring force which increases linearly with displacement from an equilibrium position is also possible in closed form. However, this solution is not as simple as the previous two cases and must be sought in terms of a power series. In fact, the eigenstates need to be expressed in terms of some complicated Hermite polynomials, and a complete exposition is not warranted by the objectives of this book.[28] Nevertheless, the results are of interest since a harmonically oscillating

---

[28] The full solution may be found, for example, in R. M. Eisberg, *Fundamentals of Modern Physics*, John Wiley & Sons, New York (1961).

charged particle gives rise to electromagnetic radiation. This is related to the problem of light emission from a laser.

Classically, the potential energy of a particle oscillating harmonically in one dimension with a frequency v is given by

$$V(x) = 2\pi^2 m v^2 x^2, \tag{3.65}$$

where m is the particle mass, and x is the displacement from equilibrium. Introducing this $V(x)$ into the Schrödinger equation and solving subject to the appropriate boundary conditions again yields eigenvalues which can be expressed as

$$E_n = (n + \tfrac{1}{2})hv, \tag{3.66}$$

where h is the Planck constant, and $n = 0, 1, 2, 3, \ldots$. Note that the bound particle has discrete energy values spaced at equal intervals of hv. From a light-emission viewpoint this means that an electron in a higher or *excited* state can drop to the ground state with the radiation of an integral number of photons of energy hv. Note too that the harmonic oscillator has a zero-point energy of $\tfrac{1}{2}hv$ which can again be shown to be a manifestation of the uncertainty principle.

3.9 THE CENTRAL-FIELD PROBLEM

Another important solution of the Schrödinger equation which can be obtained in closed form corresponds to the problem of the simplest atom, hydrogen, modeled as the motion of a single valence electron in the radial field created by one proton in a nucleus. The potential energy of the electron in the field of the proton is

$$V(r) = \frac{-q^2}{4\pi\epsilon_0 r}, \tag{3.67}$$

where q is the electronic charge, ϵ_0 the permittivity of free space, and r the distance from the central nucleus to the electron. This problem differs from the examples discussed thus far in that the three-dimensional Schrödinger equation must be solved. The latter is normally written in spherical (r, θ, ϕ) coordinates because of the spherical form of the potential-energy function.

The solution is somewhat lengthy mathematically (but beautiful), and will not be presented here.[29] However, the treatment of the one-electron atom

---

[29] For a straightforward formulation and solution of this problem see Chapter 10 of R. M. Eisberg, *Fundamentals of Modern Physics*, John Wiley & Sons, New York (1961).

forms the basis for the analysis of the many-electron atoms which comprise the electronic materials of interest to the engineer. The important results of this calculation will now be summarized, including some definitions which will be of use later in the discussion of electronic materials.

A. Results of the Solution of the One-Electron Atom Problem

Since the electron is confined to the vicinity of the nucleus (bound) by a Coulomb force, the uncertainty principle dictates that the possible electron energies must be quantized. The energy states of the electron in the hydrogen atom are given by

$$E_n = \frac{-mq^4}{8h^2\epsilon_0^2} \frac{1}{n^2},$$
(3.68)

where m is the electron mass, and h the Planck constant; the integer $n = 1, 2, 3,$ ... is usually referred to as the *principal quantum number*. These quantized energy states are derived mathematically by requiring that the wave-function solutions of the Schrödinger equation are well behaved (see Section 3.3B).

It can be shown for this case that the three-dimensional partial differential Schrödinger equation can be separated into three ordinary differential equations involving r, θ, and ϕ individually. One quantum number is derived from each of these equations, with the quantum numbers n, l, and m_l corresponding to the spherical coordinates r, θ, and ϕ. As the quantum number n refers to energy quantization, the quantum number l refers to quantization of angular momentum of the electron and is called the *orbital angular-momentum quantum number*. The quantum number m_l relates to the angular orientation of the orbital angular-momentum vector and is called the *magnetic quantum number*. Here only certain discrete directions are permitted; this constitutes *spatial quantization*.

To explain certain features in atomic optical spectra, Uhlenbeck and Goudsmit (1925)[30] postulated in an *ad hoc* fashion that the electron, besides its known orbital angular momentum, has an additional *intrinsic* angular momentum. The fact that the electron acts like a spinning solid body leads to the association of a quantum number m_s with this intrinsic angular momentum or *spin*. The Dirac formulation of the Schrödinger problem does not make an *ad hoc* introduction of electron spin necessary. Dirac (1928)[31] showed that the concept of spin quantization is derived directly from a relativistically corrected form of Schrödinger's equation. Four quantum numbers, n, l, m_l, and m_s, and their possible values are automatically shown to be necessary to fully

[30] G. E. Uhlenbeck and S. Goudsmit, "Substitution of the Hypothesis of a Non-Mechanical Force Through a Consideration of the Intrinsic Behavior of Each Electron," *Naturwiss.*, Vol. 13, p. 953 (1925).
[31] P. A. M. Dirac, "The Quantum Theory of the Electron," *Proc. Roy. Soc.* (*London*), Vol. A117, p. 610 (1928).

define the state of an electron.[32] The following is a summary of the possible values of the quantum numbers which define the possible states of an electron in any atom:

$$n = 1, 2, 3, \ldots,$$
$$l = 0, 1, 2, \ldots, (n-1),$$
$$m_l = 0, \pm 1, \pm 2, \ldots, \pm l,$$
$$m_s = \pm \tfrac{1}{2}.$$

The last number corresponds to the two spatial quantizations of the electron spin, "up" or "down." A theorem due to Pauli (1927)[33] states that each electron state can be occupied by no more than one electron. This is the famous *Pauli exclusion principle*.

PROBLEMS

3.1 Compare:

a) wave nature of the electron *and* particle nature of the electron,
b) Plank's law *and* de Broglie's law,
c) phase velocity *and* group velocity,
d) wave mechanics *and* Newtonian mechanics.

3.2 It is desired to produce x-ray radiation with a wavelength of 1 Å.

a) Through what potential difference must an electron be accelerated in vacuum so that it can, on colliding with a target, generate such a photon? (Assume all the electron's energy is transferred to the photon.)
b) What is the de Broglie wavelength of the electron of part a, just before hitting the target?
c) What is the de Broglie wavelength of a 0.10 gm mass moving at a velocity of 1 m/sec?

3.3 a) Estimate the uncertainty in the velocity of a 0.10 gm dust particle viewed with a microscope, using yellow light (6000 Å), whose objective subtends an angle of 60° with the object.
b) What is the smallest energy such a particle could have? Compare this with the kinetic energy of the mass moving at 1.0 m/sec.

3.4 An electron and a photon have the same energy. What is the value of this energy in electron-volts so that their respective wavelengths are equal?

---

[32] Spectroscopists refer to the electron states defined by $l = 0, 1, 2, \ldots$ as s, p, d, ... states, respectively. Hence the 1s quantum state refers to an energy state corresponding to $n = 1$ and $l = 0$; 2p refers to a state corresponding to $n = 2$ and $l = 1$; etc.
[33] W. Pauli, "Quantum Mechanics of the Magnetic Electrons," *Zeit. Phys.* Vol. 43, p. 601 (1927).

3.5 The work function of tungsten is 4.52 eV and cesium is 1.81 eV. The work function refers to the minimum energy necessary to remove an electron from the metal.

a) Find the maximum wavelength of light for photoelectric emission of electrons from tungsten and cesium.
b) Which, if either, of these materials can be used in designing a photocell for detecting the laser light of Example 3.1?

3.6 Assume the wave function $\Psi(x, t) = A (\sin \pi x)e^{-j\omega t}$. Find $|\Psi|^2$. Choose A such that $\int |\psi|^2 \, dx = 1$.

3.7 Show that Eq. (3.22) is a solution of the time-dependent Schrödinger equation.

3.8 Assume that $\Psi_1(x, t)$ and $\Psi_2(x, t)$ are solutions of the one-dimensional time-dependent Schrödinger equation.

a) Prove the superposition principle by showing that $\Psi_1 + \Psi_2$ is also a solution.
b) Is $\Psi_1\Psi_2$ a solution of the Schrödinger equation in general?

3.9 Suppose that the dispersion relationship between frequency and wavelength for some matter waves in a medium is given by $v = A/\lambda^3$.

a) Determine the phase velocity of these waves.
b) Determine the group velocity of these waves.

3.10 a) Show that Equation (3.34a) is a solution of the Schrödinger equation for an electron in a box, Eq. (3.29).
b) Show that the normalization constant $B = \sqrt{2/a}$.
c) Where in the box is the electron in its ground state most likely to be found?
d) By inspection of Fig. 3.5, write down an expression for the next highest vibrational mode Ψ_3, and show that this is a solution of the Schrödinger equation.
e) Where in the box is the electron in its 100th highest state most likely found?

3.11 a) Determine the eigenvalues for an electron in a three-dimensional (x, y, z) box with impenetrable walls.
b) Show that the energy separation of the higher-energy states becomes less as the energy increases.

3.12 a) Which of the terms in the solution for $\Psi(x, t)$ in the case of the step potential (Eq. 3.49) represents the "incident wave," and which the "reflected wave"?
b) Explain the difference between the standing-wave solution of Eq. (3.50) and the traveling waves of Eq. (3.49).
c) Show that the "reflection coefficient" $(|\Psi|^2_{reflected}/|\Psi|^2_{incident})$ equals $[(1 - \sqrt{1 - V_0/E})/(1 + \sqrt{1 - V_0/E})]^2$, for the case $E > V$.

d) Express the reflection coefficient numerically and explain it physically when $V_0/E \ll 1$ and $V_0/E \sim 1$.

e) Show that the solution for $E > V_0$ is a traveling wave (not a standing wave).

3.13 Consider the penetration of a potential barrier of height 3.0 eV by an electron whose energy is 1.0 eV. Determine the relative probability of finding the electron at the barrier compared to finding it

a) 10 Å beyond the barrier,

b) 100 Å beyond the barrier.

3.14 Show that the Bloch function of Eq. (3.54) is a solution of the one-dimensional Schrödinger equation for the motion of an electron in a periodic potential under the conditions represented by Eq. (3.55).

3.15 Show by the method of the Laplace transform that a solution of Eq. (3.56a) is Eq. (3.57a).

3.16 Show that Eq. (3.61) derives from Eq. (3.60) in the limit as $\gamma \to \infty$, while $V_0 b$ remains finite.

3.17 a) Write an equation which can be solved for the energy at the bottom of the lowest energy band given by the Krönig-Penney solution, in terms of the lattice scattering power P, the lattice spacing a, and the electron mass m.

b) Repeat part a for the energy at the top of the lowest band.

3.18 The Brillouin-zone concept describes the motion of the electron in k-space or p(momentum)-space.

a) What are the values of k at the boundaries of the first Brillouin zone defined by the one-dimensional Krönig-Penney solution?

b) Explain the relation of the Brillouin-zone concept to the diffraction of x-rays by a crystal.

3.19 a) How many possible 5d states exist for the electron in an atom of hydrogen?

b) Which of the following are *not* electronic states of the hydrogen atom: 1s, 10s, 112p, 2d, 1d? Only one electron can occupy each state, with the lower energy states being filled first.

c) Designate by symbols the highest state occupied by an electron in an atom of the element silicon which contains 14 electrons.

4 Introduction to the Quantum Theory of Solids

4.1 ELECTRICAL PROPERTIES OF SOLIDS

Although the Krönig-Penney treatment in Section 3.7 applies only to a hypothetical one-dimensional lattice, the basic concepts derived, which are indicated in Fig. 3.10, apply generally to the real three-dimensional crystal model. This formulation was made in order to illustrate what the Schrödinger equation has to say about the electrons in solids, in order to better explain the vast difference, for example, between the electrical and thermal properties of metals and those of insulators. The electrical resistivity of solid materials varies from about 10^{-8} Ω-m for metals to 10^{16} Ω-m for some insulators. This fantastic parameter range, more than twenty orders of magnitude, is perhaps the largest for any known measurable physical parameter. Classical physics has no explanation for this phenomenon.

What will now be developed is the quantum mechanical explanation. These ideas will make it apparent that there is an intermediate class of electrical conductors called *semiconductors*. As it turns out, these materials are even more important for the construction of electronic devices than metals or insulators. They are the "stuff" from which transistors are made!

A way of indicating the results of the Krönig-Penney model for electron waves traveling through a periodic array of ions is to plot the energy band diagram as shown in Fig. 3.10. Here are indicated the ranges of permitted and forbidden energy levels for an electron in a crystal.

A real crystal, of course, has many electrons. For example, many metal crystals have approximately one free (valence) electron per atom. Since there are approximately 10^{28} atoms/m^3 in a typical metal crystal, there are about the same number of electrons free to move and conduct current.[1] In addition, each atom in the crystal possesses many other electrons bound to the nucleus according to the Bohr-Rutherford picture of the atoms. These electrons are securely attached to their nuclei and hence do not move through the crystal lattice to conduct electric current. Now the electrons in the crystal, whether bound or free to move, may only possess energies in the allowed bands shown in Fig. 3.10. The bound electrons have large negative potential energies relative to the energy of an electron far removed from the crystal, because they are attached to their nuclei;[2] electrons physically closer to the nucleus have larger negative energies. These latter electrons represent the lower-energy electrons.

A. Energy Bands in Solids from Atomic States

It turns out that within an energy band there are a *finite number of energy levels* which electrons can occupy. This can be shown by a qualitative argument making plausible the manner in which energy bands are formed for a crystal. First consider the electron energy levels characteristic of an isolated atom. These levels are similar to those already discussed for the hydrogen atom although, in general, the location and spacing of these levels will depend upon the particular atom chosen. A result of quantum mechanics is that a discrete energy level of an atom in a perturbing electric or magnetic field will break up or split into two or more close-spaced levels due to the field. If a crystal is considered to be formed by bringing a number of atoms in close proximity to each other, the electrons of one atom will interact with the electric field of another atom (and vice versa), causing a splitting of its energy states. If a multitude of atoms interact in the crystal, an energy band is formed consisting of many such close-spaced energy levels. This is shown in Fig. 4.1, which indicates the splitting and shift of atomic energy levels as the lattice spacing is reduced. Since the separation of these multiple levels is very small (of the order of 10^{-19} eV) and their number is large, this group of allowed levels is referred to as an *energy band* of the type derived in the Krönig-Penney problem.

A theorem of quantum mechanics states that bringing atoms together leaves the total number of quantum states, corresponding to each of the atoms, unchanged. Hence in a crystal, the band formed from the s atomic state can hold 2 electrons per atom, one for each direction of spin. This latter result comes from a postulate due to Wolfgang Pauli which states that only one

---

[1] This "free electron" theory of metals was already postulated by Drude in 1900. See P. Drude, "Electron Theory of Metals," *Ann. Phys.* Vol. 1–3, p. 566, March, 1900.
[2] Energy must be supplied to remove them.

FIGURE 4.1

To the right are shown two of the energy levels characteristic of each isolated atom in a crystal. As the lattice constant or spacing between such atoms is reduced, a splitting of energy levels occurs owing to their mutual interaction. At a spacing r_0 two energy bands with a gap between them are indicated. (From W. Shockley, *Electrons and Holes in Semiconductors,* Van Nostrand Rheinhold, New York, 1950. Copyright 1950 by Litton Educational Publishing, Inc.)

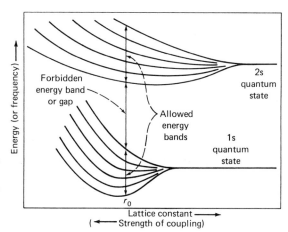

Energy (or frequency) →

Forbidden energy band or gap

Allowed energy bands

2s quantum state

1s quantum state

r_0

Lattice constant →
(← Strength of coupling)

electron may occupy each quantum state available.[3]   The s-state has two spin states, $m_s = \pm\frac{1}{2}$. This famous Pauli exclusion principle is essential for understanding the electrical properties of solids as well as the periodic table of the elements. Correspondingly, the p atomic state can contain 6 electrons per atom, and the d-state can hold 10 electrons per atom.

B. Electrical Conduction in Energy Bands

Now all the electrons in the solid may be introduced into the available energy states, the lower energy levels being filled first according to the "least energy" principle of physics. The lower energy bands are usually filled with electrons and represent the innermost (closest to the nucleus) bound electrons. *No electric current can be carried by electrons in such a filled band.* The reason for this can be seen by referring to Fig. 3.10. For any value of electron energy, because of the symmetry of the E-versus-k curve, there exist two states having wave vectors $+k$ and $-k$. This corresponds physically to equal electron momenta in the $+x$ and $-x$ directions. Hence if all energy states are filled, in thermal equilibrium the electrons are paired, with as many moving in one direction as in the other; the *net* momentum of all electrons is zero. When an electric field is applied, for a filled band there will still be just as many electrons traveling in one direction as in the opposite direction and hence the net current is zero, for every available state in the band is filled and there is no way for the field to increase the total momentum of the electrons in the direction of the field.

Of course, the field can supply energy to these electrons, conceivably raising a few across a forbidden energy gap to the next higher band, which is perhaps not completely filled.   Then the possibility does exist that these excited electrons

---

[3] This postulate was proposed prior to the Schrödinger theory, just after the Bohr hydrogen-atom model was developed. The quantum states here refer to the spatial quantum numbers as well as the spin quantum number (see Section 3.9).

FIGURE 4.2

The band structure of (a) a typical insulator, (b) a metal, and (c) a semiconductor according to the energy band theory of solids. The cross-hatched areas represent bands filled with electrons. The Fermi energy E_F is indicated for all three cases. Its significance will be discussed soon.

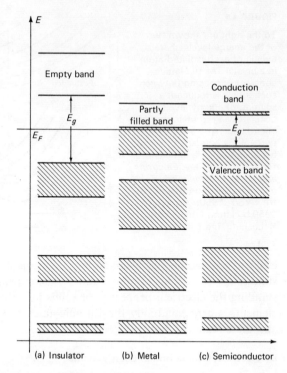

(a) Insulator   (b) Metal   (c) Semiconductor

may enter into electrical conduction. This, in fact, does happen under very high field conditions (10^8 V/m) under special conditions in semiconductor diodes and is known as the *Zener effect*.[4] Normally, however, electrical conduction will occur only if an energy band is partially filled. This is the case in a number of metals such as copper, silver, sodium, and aluminum, in which the highest energy band containing electrons is about half filled.

C. Electrical Conduction in Metals and Insulators

Figure 4.2 indicates the distinction between a metal and an insulator in terms of the energy band theory of solids. In an insulator (Fig. 4.2a) the number of electrons present is just sufficient to fill some lowermost energy bands. There is an empty band just above the filled bands where electron conduction can take place. A forbidden energy gap of 3 to 5 eV (E_g) separates the highest filled band from the next empty band. Perhaps at very high temperatures sufficient thermal energy may be present to cause a few electrons to be raised into conduction. However, the probability that an electron in the filled band will make a transition upward in energy to a conduction band is very small at room temperature. Hence at low temperatures insulators like mica,

[4] C. Zener, "Non-Adiabatic Crossing of Energy Levels," *Proc. Roy. Soc.*, Vol. A137, p. 696 (1932).

quartz, and diamond have electrical resistivities of greater than 10^{14} Ω-m (see Table 4.1).

This is in contrast to the resistivities of about 10^{-8} Ω-m exhibited by some metals (also shown in Table 4.1). Here the uppermost band is about half filled with electrons (Fig. 4.2b), accounting for the high electrical conductivity. (The lower filled bands correspond to tightly bound electrons near the atomic nucleus, whereas the conducting electrons correspond to the outer electrons normally responsible for chemical valence.)

D. Electrical Conductivity of Semiconductor Materials

There is a class of materials called *semiconductors* which appear quite similar in band structure to insulators. In fact, the major difference is the width of the energy gap above the uppermost filled band. Figure 4.2c is a sketch of the band structure of an intrinsic (pure) semiconductor. The uppermost filled band is called the *valence band*. The energy gap E_g is usually between 0.1 and 2 eV. Hence even at room temperature some electrons are thermally excited into the next empty band, called the *conduction band*. These materials have intrinsic resistivities intermediate between insulators and metals or about 1 to 10^5 Ω-m. Some values for the electrical resistivities of typical semiconductors are given in Table 4.1.

The value of energy gap that distinguishes insulators from semiconductors is rather arbitrary. However, active electronic devices primarily utilize semi-conductor materials with about 1 eV gaps. The reason for this is twofold: (1) the electrical conductivity is then sufficient so that the application of a few volts yields some milliamperes of current, and (2) this conductivity can then be

TABLE 4.1

Electrical resistivity of metals, semiconductors, and insulators at room temperature (from R. E. Bolz and G. L. Tuve, *Handbook of Tables for Applied Engineering Science*, Chemical Rubber Company, 1970)

Material	Resistivity, Ω-m
Metals	
Ag	1.59×10^{-8}
Al	2.66×10^{-8}
Au	2.35×10^{-8}
Cu	1.67×10^{-8}
Fe	9.70×10^{-8}
Hg	98.4×10^{-8}
Ni	6.85×10^{-8}
Na	4.1×10^{-8}
Pb	20.6×10^{-8}
Semiconductors	
Ge	0.53
Si	2.3×10^3
GaAs	$> 1 \times 10^6$
Insulators	
Mica	9×10^{14}
Quartz	3×10^{14}
Diamond	10^{14}

modified by the introduction of impurity atoms into the crystal. The latter modify the band structure somewhat by introducing energy levels into the forbidden energy gap, thereby enhancing electrical conduction. But this will be discussed in some detail in the later sections on semiconductors. At that time it will be pointed out that there exist two types of *impurity* (as distinct from intrinsic) semiconductor materials; one conducts electricity via negative charges while the other conducts via positive charges. This fact is of primary importance in operation of bipolar (two-carrier) junction transistors of the type discussed briefly in Chapter 1.

4.2 INTRODUCTION TO THE QUANTUM PHYSICS OF METALS

The description of the motion of an electron according to the Krönig-Penney quantum mechanical model is rather complex, even in the one-dimensional case. Solutions of the three-dimensional problem, such as are needed for real elemental crystals like sodium or germanium, cannot be obtained in closed form. Approximation techniques are used, and calculations need to be carried out with the aid of computers. However, as previously discussed, the general aspects of the three-dimensional solutions are indicated by one-dimensional analysis. In fact, many of the electrical and thermal properties of crystal materials can be described from a quantum mechanical point of view, in simple terms, if appropriate approximations are made. The trick is to select an approximation which is valid under the conditions of the problem under discussion. For example, the Krönig-Penney solution predicts a complicated relationship between the energy of the electron in a crystal lattice and its wave vector (wavelength or momentum), that is, $E = E(k)$. However, when this function is plotted (as in Fig. 3.10), it is seen that for a small range of energies near where discontinuities of energy occur or at $k = 0$, ΔE is essentially proportional to k^2. This parabolic relationship suggests that the electron with wavelength other than that corresponding to the Bragg condition (see Eq. 3.63) is essentially free. This nearly free electron at $k = 0$, or with energy close to a discontinuity at E_0, will have energy of motion given by

$$|E - E_0| = \frac{1}{2}mv^2 = \frac{p^2}{2m} = \frac{\hbar^2 k^2}{2m}, \qquad (4.1)$$

which indicates that this simple expression can be applied, but only where strong interaction with (or scatter by) the lattice atoms is negligible. This results in the *free electron* theory, which can be successfully applied to several univalent metals like copper, silver, and sodium. This theory will lead us to a definition of the Fermi energy, which is of importance in describing many of the properties of both metals and semiconductors.

Next the concept of the Brillouin zone will be introduced in order to explain the behavior of divalent metals such as cadmium, zinc, and calcium. This will lead directly to the concept of *holes* as positively charged electric carriers, present in some metals and particularly in semiconductors. Finally, a derivation of Ohm's law will be given from a microscopic point of view to describe the transport of charged carriers in metals and semiconductors.

A. Free Electron Theory of Metals: The Fermi Energy

A number of important physical properties of monovalent metals can be described in terms of a free electron gas made up of the outer electrons which are loosely bound to the metal-atom nuclei. These valence electrons are known as *conduction* electrons. For a detailed understanding of the electrical properties of these metals, it is necessary to determine the number of energy states available to electrons in the conduction band. This will now be pursued through the concept of *momentum space* and the uncertainty principle.

Let us extend Eq. (4.1) to three dimensions by writing the electron kinetic energy as

$$E = \frac{p_x^2 + p_y^2 + p_z^2}{2m},$$

(4.2)

FIGURE 4.3

a) An octant in three-dimensional momentum space. The spherical surface is one of constant energy for free electrons.
b) An elemental volume in momentum space whose size is h^3/L^3.

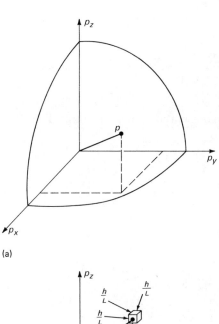

(a)

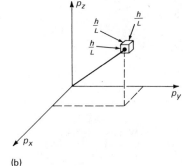

(b)

where p_x, p_y, and p_z are the x, y, and z components of momentum, m is the electron mass, and E_0 is arbitrarily taken as zero. It is convenient now to define a momentum space in a manner similar to the way configurational (x, y, z) space is normally defined. That is, consider three mutually perpendicular orthogonal axes labeled p_x, p_y, and p_z. Figure 4.3 shows an octant in this three-dimensional space. Also shown is a spherical surface in this space which is a surface of constant energy. This follows from Eq. (4.2), compared with the well-known analogous equation of a spherical surface in configurational space, $x^2 + y^2 + z^2 = r^2$, where r is the sphere radius. The radius of the sphere in momentum space is $\sqrt{2mE}$.

In classical theory a continuous set of energy values would be available to the essentially free electrons. However, the uncertainty principle restricts our ability to define both the electron momentum and its position with arbitrary accuracy; this leads to a discrete number of energy values in a crystal of finite dimensions. Let us now determine the number of electron states possible in a crystal of a given size.[5]

If the metal crystal in which we are considering these electrons is assumed to be a cube whose sides are of length L, the electron's location is given by

$$0 < x < L,$$
$$0 < y < L, \qquad\qquad (4.3)$$
$$0 < z < L.$$

Hence $\Delta x = \Delta y = \Delta z = L$. The minimum uncertainty (fuzziness) of the electron momenta in each of the three directions is then

$$\Delta p_x = \frac{h}{L},$$

$$\Delta p_y = \frac{h}{L}, \qquad\qquad (4.4)$$

$$\Delta p_z = \frac{h}{L}.$$

This follows from Eq. (3.3). The product $\Delta p_x \, \Delta p_y \, \Delta p_z$ defines an elemental volume in momentum space (see Fig. 4.3b) and represents a discrete electronic state. That is, it is not possible to define or resolve any electron state of momentum or energy any more finely than by this volume in momentum space. Hence, from Eq. (4.4), this elementary quantum state is defined by

$$\Delta p_x \, \Delta p_y \, \Delta p_z = \frac{h^3}{L^3}. \qquad\qquad (4.5)$$

---

[5] This problem may be treated by solving the Schrödinger equation for a free electron in a three-dimensional box; such a formulation *automatically* yields energy quantization corresponding to the uncertainty principle!

These cubes in p-space correspond to electronic states with energies below some maximum, which we will call the *Fermi energy*, $E_F = E_{max}$. Hence all electron states are contained within a sphere of radius $\sqrt{2mE_F}$.

Consider the above description to refer to the electrons in the conduction band in a monovalent metal, as indicated in Fig. 4.2b, which is filled up to a certain maximum energy value which we now refer to as the Fermi energy. The exclusion and minimum energy principles dictate that the N electrons present will occupy all energy states up to a value E_F. Since two electrons (with opposite spin) can be accommodated in each elemental momentum space volume, the number of these electrons can be expressed as twice the total volume under E_F divided by the elemental volume or

$$N = 2 \frac{(4\pi/3)(2mE_F)^{3/2}}{(h/L)^3}. \tag{4.6}$$

The Fermi energy then can be obtained by rewriting Eq. (4.6) as

$$E_F = \frac{h^2}{8m}\left[\frac{3}{\pi}\frac{N}{L^3}\right]^{2/3}. \tag{4.7}$$

Hence the Fermi energy of a crystal of volume L^3 depends only on the number of electrons per unit volume of metal crystal. Table 4.2 gives some values of Fermi energy calculated for a number of monovalent metals.

EXAMPLE 4.1

The density of sodium is 970 kg/m³, and its atomic weight is 2.3×10^{-2} kg/gm-atom. Assuming that the free electron theory of metals applies to sodium, compute the Fermi energy of this metal.

SOLUTION

$$\text{Number of Na atoms/m}^3 = \frac{970 \text{ kg/m}^3(6.02 \times 10^{23} \text{ atoms/gm-atom})}{2.3 \times 10^{-2} \text{ kg/gm-atom}}$$
$$= 2.5 \times 10^{28} \text{ atoms/m}^3$$

Assuming that one electron per atom contributes to electrical conduction, $N/L^3 = 2.5 \times 10^{28}$ electrons/m³ and from Eq. (4.7)

$$E_F = \frac{h^2}{8m}\left(\frac{3}{\pi}\frac{N}{L^3}\right)^{2/3}$$

$$= \frac{(6.62 \times 10^{-34})^2 \text{ J}^2\text{-sec}^2}{8(9.1 \times 10^{-31}) \text{ kg}}\left(\frac{3}{\pi} \times 2.5 \times 10^{28} \text{ electrons/m}^3\right)^{2/3}$$

$$= 4.9 \times 10^{-19} \text{ J} \frac{1 \text{ eV}}{1.6 \times 10^{-19} \text{ J}} = 3.1 \text{ eV}.$$

TABLE 4.2

Electron concentration
and Fermi energy in
metals (it is assumed that
each atom contributes
one conduction electron)

Element	Electron concentration N/V, m^{-3}	Energy E_F, eV
Li	4.6×10^{28}	4.7
Na	2.5×10^{28}	3.1
K	1.34×10^{28}	2.1
Rb	1.08×10^{28}	1.8
Cs	0.86×10^{28}	1.5
Cu	8.50×10^{28}	7.0
Ag	5.76×10^{28}	5.5
Au	5.90×10^{28}	5.5

The Fermi energy as derived here simply represents the maximum kinetic energy of electrons in a monovalent metal. The concept is much more general, though, and is of extreme importance in the description of the electronic properties of semiconductors. The Fermi energy is related to the chemical potential, which is a thermodynamic parameter that defines the state of a thermodynamic system, such as a semiconductor, in equilibrium. The Fermi energy can be measured experimentally for a metal since it is related to the *work function* of the metal. (The latter is defined as the energy necessary to remove an electron from a metal to a point infinitely far removed.) Thermionic cathode emission experiments can be used to measure work function and hence Fermi energy.[6]

In the previous derivation of the Fermi energy the electron mass was considered as that of an electron in free space, m. However, in our present problem the electron is actually moving through a crystal lattice. It is found convenient to take into account the effect of the lattice ions on the motion of the nearly free electrons by assigning them an effective mass m^*. (The broader significance of this concept will be discussed in the next chapter.) If this is done then the motion of the electrons, for example in an applied electric field, can be given a pseudo-classical description in terms of Newton's law, $F = m^*a$. It can be seen that the degree of binding of the electron to the lattice ions will determine its effective mass. Strong binding would be indicated by a larger effective mass, as the ionic electric field slows down the electron. However, the picture is not quite so simple, and it is found from electron conductivity measurements that the ionic field in some cases actually aids electron motion, or that the effective mass is somewhat lower than the free-space mass. A more rigorous and general definition of the effective mass will be given later in a more detailed discussion of electron transport. Table 4.3 gives values of the effective mass of electrons in some metals as calculated from Eq. (4.7), using measured values of Fermi energy.

[6] For details see A. J. Dekker, *Solid State Physics*, p. 212, Prentice-Hall, Inc., Englewood Cliffs, N. J. (1965).

TABLE 4.3

Ratio of the effective mass of the electron to the free electron mass (estimated from experimental Fermi-level measurements)

Metal	m^*/m
Al	0.97
Ca	1.4
Cu	1.0
Li	1.2
Na	1.2
K	1.1
Be	1.6

B. Density of Electronic States in Metals

To explain the high electrical conductivity of monovalent metals such as sodium, as compared with that of a divalent metal such as cadmium, it is necessary to calculate the number of energy states which electrons can occupy in any given energy interval. This *density of states* will be calculated assuming a nearly free electron model. The results of this calculation are quite suitable for a monovalent metal. However, it will be necessary to introduce the concept of the Brillouin zone to modify the simple theory in the case of divalent metals. This modification must not only explain the lower electrical conductivities usually observed in divalent metals, but also why, in the case of cadmium for example, the charge carriers are observed to be *positive*.

Let us return to Eq. (4.6) and write down the number of electronic states with energy between some value E and $E + dE$ (see Fig. 4.4). This number may be obtained by determining the elemental volume in momentum space between the spherical shells corresponding to energies E and $E + dE$, and dividing by the elemental volume of one electronic state, h^3/L^3. Letting dE go to 0 will yield the desired number. Mathematically this operation may be performed by

FIGURE 4.4

A figure useful for the determination of the density of electronic states versus energy of nearly free electrons in a metal. Two spherical surfaces of constant energy in momentum space, differing only infinitesimally in energy, are shown.

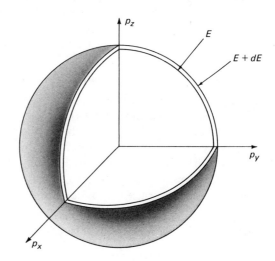

FIGURE 4.5

Density of filled electronic
states versus energy for
sodium obtained by observing x-ray emission on electron bombardment of sodium.
(From H. Skinner, *Prog. in
Phys.*, Vol. 5, 1938.) Also
indicated is a plot (dashed)
of the theoretical prediction
of Eq. (4.9).

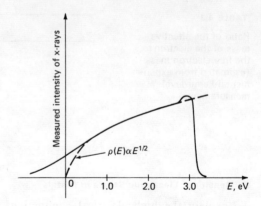

taking the differential of the total number of states N with energy less than E, which from Eq. (4.6) is $N(E) = 8\pi(2m^*E)^{3/2}L^3/3h^3$. This gives

$$dN = \frac{4\pi(2m^*)^{3/2}L^3}{h^3} E^{1/2}\, dE. \qquad (4.8)$$

Hence the energy density-of-states function, $\rho(E) = (dN/dE)/L^3$, is

$$\rho(E) = \frac{8\sqrt{2}\pi(m^*)^{3/2}}{h^3} E^{1/2}, \qquad (4.9)$$

which represents the number of quantum states at any energy per unit energy per unit volume; it varies as the square root of the energy. These energy states exist in the energy band and are available to electrons but are not necessarily occupied.

The form of this density-of-states function and the filling of the energy states to some maximum energy value (Fermi level) was confirmed experimentally for the metal sodium by H. Skinner in 1938.[7] His results, shown in Fig. 4.5, were obtained by observing the x-ray emission from sodium bombarded by high-energy electrons. This type of experiment gives information about the electrons occupying energy levels in metals in a manner similar to the method of studying energy levels in atoms in a gas by electric discharge excitation.

4.3   THE FERMI-DIRAC DISTRIBUTION FUNCTION

The density of electronic energy states in an energy band has been derived. It has been stated that in a metal these states are generally filled up to some maximum energy value E_F. Another way of expressing this result is that the probability of occupancy of energy levels by electrons is 1 below E_F, and 0 above E_F. The fact is that the latter statement is an approximation, only true at a temperature of absolute zero. At any other temperature the thermal energy present will raise (or excite) electrons to higher energy levels, leaving

[7] H. Skinner, "Soft X-Ray Spectroscopy of Solid State," *Progress in Phys.* Vol. 5, p. 257 (1938).

FIGURE 4.6

The Fermi function $f(E)$ for two temperatures: (a) $T = 0°K$, and (b) $T > 0°K$. At $T = 0°K$ the probability of occupancy of energy levels by electrons is absolute (1) for energy values less than the Fermi energy E_F, and zero for $E > E_F$. At $T > 0°K$, electrons may occupy levels above E_F. At $E = E_F$ the occupation probability is always $1/2$.

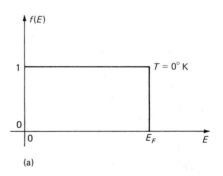

(a)

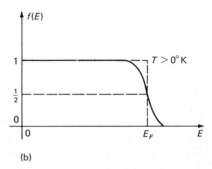

(b)

some levels below empty. A function for expressing the probability of an electron occupying any particular energy level E at an absolute temperature T is given by

$$f(E) = \frac{1}{e^{(E-E_F)/kT} + 1}.$$

(4.10)

This is the famous *Fermi-Dirac distribution function* which was derived by Enrico Fermi in 1926[8] for electrons obeying the Pauli exclusion principle. This derivation was accomplished by asking the question: Given a set of energy states, what is statistically the most probable distribution of electrons among these states, assuming no *a priori* preferable state and no more than one electron per state?

The function is plotted for $T = 0°K$ in Fig. 4.6a and indicates that all energy levels are occupied [$f(E) = 1$] for energy values below the Fermi energy E_F, and empty [$f(E) = 0$] for energy values above E_F. This is no longer true at $T > 0°K$ as is shown in Fig. 4.6b. Here the effect of temperature is to make it possible for some electrons to have energies above E_F. In fact, there is a reasonable probability that some electrons have kT units of energy in excess of E_F, but

[8] This function was derived by Fermi and, independently, by P. A. M. Dirac in the same year. See E. Fermi, "Quantization of the Ideal Monatomic Gas," *Zeit. Phys.* Vol. 36, p. 902 (1926).

FIGURE 4.7

Graph of the number of electrons per unit volume in any energy state between E and $E + dE$ at a temperature $T > 0°$K. The Fermi energy at which the occupation probability is $1/2$ is indicated.

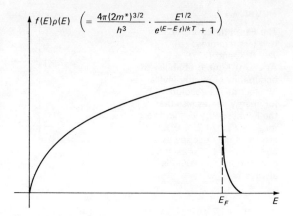

$$f(E)\rho(E) \quad \left(= \frac{4\pi(2m^*)^{3/2}}{h^3} \cdot \frac{E^{1/2}}{e^{(E-E_f)/kT} + 1} \right)$$

not much more. This result may be obtained by referring to Eq. (4.10). Since the Fermi energy is of the order of 5 eV for metals and kT is only 0.026 eV at room temperature, the statement that all levels are occupied below E_F is a very good approximation even at temperatures above $0°$K.

Figure 4.6b indicates a very useful definition of the Fermi energy level. *The Fermi level is that energy for which the probability of occupancy by an electron is $\frac{1}{2}$.* This concept is universally true for all electrons which obey the Fermi statistics and will be especially helpful in our discussion of semiconductor materials.

Expressions have now been given for the density of states and the probability of occupancy of these states by electrons. The product of the quantities given by Eqs. (4.9) and (4.10) yields an equation specifying the number of electrons per unit volume occupying an energy state between E and $E + dE$ at a particular temperature. A plot of this function is shown in Fig. 4.7. This concept will be particularly useful in the discussion of the electrical properties of semiconductors in the next chapter.

EXAMPLE 4.2

Determine the probability that an energy level in sodium is occupied by an electron at $300°$K if it is located above the Fermi level by (a) 0.026 eV ($=kT$), (b) 0.078 eV ($=3kT$). The Fermi level of sodium is located 3.1 eV above the bottom of the conduction band.

SOLUTION From Eq. (4.10),

a) $f(E)|_{0.026\,\text{eV}} = \dfrac{1}{e^{kT/kT} + 1} = 0.27.$

b) $f(E)|_{0.078\,\text{eV}} = \dfrac{1}{e^{3kT/kT} + 1} = 0.05.$

The latter result indicates that at room temperature, an energy level (0.078 eV/ 3.1 eV) × 100, or 2.5 percent higher than the Fermi level in sodium, has a chance of only 1 in 20 of being occupied. At the Fermi energy this chance of occupancy is 1 in 2.

4.4 BRILLOUIN ZONES: HOLES

Knowing something about the density and occupation of states in a metal permits us to come back to the question of the difference in electrical conductivity between monovalent and divalent metals. It has already been indicated that the conduction band in a metal has finite width and hence a limited number of electronic states. The number of nearly free electrons available for conduction in a monovalent metal has already been indicated to be about one per atom of the crystal. It would appear that a divalent metal would have about twice that number. It can be shown[9] that the conduction band in a metal which has one atom per primitive cell contains two energy states per atom. Hence the monovalent conduction band would be only half full while the divalent metal conduction band would be nearly full. Since no electrical conductivity can be obtained in principle when a band is filled, the divalent metal should be a poor conductor or even an insulator. This is not the total picture, and these divalent metals are not insulators. However, generally, divalent metals such as cadmium and zinc are poorer conductors than monovalent copper or silver, even though they contain twice the number of valence electrons.

The density of energy states as formulated in Eq. (4.9) and shown in Fig. 4.5 applies only to the case of nearly free electrons in monovalent metals such as copper or silver. The available electrons are placed into energy states according to the principle of least energy and the exclusion principle, up to the Fermi level. If more electrons are available,[10] these will have to occupy higher energy states corresponding to higher momenta. By the de Broglie relation this corresponds to electrons of shorter wavelength or larger wave vector k. When k becomes comparable to the atomic spacing a, the Krönig-Penney formulation of this problem indicates intense interaction with the binding field of the nuclei. These electrons can no longer be considered as free; hence a modification must be made in the simple theory presented thus far. The technique follows from a technique for analyzing periodic structures developed by Leon Brillouin.[11] For this purpose the concept of the Brillouin zone will be introduced.

To define the first Brillouin zone refer to Fig. 3.10 and the graph of E versus k. Note that the curve is nearly parabolic ($E \propto k^2$) for small values of k. Since the electron momentum is $p = \hbar k$, the electrons in these lower energy

---

[9] See C. Kittel, *Introduction to Solid-State Physics*, 4th edition, John Wiley & Sons, New York (1971).

[10] These come from the second valence electron attached to each divalent metal atom.

[11] See L. Brillouin, *Wave Propagation in Periodic Structures: Electric Filters and Crystal Lattices*, 2nd ed., Dover, New York (1953).

FIGURE 4.8

Cubic first Brillouin zone for
a simple cubic lattice. If electrons fill the sphere shown
within this zone, the energy
corresponding to the surface
of this sphere is the Fermi
energy.

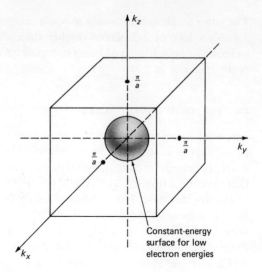

Constant-energy
surface for low
electron energies

states are nearly free. However, for values of k approaching $\pm\pi/a$ the curve
no longer follows this parabola, and in fact, has a point of inflection. At
$k = \pm\pi/a$ there is an energy discontinuity indicating the top of the energy
band and the beginning of a forbidden energy gap. This is a one-dimensional
analysis, but the concept can be extrapolated to three dimensions to represent
the case of a simple cubic three-dimensional crystal. Then the scattering
boundaries are denoted by planes given by $k_x = \pm\pi/a$, $k_y = \pm\pi/a$, and
$k_z = \pm\pi/a$, which define a cube.[12] This is shown in Fig. 4.8; the volume within
the cube is known as the *first* Brillouin zone. The second, third, etc., Brillouin
zones derive from the second, third, etc., energy bands shown in Fig. 3.10.

A filled energy band means that all the energy states[13] contained in the
Brillouin zone are occupied by electrons. A material for which this were true
would be an electrical insulator if the energy gap above this band were large
(3 to 5 eV). Consider the filling of the first Brillouin zone with electrons suc-
cessively. Initially these lie within the volume of a sphere in k-space or momentum space, the surface of the sphere representing constant energy. As more
electrons are introduced, the sphere radius increases until it approaches the
walls of the Brillouin zone. Then near the zone walls, the surface deviates
from a spherical form owing to interaction with the periodic field of the ions.
A sketch of possible surfaces of this type is shown in Fig. 4.9, drawn in two
dimensions for simplicity. (Here the loci of constant energy are contour
lines.) Note that when sufficient electrons are available to fill states up to

[12] Brillouin-zone geometries are much more complicated in most other crystal structures,
such as bcc, fcc, hcp.
[13] An elemental energy state in this k-space has a volume $(2\pi/L)^2$ which follows from the
discussion of momentum space and the uncertainty principle in Section 4.2.

energy E_3, the constant-energy line is no longer circular and, in fact, becomes disconnected. When more electrons are added up to energy E_4, the constant-energy surface breaks up into four symmetrical parts, each of which is nearly circular but with a center now at $(k_x, k_y) = (\pm\pi/a, \pm\pi/a)$, rather than at $(k_x, k_y) = (0, 0)$ as for low-energy surfaces. This indicates that electrons with wave vectors and energies corresponding to the corners of the Brillouin-zone cube (square in two dimensions) have energies given by

$$E = E_{max} - \frac{h^2(k')^2}{8\pi^2 m^*} = E_{max} - \frac{h^2}{8\pi^2 m^*}\left[\left(k_x \pm \frac{\pi}{a}\right)^2 + \left(k_y \pm \frac{\pi}{a}\right)^2 + \left(k_z \pm \frac{\pi}{a}\right)^2\right], \quad (4.11)$$

where E_{max} is the energy corresponding to a corner of the Brillouin zone.

The energy relation for these electrons is nearly parabolic $[|(E_{max} - E)| \propto (k')^2]$ as for free electrons, and in many respects can be analyzed in a similar manner. (It must be borne in mind that these electric carriers are certainly not free, but in some respects can be treated as such.) Note that for such electrons the *rate* of energy increase *decreases* as its wave vector (momentum) *increases*. (See Eq. 4.11.) It is this peculiar property that indicates that these electrons exhibit a *negative* effective mass. It is often important to account for electrical conduction in a band completely filled with electrons except for a few empty states at the corners of the Brillouin zone. The conduction properties of this nearly filled band are most conveniently accounted for by treating the electrical conduction due to the absence of electrons (with negative charge and negative mass) in these few empty states. This may be accomplished by considering instead that these states are filled with particles of positive charge and positive mass called *holes*.

FIGURE 4.9

Schematic representation of constant-energy contours for a two-dimensional square lattice in the first Brillouin zone. E_1 and E_2 are surfaces corresponding to nearly free electrons. E_3 is a case of strong interaction with the crystal lattice. E_4 represents the case of "holes."

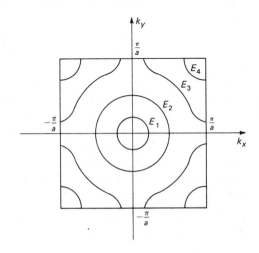

FIGURE 4.10

The density of state functions for (a) an energy band in a simple cubic crystal; (b) the conduction band of a simple monovalent metal; (c) an insulator; (d) the conduction band of a simple divalent metal or the valence band of a semiconductor; (e) a metal with overlapping bands.

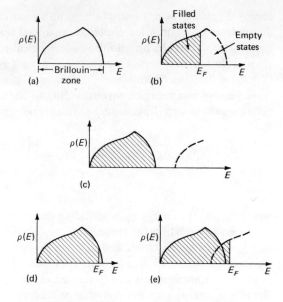

Returning now to the density-of-states function for the complete first Brillouin zone, we arrive at a picture exactly equal to that shown in Fig. 4.8 for low values of energy. However, as the energy increases and the constant-energy surfaces distort and enter the corners of the Brillouin cube, the energy density of states begins to decrease. This results from the fact that for an energy increase dE, the volume in k-space between E and $E + dE$ becomes proportionately less as only a small volume is left in the corners of the Brillouin cube. (See the two-dimensional version indicated in Fig. 4.9.) The complete density-of-states function so derived for the full energy band is shown in Fig. 4.10a.

It is now possible to represent the electron occupation of energy states for monovalent and divalent metals and make predictions about their respective electrical conductivities. Figure 4.10b shows a sketch of the occupation of energy states for a simple monovalent metal. Note that the states are about half filled, so energy supplied by an electric field can accelerate the electrons, raising them to higher available energy states. This is the case of high electrical conductivity. Figure 4.10c is the case of an insulator where no empty energy states are available except for a very few made accessible by electron excitation across the wide forbidden energy gap to the next empty energy band.[14] This is the case of very low electrical conductivity. The extremely low probability for excitation across the energy gap accounts for the conductivity difference of 20 orders of magnitude or more between metals and insulators. Again, if the energy gap is smaller (~ 1 eV) than for the insulator we get semiconductor behavior with electrical conductivity somewhere between metal and insulator.

[14] The excitation probability is proportional to $\exp(-AE_{gap})$ and hence is extremely sensitive to small changes in energy gap.

TABLE 4.4

Electrical resistivity of monovalent and divalent metals at 300°K and the the polarities of their electric carriers

Metal	Charge	Type	Resis-tivity, Ω-m	Electrical carrier charge
Al	—	Mono-valent	2.66×10^{-8}	Negative
Ag	—	Mono-valent	1.59×10^{-8}	Negative
Cu	—	Mono-valent	1.67×10^{-8}	Negative
Cd	+	Divalent	7.40×10^{-8}	Positive
Mg	—	Divalent	4.45×10^{-8}	Negative
Zn	+	Divalent	5.92×10^{-8}	Positive

Finally what sometimes occurs is that some empty energy states are available at the top of the energy band (corners of the Brillouin zone). This is illustrated by Fig. 4.10d. Since some empty states are available, electric conduction is still relatively high, although somewhat lower than that of a simple monovalent metal, owing to the lower density of states in the corner of the zone. The conductivities of monovalent and divalent metals are compared in Table 4.4. The striking aspect of these data is not the numerical differences in electrical conductivity, but the fact that some divalent metals such as Zn and Cd conduct via *positively* charged carriers or holes. This finding is consistent with the behavior expected of electrons in nearly filled energy bands, already discussed for divalent metals.

Semiconductor crystals are particularly interesting from this viewpoint. Their uppermost filled band is not very far removed in energy from the next highest empty band. At absolute zero no conduction takes place in the filled band or, of course, in the empty band; they act as insulators. However, at elevated temperatures thermal excitation raises a few electrons just to the bottom of the empty band, leaving some empty states at the top of the filled band. Hence limited electric conduction takes place not only via electrons in the nearly empty band but also via holes in the nearly filled band. The fact that semiconductor materials can conduct current via both negatively and positively charged carriers makes bipolar, two-carrier transistors possible. This will be discussed more extensively in a later section on semiconductor materials.

Sometimes divalent metals have two consecutive overlapping bands as is shown in Fig. 4.10e. This may cause a material which could be expected to have only a few empty states at the top of an energy band to have many available energy states and hence exhibit strong electron conduction.

4.5 ELECTRONIC TRANSPORT IN METALS

The electronic transport of current carriers in metals will now be considered from a quantum mechanical viewpoint, leading to a simple derivation of

FIGURE 4.11

(a) Energy, (b) velocity, and
(c) effective mass as a function of the electron wave
vector k for an electron
moving in a crystal.

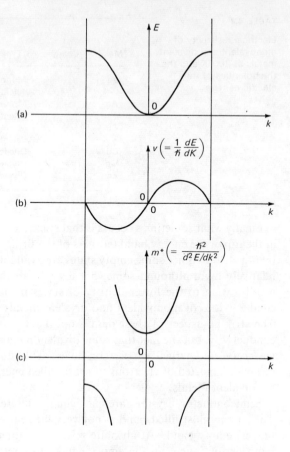

Ohm's law. The previous section discussed the availability of electrons for conduction. This section considers the speed with which they move. These two factors together determine the electrical conductivity of a crystal material.

A. The Electron Velocity

To discuss the motion of an electron in a crystal, let us consider again the energy band structure for a simple crystal. The E-versus-k curve for the first Brillouin zone is redrawn in Fig. 4.11a. To consider the velocity of an electron in such a periodic structure, we invoke the concept of the electron as a wave packet which travels with a group velocity given by $v_g = d\omega/dk$ (see Eq. 3.28). Since quantum mechanics postulates that $E = h\nu = \hbar\omega$, the group velocity of the electron in the crystal is seen by differentiation to be

$$v = \frac{1}{\hbar}\frac{dE}{dk}.$$

$$(4.12)$$

Hence the electron velocity depends on the slope of the E-versus-k curve[15] as indicated in Fig. 4.11b. Note that the electron velocity approaches zero at both the top and the bottom of the band; this must always be true, regardless of the problem, because of the periodic nature of the solution using the Bloch function (see Eq. 3.54). The zero velocity of electrons at the top and bottom of the band can be explained in physical terms using the wave concept of the electron. At the edges of the band the Bragg condition applies (see Eq. 3.63). Hence for electrons of this wave vector (wavelength), standing waves result so that no such electron waves can propagate through the lattice. Note, too, that the electron velocity reaches a maximum somewhere near the middle of the energy band and then decreases for increasing values of energy. This type of behavior is certainly drastically different from the motion of electrons in free space; it is caused by the electrical interaction of the electrons with the ions of the crystal lattice.

B. The Effective Electron Mass

The motion of an electron in an accelerating electric field $\mathcal{E}$ applied to a crystal will now be discussed. Consider the energy gained by the electron when acted on for a short time dt by the electric field. This is given classically by the product of the force F and the distance $dx = v\, dt$, or

$$dE = -q\mathcal{E}v\, dt. \tag{4.13}$$

If we introduce the electron velocity in the crystal given by Eq. (4.12), then (4.13) becomes

$$dE = \frac{q\mathcal{E}}{\hbar}\frac{dE}{dk}\, dt. \tag{4.14}$$

By noting that $dE = (dE/dk)\, dk$, we can solve this equation for dk/dt as

$$\frac{dk}{dt} = -\frac{q\mathcal{E}}{\hbar} = \frac{F}{\hbar}, \tag{4.15}$$

where F is the external force exerted on the electron by the field. From Eq. (4.12) for velocity we can obtain the electron acceleration by differentiation as

$$a = \frac{dv}{dt} = \frac{1}{\hbar}\frac{d^2E}{dk^2}\frac{dk}{dt}. \tag{4.16}$$

Introducing dk/dt from Eq. (4.15) yields

$$a = \frac{-q\mathcal{E}}{\hbar^2}\frac{d^2E}{dk^2} = \frac{F}{\hbar^2}\frac{d^2E}{dk^2}. \tag{4.17}$$

[15] The variation of E with k is always sought by means of solution of the Schrödinger equation for a particular crystal material, no matter how complex in structure.

The force F here is that due only to the electric field applied, excluding the forces of the crystal lattice ions. The interaction of the electrons with the lattice ions is taken into account by the introduction of the concept of an effective mass. The expression for electron acceleration, Eq. (4.17), is strikingly different from the Newtonian classical law of motion, $F = ma$. However, it is convenient to discuss the motion of an electron in a crystal in a manner similar to the way we describe a ball rolling down an inclined plane. This is possible if we compare Eq. (4.17) with Newton's law and define an effective electron mass as

$$m^* = \frac{\hbar^2}{d^2E/dk^2}. \tag{4.18}$$

Now mathematically the second derivative of E with respect to k represents the curvature of the E-versus-k curve of Fig. 4.11a. Zero curvature occurs at points of inflection; owing to the reciprocal relationship between effective mass and curvature, the effective mass becomes infinite there. This is shown in Fig. 4.10c. Note, too, that the effective mass has a minimum absolute value at both the bottom and the top of the energy band. However, this mass is *positive* at the bottom of the band but *negative* at the top. Hence, for a quantum mechanical description of the motion of an electron in a crystal lattice, the mass can no longer be considered a constant in the classical sense. To provide a pseudo-classical description, the particle is considered to move as a free particle according to Newton's laws, and the effect of the crystal lattice is taken into account by defining an essentially constant effective mass. This is only possible for particles with energy near the bottom or top of a band. In this way it is possible in many cases to consider the electron as nearly free and to introduce the effect of the periodic potential by substituting a constant m^* for m, since the electrons are often near an extremum in $E(k)$.

The significance of the negative mass assigned to electrons at the top of the band is related to the discussion of electron states at the top of a nearly filled energy band in Section 4.4. There the concept of the hole was defined, and it was indicated that the reduced rate of increase of energy with increasing momentum would lead to the concept of a negative electron mass. It was then noted that for convenience the motions of particles in a nearly filled band could be described by assigning to the empty states pseudo-particles (holes) with a positive mass and a positive charge. The concept of holes and electrons is particularly important in the description of semiconductor materials and will be discussed again in the next chapter.

A method of measuring the effective mass experimentally is by solid-state cyclotron resonance. The electrons are made to rotate in circles by subjecting them to a magnetic field. In analogy with the atomic-physics experiment using electrons in free space, the rotational (cyclotron) frequency of the electrons is

$$f = \frac{qB}{2\pi m^*} \quad \text{or} \quad \omega = \frac{qB}{m^*}, \tag{4.19}$$

where B is the magnetic induction, q the electronic charge, and m^* the effective mass. If the material under test is introduced into a tuned circuit and a microwave signal is impressed, a resonance or absorption of energy will occur at the specific frequency given by Eq. (4.19). This indicates that at this signal frequency energy is being absorbed by the rotating electrons, and this then permits calculation of the effective mass. This technique is used most effectively for semiconductor materials. Skin effect complicates the measurement in the case of metals. Then, the carrier effective mass may be calculated from the thermal conductivity since electrons are responsible for heat conduction in metals.

C. Ohm's Law

The task which still remains is to calculate Ohm's law for electrical conduction in solids. The formulation is due to Drude and was made in 1900, long before quantum theory. This treatment results in a correct phenomenological equation. However, the details of the parameters in Drude's theory are only now calculable using a microscopic quantum mechanical approach.

Under the influence of an electric field $\mathcal{E}$, a randomly moving free electron would have an acceleration $a = q\mathcal{E}/m$ in a direction opposite to the field. Its velocity would continue to increase with time. The electron in the crystal lattice, however, will collide with some lattice ion a certain time after having been put in motion; the average time between random collisions is taken as τ, which may be a function of the random electron velocity. The result of a collision is to randomize the motion or remove all the electron velocity due to the accelerating field.[16] Hence after a collision the electron "relaxes" to its condition of random velocity as before acceleration. The time τ is called the *relaxation time*. The average velocity increase of the electrons between collisions due to the field (drift velocity) is given by

$$\bar{v}_d = \frac{1}{2} at = \frac{1}{2} \frac{q\mathcal{E}\tau}{m^*}, \qquad (4.20)$$

where the effective mass is introduced to take into account the fact that the electron is moving in a crystal lattice. Now the density of electrons n moving per second in a particular direction through 1 m^2 of area perpendicular to that direction is $n\bar{v}_d$, the electron flux. The rate of charge flow is the particle charge times this flux, which constitutes the electric current density given by

$$J = nq\bar{v}_d. \qquad (4.21)$$

Introducing Eq. (4.20) into (4.21) gives

$$J = \frac{nq^2\tau}{2m^*} \mathcal{E}. \qquad (4.22)$$

[16] Note that the electron does not start from rest. In the absence of an externally applied electric field, the electron will have a velocity, as previously described, according to its state in the energy band.

This is the correct form of Ohm's law except for the factor of 2 in the denominator. The latter is eliminated if τ is averaged over the distribution of electron velocities.

Since the usual statement of Ohm's law is $J = \sigma \mathcal{E}$, where σ is the electrical conductivity of the crystal, this conductivity is

$$\sigma = \frac{nq^2\tau}{m^*}.$$ (4.23)

D. The Electron Mobility

It is convenient as a measure of the speed of an electron in a field to define an electron carrier *mobility* μ which is defined as the velocity per unit applied electric field; that is,

$$\mu = \frac{\bar{v}_d}{\mathcal{E}} = \frac{q\tau}{m^*}.$$ (4.24)

Then the conductivity can be written as

$$\sigma = nq\mu.$$ (4.25)

EXAMPLE 4.3

Assuming the free electron theory of metals to apply to silver and using the data of Tables 4.2 and 4.4, determine the mobility of the electron current carriers in silver at 300°K. Also determine their collision relaxation time.

SOLUTION   Assume one conduction electron per atom as in Table 4.2 and note that the electrical conductivity is the reciprocal of the resistivity as given in Table 4.4. Then Eq. (4.25) gives for the electron mobility

$$\mu = \frac{\sigma}{nq}$$

$$= \frac{(1.59 \times 10^{-8})^{-1}(\Omega\text{-m})^{-1}}{(5.76 \times 10^{28} \text{ electrons/m}^3)(1.6 \times 10^{-19} \text{ C})}$$

$$= 6.8 \times 10^{-3} \text{ m}^2/\text{V-sec}.$$

The relaxation time is obtained from Eq. (4.24) as

$$\tau = \frac{\mu m^*}{q}$$

$$= \frac{6.8 \times 10^{-3} \text{ m}^2/\text{V-sec } (9.1 \times 10^{-31} \text{ kg}}{1.6 \times 10^{-19} \text{ C}}$$

$$= 3.9 \times 10^{-14} \text{ sec}.$$

Drude's theory does not permit evaluation of the parameters τ and m^*. It is the task of quantum mechanics to do so. If the crystal structure of the material is known, the ionic potential field can be estimated and introduced into Schrödinger's equation to yield an E-versus-k relation. Now the effective mass can be calculated from Eq. (4.18) and is a property of the material whether or not there is an applied field. At absolute zero temperature, where the ions in the crystal lattice have no thermal energy, in principle the electrons would move through the lattice without any net scattering. That is, the lattice ions would periodically supply equal amounts of acceleration and deceleration, and the electrical conductivity would tend toward infinity. However, at any finite temperature the lattice vibrations would tend to scatter the electrons in motion since all ions at any particular time would not be in their precisely periodic locations. This has already been described as a type of crystal irregularity or crystal defect. The electrical conductivity of a metal is determined by adding the effects of all crystal defects. For example, at low temperatures the conductivity of an elemental metal is controlled by trace impurity atoms and other crystal imperfections which contribute to electron scatter.

The quantum mechanical calculation of the lattice relaxation time by consideration of these scattering mechanisms is quite difficult. It is often much easier to measure the electron mobility in an electric field, measure the effective mass by a method such as cyclotron resonance, and then calculate the relaxation time from Eq. (4.24). Sometimes (as in the case of semiconductor materials) it is possible to measure the carrier mobility directly. However, this has not been done in metals, so the mobility has to be inferred from Eq. (4.25). In this equation, however, the carrier density n is not known but can be determined by Hall measurement. This phenomenon is discussed next.

E. The Hall Effect

The carrier density can be determined directly by measuring the Hall effect.[17] Here a voltage is measured in the y-direction in a sample while an electric field is applied in the x-direction and a magnetic field is applied in the z-direction, as shown in Fig. 4.12. The Hall voltage measured in the y-direction is given by

$$V_H = \frac{IB}{Wnq}, \qquad (4.26)$$

where I is the electric current in the x-direction, B is the magnetic induction in the z-direction, W is the sample thickness in the z-direction, n is the carrier density, and q is the carrier charge. If the Hall voltage, the current, and the magnetic field are measured, then the carrier density can be calculated directly. By taking into account the polarity of the voltage, the sign or polarity of the

[17] E. H. Hall, "The Hall Effect," *Science*, Vol. 5, p. 249 (1885).

FIGURE 4.12

The directions of current, magnetic field, and Hall field for determining the carrier density in a rectangular sample.

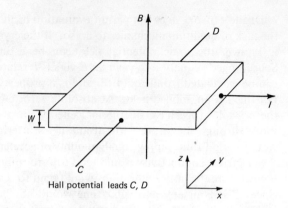

Hall potential leads C, D

TABLE 4.5

Physical properties of semiconductors and metals at 300°K (values from W. C. Dunlap, *Encyclopedia of Physics*, Reinhold Publishing Corporation, New York, 1966)

Material	Band gap, eV	Electron mobility, $m^2 V^{-1} sec^{-1}$	Hole mobility, $m^2 V^{-1} sec^{-1}$
Si	1.15	0.14	0.048
Ge	0.66	0.38	0.18
GaAs	1.43	0.85	0.04
GaSb	0.69	0.40	0.065
InSb	0.17	7.0	0.1
SiC	3.0	0.006	0.0008
PbS	0.37	0.08	0.1
ZnO	3.2	0.019	
CdS	2.4	0.02	
HgTe	0.2	0.22	0.016
Al		0.0012	
Cu		0.0032	
Li		0.0018	
Na		0.0053	
Zn			0.0056
Cd			0.0081
Co			0.0026

charged carriers in the sample can be determined, for the Hall voltage derives from the magnetic-field deflection of the charge carriers in the y-direction. They are deflected to one edge of the sample; this charge pile-up causes an electric field. The direction of the field depends on the polarity of the carrier charge. A more complete analysis of the Hall effect is given in Kittel's book.[18]

Measurement of the Hall voltage for a material yields a value for the carrier density n. This, coupled with an electrical conductivity measurement and using Eq. (4.25), yields a value of the so-called Hall mobility. Some typical values for the mobility of carriers in metals and semiconductors are given in Table 4.5.

[18] C. Kittel, *Introduction to Solid-State Physics*, 4th edition, John Wiley & Sons, New York (1971).

PROBLEMS

4.1 Assume the conduction electrons in a metal such as sodium to be essentially free.

a) Calculate the total number of energy states below the Fermi energy of 3.1 eV, in a volume of 1.0×10^{-5} m^3.
b) Estimate the energy separation, in joules, between these states.

4.2 Show that at $T = 0°$K the average kinetic energy of the nearly free electrons in a metal is $\frac{3}{5}E_F$, where E_F is the Fermi energy. (*Hint*: $E_{ave} = \int E\, dN / \int dN$.)

4.3 Assume that one nearly free electron (4s) is associated with each atom of copper. The density of copper is 8920 kg/m^3 and its atomic weight is 63.54 gm/gm-atom.

a) Determine the Fermi energy at $T = 0°$K for copper.
b) For what temperature would kT be equal to the Fermi energy for copper $(k = 1.38 \times 10^{-23}$ J/$°$K)?
c) Calculate the linear velocity of electrons at the Fermi energy of copper at $0°$K.
d) Calculate the de Broglie wavelength of the electrons in part c.

4.4 The work function of the metal potassium is 2.3 eV, the density is 860 kg/m^3, and the atomic weight is 39.1 gm/gm-atom. The potential energy of an electron at rest inside the metal is 4.8 eV. Calculate the atomic density in a bar of potassium:

a) from the work function, assuming the free electron theory,
b) using Avogadro's number.

4.5 The Fermi energy for tungsten at $300°$K is 4.5 eV. The electrons in tungsten follow the Fermi energy distribution function.

a) Find the probability of an energy level with energy value 10 percent below the Fermi energy being occupied.
b) Repeat part a at $3000°$K, assuming the Fermi energy is relatively temperature independent.
c) What is the occupation probability at $2kT$ units of energy above the Fermi energy?
d) What is the occupation probability of an energy level at the Fermi energy at $300°$K? At $3000°$K?

4.6 a) Calculate the highest wave-vector (k) value for an electron in the conduction band of copper.
b) Determine an order-of-magnitude value for the k vector at the wall of the first Brillouin zone of copper by assuming a simple lattice structure for the crystal.

c) What do parts a and b indicate about the free-electron approximation for copper?

4.7 The energy of an electron at the bottom of the conduction band of a metal is given by $E = Ak^2$, where k is the wave vector of the electron.

a) Plot E versus k if A is a constant, independent of k. On the same graph plot E versus k for a value of A three times as large ($3A$).
b) What is the effective mass for electrons for A and $3A$?
c) Repeat part a for holes at the top of a conduction band where $E = E_{max} - Bk^2$, where E_{max} is the energy at the top of the band.
d) What is the effective mass for holes at the top of the band?

4.8 a) Calculate the average drift velocity of the electrons in a field of 1000 V/m for copper, according to the free-electron theory of monovalent metals.
b) Repeat part a for silver.
c) What is the resistance of a piece of copper wire 0.10 mm in diameter and 1.0 cm long?

4.9 For copper, starting with the known value of electrical conductivity, calculate:

a) The electron relaxation time τ at 300°K.
b) The mean free path λ (average distance traveled between collisions) at 300°K. (Note that $\lambda = v_F \tau$, where v_F is the electron velocity near the Fermi energy.)
c) The electron drift velocity in a field of 100 V/m. About how far does an electron move in a lamp filament 1 m long in a half cycle of 110 V, 60 Hz alternating current?
d) The drift velocity is much smaller than the speed of light, yet current pulses are known to travel down wires at speeds approaching that of light. Explain.
e) At what signal frequency does the derivation of Ohm's law break down for copper?

4.10 The mean free path of a conduction electron in copper is about 10^{-4} m at liquid helium temperature (4.2°K). (See Problem 4.9b for the definition of mean free path.)

a) Approximately what is the lowest frequency that could be used in a cyclotron resonance experiment at 4.2°K to determine the effective mass of an electron in copper?
b) What magnetic induction B is required for this experiment?

4.11 The resonant frequency for the cyclotron resonance of electrons in sodium is found to be 3000 MHz. The magnetic induction applied is 0.10 webers/m². Determine the effective mass of electrons in sodium.

4.12 Assuming that the electrons in sodium are essentially free:

a) Find the electron relaxation time τ.
b) Calculate the mobility of electrons in sodium.

4.13 The Hall voltage for the metal sodium is 0.001 mV, measured at $I = 100$ mA, $B = 2.0$ webers/m^2, and $W = 0.05$ mm.

a) Calculate the number of carriers per cubic meter in sodium.
b) Calculate the mobility of the electrons in sodium, using its known value of electrical conductivity.

5 Introduction to Semiconductor Physics

5.1 PROPERTIES OF SEMICONDUCTORS

Many of the properties of semiconductors must be explained in terms of the energy band concept which derives from the quantum theory of solids. In fact, the explanation of the electrical behavior of semiconductor materials is one of the great triumphs of quantum theory. This chapter presents a quantitative description of pure and impurity-containing semiconductor crystals. Much of the material in Chapters 3 and 4 on quantum mechanics and Chapter 2 on crystal structure will be referred to in this chapter.

A. Semiconductor Crystal Materials

The two most important materials used today in the manufacture of electronic devices are the semiconductors silicon and germanium. Silicon came into general use during World War II (1942) when the point-contact (metal-whisker) silicon-diode rectifier was used in the detection of radar signals. One of the most important electronic technological discoveries of our era was made in 1948 when the point-contact transistor was developed using germanium. However, the introduction of many of the advanced electronic devices used today, such as integrated circuits, had to await a further development in the understanding of the materials used in the fabrication of these devices. Semiconductor materials with purities exceeding that of any known substance were developed during the 1950's. Today the element silicon with an impurity content of less than 1 part in 10^{10} is available in large quantities and at reasonable prices. This is beyond the limits of normal chemical purity, and *semiconductor purity* is the term used to denote this degree of impurity control. In fact, electrical rather than chemical measurements are used to determine the purity level of semiconductor materials. Silicon and germanium can be purified to such an extent owing to their being elements and the fact that electrically important impurities are only slightly soluble in these materials. As we shall see, the covalent bond which is typical of silicon and germanium crystals is another important factor which accounts for their suitability as electronic-device starting materials. In addition, techniques have been developed to grow single crystals of Si and Ge containing remarkably few crystal imperfections.

5.2 INTRINSIC SEMICONDUCTORS

Consider first the properties of pure semiconductor materials which contain practically no extraneous impurities. These are called *intrinsic* semiconductors. An extremely small number of certain types of impurity atoms produce a large effect on the electrical properties of semiconductors; this subject, important in a discussion of semiconductor devices, will be deferred until later.

Silicon and germanium crystals are formed by covalent bonds of atoms in a diamond structure. A three-dimensional sketch of this structure is shown in Fig. 2.10a. A two-dimensional (for simplicity) representation for silicon is given in Fig. 5.1. Here the covalent bonds result in the sharing by a central atom of one electron from each of four surrounding atoms. Since the central silicon atom has four electrons in an outer shell, the sharing of these electrons tends to complement the original four, resulting in a completed shell of eight electrons. This is a very stable structure and, owing to the strength of the covalent bond, practically no electrons can move through the crystal lattice at low temperatures. The material is an electric insulator at absolute zero temperature. However, as the temperature is raised, thermal energy tends to break a few electrons loose from their bonds, and some electric conduction can take place.

FIGURE 5.1

Two-dimensional schematic diagram of a silicon crystal. Each pair of lines represents a two-electron covalent bond.

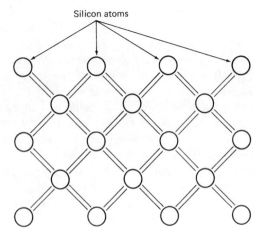

Silicon atoms

This is a qualitative description of the variation of the electrical behavior of an intrinsic semiconductor with temperature. For a quantitative discussion which will lead to a calculation of the electrical conductivity of silicon or germanium as a function of temperature, it is necessary to refer to the energy band description of these crystals.

A. Energy Bands in Intrinsic Semiconductors

A solution of Schrödinger's equation for the energy states of an electron in a three-dimensional periodic diamond lattice typical of Si and Ge has been obtained. The results predict a series of energy bands and energy gaps along the lines of the one-dimensional Krönig-Penney solution. However, owing to the three-dimensional nature of the semiconductor crystal problem, the solution is vastly more difficult. Some of the more complicated features of this solution will be presented later. It is sufficient for the description of a number of semiconductor phenomena to consider in detail the uppermost nearly filled energy band and the next highest nearly empty band separated from it by a forbidden energy gap.[1] At absolute zero temperature the lower band is completely filled, with all electron energy states occupied corresponding to complete electronic shells; the upper band is completely empty. This is shown in Fig. 5.2a. This lower filled band is referred to as the valence band since it corresponds to the energy states of the valence electrons of the atoms. The empty band, called the conduction band, represents the many energy states available for electrons to be excited to, by gaining energy from thermal excitation or a very high electric field. At absolute zero temperature Si and Ge are perfect insulators since the filled band can contribute nothing to electric conduction and the empty band contains no current carriers. When the applied electric field is moderate, there is not sufficient energy available to raise electrons across the forbidden energy gap into the conduction band, and no electric conduction takes place. As the temperature is raised, however, there is a small but definite probability that a given electron will be excited into the conduction band. Since there are many electrons in the valence band, a large number will go into conduction. This is shown in Fig. 5.2b. When this occurs an electric field will be able to supply energy, and hence momentum, to the electrons in the conduction band, resulting in electric conduction. Also, some states in the lower valence band which have been emptied by this excitation are available, so that other electrons in the valence band may be raised in energy. Hence conduction also takes place in the valence band owing to the availability of a few empty states. It was shown previously that these empty states in the valence band may be considered to conduct electricity as positively charged particles called "holes." For an intrinsic semiconductor it follows that the number of electrons in conduction equals the number of holes.

---

[1] There are several other filled bands below these bands which correspond to complete inner electronic shells of the semiconductor atoms.

FIGURE 5.2

Conduction and valence
bands of a pure semi-
conductor (a) at absolute
zero and (b) at room tem-
perature, showing thermally
excited electrons and holes.

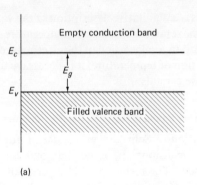

(a)

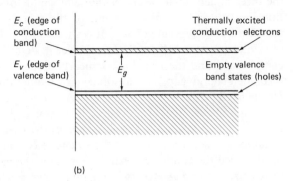

(b)

The forbidden energy gap at room temperature is about 1.15 eV for silicon
and about 0.66 eV for germanium. The element carbon forms a crystalline
substance known as diamond by tetrahedral covalent bonds; this material,
however, is referred to as an insulator since the width of its energy gap is
about 5.3 eV at room temperature. With this wide gap, thermal energy at
normal temperatures is insufficient to excite any substantial number of electrons
into conduction, and the material has extremely low electrical conductivity. A
result of the Krönig-Penney solution predicts that the energy gap width depends
inversely on the interatomic spacing. It is interesting to note that the dimensions
of the conventional unit cell[2] for diamond, silicon, and germanium are, respectively, 3.567, 5.428, and 5.648 Å and the gaps are 5.3, 1.15, and 0.66 eV. Of
course, this problem is quite complex, and factors such as the electron shell
structure of the atoms enter as well.

B. The Density of Electrons and Holes in Semiconductors

The task is now to calculate the number of conducting electrons and holes in an
intrinsic semiconductor (such as silicon) to be able to express the conductivity
of the substance in terms of this and other physical parameters, such as carrier

---

[2] These are known as the *lattice constants* and are determined by x-ray diffraction studies.

collision relaxation time and effective mass. In this way the differences between the conducting properties of silicon, germanium, and even gallium arsenide can be understood. It turns out to be convenient to give the number of current carriers in terms of a parameter which expresses the state of a material—the Fermi energy.

The density of electrons and holes in an intrinsic semiconductor can be stated in terms of the density of states in the conduction and valence bands plus the Fermi-Dirac distribution function for occupancy of these states. It has already been pointed out in Section 4.4 that the electrons at the very top and bottom of an energy band can be treated as nearly free since, essentially, $|E - E_0| \propto k^2$. Hence the density-of-states function per unit energy, E, per unit volume near the bottom of a conduction band has the form (see Eq. 4.9)

$$\rho_c(E) \, dE = \frac{8\sqrt{2\pi}}{h^3} m_n^{*3/2}(E - E_c)^{1/2} \, dE. \tag{5.1}$$

The density of states near the top of a valence band has the form

$$\rho_v(E) \, dE = \frac{8\sqrt{2\pi}}{h^3} m_p^{*3/2}(E_v - E)^{1/2} \, dE. \tag{5.2}$$

Hence $\rho_c(E)$ and $\rho_v(E)$ represent the density of states per unit energy per unit volume near the bottom of the conduction band and near the top of the valence band respectively, of primary interest for an intrinsic semiconductor; m_n^* and m_p^* refer to the effective mass of electrons (negative charge) and holes (positive

FIGURE 5.3

Calculation of the density of electrons and holes in an intrinsic semiconductor.
a) Density-of-states function at the bottom of the conduction band and top of the valence band.
b) The Fermi function with $f(E)$ greatly magnified at the edges of the bands and shown in the circles.
c) Density of electrons and holes.

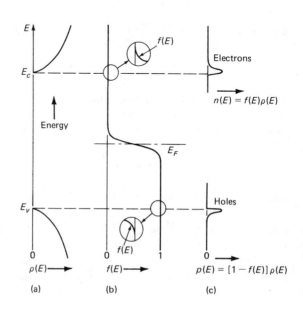

charge) respectively; and E_c and E_v correspond to the energy defining the bottom of the conduction band and top of the valence band. This is shown in Fig. 5.3.

Next we must consider the actual number of electrons occupying these states. This information is provided by the Fermi-Dirac distribution function of Section 4.3. It states that at a temperature of absolute zero any energy level below the Fermi energy is occupied whereas all levels above the Fermi energy are empty. Since it is known that at absolute zero temperature the valence band in a semiconductor is completely full and the conduction band completely empty, this locates the Fermi energy as somewhere between E_c and E_v. At any temperature above absolute zero some electrons fill energy levels in the conduction band corresponding to electrons which have been excited there from the valence band, leaving an equal number of empty states (holes) behind. The statement that the number of electrons within dE of E_c equals the number of holes within dE of E_v is expressed as

$$\rho_c(E)f(E)\,dE = \rho_v(E)[1 - f(E)]\,dE, \tag{5.3}$$

where $1 - f(E)$ represents the probability of an energy level in the valence band being empty. If the effective masses of electrons and holes are approximately equal, then $\rho_c(E) \simeq \rho_v(E)$ and, using Eq. (4.10) for the Fermi-Dirac distribution function, we get for electrons and holes just at the band edges E_c and E_v,

$$\frac{1}{e^{(E_c - E_F)kT} + 1} = 1 - \frac{1}{e^{(E_v - E_F)/kT} + 1}. \tag{5.4}$$

After some algebraic manipulations, we get for E_F

$$E_F = \frac{E_c + E_v}{2}, \tag{5.5}$$

and since $E_c - E_v = E_g$, the forbidden energy gap, we have

$$E_F = E_v + \frac{E_g}{2}. \tag{5.6}$$

Hence the Fermi energy level for an intrinsic semiconductor is at the middle of the energy gap if the electron and hole effective masses are equal. This is essentially true even if m_n^* doesn't equal m_p^*, for then

$$E_F = E_v + \frac{E_g}{2} + kT\ln\left(\frac{m_p^*}{m_n^*}\right)^{3/4}, \tag{5.7}$$

and the additional term is usually small compared to $E_g/2$.

To calculate the density of conduction electrons in an intrinsic semiconductor, consider first the Fermi-Dirac distribution function under the condition that $E_c - E_F > kT$. Near room temperature kT is about 0.026 eV and $E_c - E_F > kT$ means that the Fermi energy is at least a multiple of kT removed from E_c. In practical cases involving silicon, E_F is at least one-tenth of an electron-volt below the conduction band edge E_c. Then the Fermi-Dirac function reduces to

$$f(E) \cong e^{-(E-E_F)/kT}.$$

With this approximation, the number of conduction electrons per unit volume at a temperature T is obtained, by integrating the product of the occupation probability and the density of available states over the conduction band, as

$$n(T) = \int_{E_c}^{\infty} f(E)\rho(E) \, dE = \int_{E_c}^{\infty} \frac{8\sqrt{2}\pi}{h^3} m_n^{*3/2} \, e^{(E_F-E)/kT}(E - E_c)^{1/2} \, dE.$$

$$(5.8)$$

Actually the top of the conduction band has a finite energy value but, since the Fermi-Dirac function falls off so rapidly as energy increases, extending the integral limit to infinity contributes little error and does facilitate the integration. This means that electrons only occupy states near the bottom of the conduction band.

Since energy is the only variable in Eq. (5.8), it may be rewritten as

$$n(T) = \frac{8\sqrt{2}\pi}{h^3} m_n^{*3/2} e^{E_F/kT} \int_{E_c}^{\infty} (E - E_c)^{1/2} e^{-E/kT} \, dE. \qquad (5.9)$$

Changing the variable of integration to $x = (E - E_c)/kT$ yields

$$n = \frac{8\sqrt{2}\pi}{h^3} (m_n^* kT)^{3/2} e^{(E_F-E_c)/kT} \int_0^{\infty} x^{1/2} e^{-x} \, dx. \qquad (5.10)$$

Integrals of the form occurring in Eq. (5.10) are tabulated, and this one equals $\sqrt{\pi}/2$. Hence we get

$$n = 2(2\pi m_n^* kT/h^2)^{3/2} e^{-(E_c-E_F)/kT}. \qquad (5.11)$$

In a very similar manner the number of empty states near the top of the valence band can be calculated. Here we must use the Fermi-Dirac function for *unoccupied* states, which is defined as

$$f'(E) = 1 - f(E) = 1 - \frac{1}{e^{(E-E_F)/kT} + 1} \qquad (5.12)$$

or

$$f'(E) = \frac{1}{e^{(E_F-E)/kT} + 1}. \qquad (5.13)$$

By an argument similar to that in the case of electrons in the conduction band, assume $E_F - E_v > kT$. That is, the Fermi energy is at least a few kT energy units above the top of the valence band. Then Eq. (5.13) reduces to

$$f'(E) \cong e^{-(E_F-E)/kT}. \qquad (5.14)$$

The total number of empty energy states in the valence band per unit volume then becomes

$$p(T) = \frac{8\sqrt{2}\pi}{h^3} m_p^{*3/2} e^{-E_F/kT} \int_{-\infty}^{E_v} (E_v - E)^{1/2} e^{E/kT} \, dE. \tag{5.15}$$

Again for mathematical convenience the lower integration limit is chosen as minus infinity, although there exists a finite lowest energy in this valence band. The rationale again is the rapid falloff of the Fermi function for holes as the energy decreases. Introducing $y = (E_v - E)/kT$ as the integration variable in Eq. (5.15) yields

$$p = \frac{8\sqrt{2}\pi}{h^3} (m_p^* kT)^{3/2} e^{-(E_F - E_v)/kT} \int_0^{\infty} y^{1/2} e^{-y} \, dy. \tag{5.16}$$

Writing in the value of the integral as before gives

$$p = 2(2\pi m_p^* kT/h^2)^{3/2} e^{-(E_F - E_v)/kT} \tag{5.17}$$

for the density of holes in the valence band. As was already discussed, in interpreting the conduction properties of semiconductors these empty states or holes can be considered as positively charged current carriers with a positive effective mass m_p^*. It was pointed out previously that this is more convenient than treating the electrical conduction due to all the occupied electron states in the conduction band with the exception of a few empty states at the top of the band (see Section 4.4).

C. The Electron-Hole Product

Multiplying Eqs. (5.11) and (5.17) together yields an extremely useful expression, the electron-hole product, which is

$$np = 4\left[\frac{2\pi(m_p^* m_n^*)^{1/2} kT}{h^2}\right]^3 e^{-E_g/kT}. \tag{5.18}$$

This is the so-called *law of mass action* of electrons and holes in a semiconductor. It states that the product of electrons and holes in a semiconductor in thermal equilibrium is a function only of temperature since the hole and electron effective masses and the energy gap are relatively temperature independent.

The fact is that foreign atoms introduced substitutionally into a pure semiconductor crystal can contribute current carriers. Such a material will be referred to as an *impurity* semiconductor in contrast to an intrinsic semiconductor. In this case, the numbers of electrons and holes are no longer necessarily equal. Nevertheless, the assumptions made in the above derivation of the np product are also true for the impurity semiconductor. Hence Eq. (5.18) applies for both intrinsic and impurity semiconductors. Therefore, given the number of electrons in a semiconductor of the type being discussed, the number of holes is automatically determined if m_n^*, m_p^*, and E_g/kT are known.

This fact is extremely useful in explaining the behavior of semiconductor devices and will be used later.

In the intrinsic semiconductor $n = p$ and this density of carriers is usually referred to as the intrinsic number of carriers per unit volume, n_i, which is given by

$$n_i(T) = 2 \left[\frac{2\pi (m_p^* m_n^*)^{1/2} kT}{h^2} \right]^{3/2} e^{-E_g/2kT}. \qquad (5.19)$$

For calculation purposes it is useful to note that

$$2 \left(\frac{2\pi m_n^* kT}{h^2} \right)^{3/2} \equiv 2 \left(\frac{2\pi mkT}{h^2} \right)^{3/2} \left(\frac{m_n^*}{m} \right)^{3/2}$$

$$= 4.83 \times 10^{21} \left(\frac{m_n^*}{m} \right)^{3/2} T^{3/2} \quad \text{electrons/m}^3 \qquad (5.20a)$$

and

$$2 \left(\frac{2\pi m_p^* kT}{h^2} \right)^{3/2} \equiv 2 \left(\frac{2\pi mkT}{h^2} \right)^{3/2} \left(\frac{m_p^*}{m} \right)^{3/2} T^{3/2}$$

$$= 4.83 \times 10^{21} \left(\frac{m_p^*}{m} \right)^{3/2} T^{3/2} \quad \text{holes/m}^3. \qquad (5.20b)$$

At $300°K$, n_i is 1.5×10^{16} m^{-3} for Si and 2.5×10^{19} m^{-3} for Ge. Now the electrical conductivity of a semiconductor may be written as

$$\sigma = nq\mu_n + pq\mu_p, \qquad (5.21)$$

where μ_n and μ_p correspond to the electron mobility of electrons and holes respectively. If the semiconductor is intrinsic (pure), we can then write

$$\sigma = 4.83 \times 10^{21} T^{3/2} q(\mu_p + \mu_n) e^{-E_g/2kT} \left(\frac{m_n^* m_p^*}{m^2} \right)^{3/4}, \qquad (5.22)$$

taking into account that the numbers of electrons and holes are equal.

EXAMPLE 5.1

Intrinsic semiconductor material A has an energy gap of 0.36 eV while material B has an energy gap of 0.72 eV. Compare the intrinsic density of carriers in these two semiconductor materials at $300°K$. Assume the effective masses of all the electrons and holes are equal to the free electron mass.

SOLUTION Using Eq. (5.19), we have

$$\frac{(n_i)_A}{(n_i)_B} = \frac{e^{-(E_g)_A/2kT}}{e^{-(E_g)_B/2kT}} = e^{[(E_g)_B - (E_g)_A]/2kT}$$

$$= e^{(0.72 - 0.36) eV/0.052 eV}$$

$$= 1000.$$

Hence, although the energy gap of these two intrinsic semiconductors differs by only a factor of two, the intrinsic density of carriers in the narrower-gap material is 1000 times greater than that in the wider-gap semiconductor.

5.3 THE IMPURITY SEMICONDUCTOR

The introduction of foreign atoms into the lattice of a semiconductor crystal significantly affects the electrical conductivity of that material. Atoms of the elements antimony, arsenic, boron, or gallium can be introduced into a silicon or germanium crystal lattice by heating the crystal in contact with a vapor of these elements at a temperature near 1000°C. Because of the similarity of the atomic radius of these impurity atoms and that of the semiconductor, they enter the lattice in place of silicon atoms, *substitutionally*.

A. Donors

An arsenic atom can be made to occupy an empty lattice site normally occupied by a silicon atom (vacancy) in a pure silicon crystal. Now the unique semiconducting properties of silicon and germanium result from their electron valency of four. When an arsenic atom substitutes for a silicon atom, it not only supplies the four electrons for bonding normally supplied by the silicon, but introduces an extra electron owing to its valency of five. The four electrons are strongly bound in the silicon lattice in the normal tetrahedral covalent bonds. However, the extra electron is very loosely bound to its nucleus, much like the valence electron in metals. At some moderate temperature this extra electron contributes to the electrical conductivity of the semiconductor, above and beyond that characteristic of pure silicon. This is shown schematically in Fig. 5.4.

FIGURE 5.4

Free electron arising from ionization of a substitutional arsenic impurity atom which becomes positively ionized.

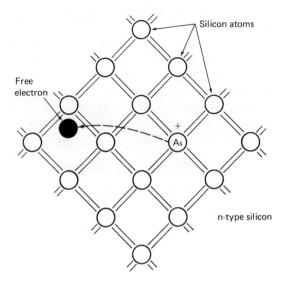

FIGURE 5.5

Energy levels of some impurity states in the forbidden energy gap of silicon. The levels are drawn as short dashes near the particular impurity to signify the localized nature of an electron in this energy level. The donor and acceptor energy values are measured from the conduction and valence band edges, respectively.

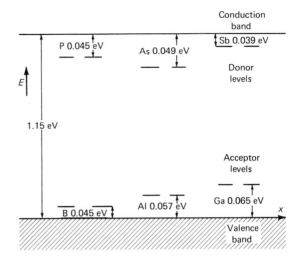

For convenience in dealing quantitatively with these electron-contributing or *donor-type* impurities, some technique must be devised for fitting these impurity atoms into the energy band scheme. Let us assume (as is usually the case) that there are only relatively few of these impurities in a silicon lattice.[3] Then the extra donor electron of the arsenic atom is attracted by the unit net positive charge of its nucleus (the atom must be electrically neutral). This is taken as similar to the problem of the electron attached to the hydrogen-atom nucleus, with the exception that the arsenic atom is embedded in a matrix of silicon whereas the hydrogen atom is in vacuum. Now the binding energy of an electron to a hydrogen nucleus is 13.6 eV. This is obtained by use of Eq. (3.68):

$$E_B = \frac{-mq^4}{8\epsilon_0^2 h^2} = -13.6 \text{ eV}. \tag{5.23}$$

The energy binding the electron to the arsenic nucleus in a silicon lattice can be roughly estimated by modifying this expression somewhat. The dielectric constant now should be 12 as for silicon, compared to 1 for vacuum. Since the dielectric constant enters into the energy expression as an inverse square, the donor electron binding energy becomes $13.6/(12)^2 \simeq 0.1$ eV.[4] This indicates that a donor electron requires about 0.1 eV of energy to remove it from the influence of its nucleus so that it can participate in electronic conduction. Hence the donor energy state must be about 0.1 eV less than the conduction-band edge of silicon. This can be indicated on the normal energy band diagram

---

[3] Common impurity levels range from 1 to 1 million foreign atoms per billion atoms of silicon.

[4] The actual calculated value should be even less since the "effective" electron mass in silicon is less than the free electron mass.

FIGURE 5.6

Free hole arising from
ionization of a substitutional
boron impurity atom which
captures an electron,
becoming a negative ion.

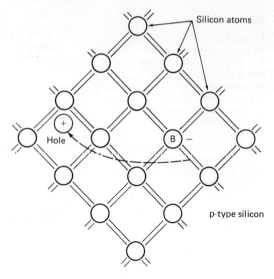

for intrinsic silicon as shown in Fig. 5.5. The figure shows that about 0.049 eV
of energy is required to raise an electron on an arsenic donor level just into
the conduction band. When this occurs the donor atom takes on a net positive
charge. The 0.049 eV is a measured rather than calculated value. Note that the
impurity levels are localized in space, as indicated by short lines, whereas the
edges of the conduction and valence bands are not. This corresponds to the fact
that the relatively few impurity atoms present occur only occasionally in the
silicon lattice.

B. Acceptors

A corresponding set of arguments can be given now for the case of an impurity
atom like boron introduced substitutionally into a silicon crystal. The boron
atom has a valence of only three and hence lacks one electron from satisfying
the covalent silicon atom bonding structure. It is sometimes convenient to
view this as an acceptor nucleus with a positively charged hole orbiting around
it. Should a free electron somehow be made available, it would tend to be
captured on this boron acceptor site, completing the tetrahedral bond and
ionizing the acceptor by giving it a negative charge. This is shown in Fig. 5.6.
The electron may actually be one which comes from one of the valence electrons
of some nearby atom. This provides an empty energy state or hole in the
valence band structure of silicon.[5] Energy must be supplied to remove the
electron from the silicon covalent tetrahedral bond so that it can be captured by
an acceptor atom. This energy is also of the order of 0.1 eV as for the donor

---

[5] An ionized acceptor contributes a positive hole to conduction in the valence band. This is
identically equivalent to stating that the acceptor energy level has an electron excited into it
from the valence band, leaving a mobile hole behind.

FIGURE 5.7

A free hole arising from ionization of a substitutional boron impurity atom is accelerated to the right by the electric field. Correspondingly, a valence electron moves to the left to fill the empty state, leaving an empty valence state behind.

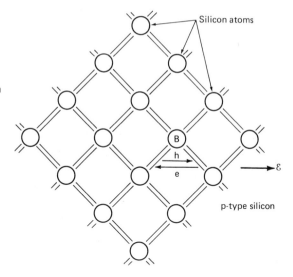

electron ionization, and hence the acceptor level appears about 0.1 eV above the edge of the valence bond. Some such acceptor levels are indicated in Fig. 5.5.

Electric conduction in the valence band of a semiconductor containing acceptor atoms can now be described. Under thermal equilibrium conditions in the crystal lattice, some valence electrons of silicon atoms come free from their normal positions and are immobilized or trapped on acceptor impurity sites. This action leaves behind an empty energy level in the semiconductor valence band. Now electric conduction can take place in the valence band by way of another valence electron being raised in energy by an electric field so as to fill this empty hole. The electron so raised in energy itself creates a hole, and another valence electron can be raised into this energy level by the field. This is pictured schematically in Fig. 5.7. If the excited valence electron moves to the left to fill an empty valence state there, leaving behind a hole, the empty state has apparently moved to the right. This leads to a simple but not strictly correct way[6] of visualizing the motion of these bound electrons by considering the corresponding hole moving in a direction opposite that of the valence electrons. Thus in an electric field the holes in the valence band will move in a direction opposite to electrons in the conduction band—and hence the concept of the hole as a particle with a positive charge. This is not strictly correct since it is the motion of the enormous number of electrons present in a nearly filled valence band that must be considered, rather than the motion of the few holes. The concept of the mobile charge carrier called the hole is an

[6] See R. B. Adler et al., *Introduction to Semiconductor Physics*, Section 1.3.2, S.E.E.C. Vol. I, John Wiley & Sons, New York (1964).

expedient in discussing the electrical conductivity exhibited by all the electrons in the valence band of a semiconductor; this is similar to the way the effective mass permits the use of a nearly free electron analysis for the motion of an electron in the crystal field of a solid.

C. Impurity Semiconductor Doping

Note that the elements from group III in the periodic table (B, Al, Ga, and In) are acceptors in Ge and Si and produce conduction by holes. Ge and Si with the addition of these materials exhibit positive Hall effect and are called p-type (see Section 4.5). The elements from group V in the periodic table (P, As, and Sb) are donors, produce conduction by electrons, exhibit negative Hall effect, and are referred to as n-type. The purposeful addition of p- and n-type impurities to a semiconductor material is called doping.

A semiconductor material contains both n- and p-type impurities. If the impurities are primarily acceptors the material is p-type; if they are mainly donors it is n-type. Sometimes a semiconductor material can contain nearly equal quantities of donors and acceptors; it is then said to be *compensated*. When this occurs the extra electrons from the donors become trapped on the acceptor sites, and hence these two impurities compensate each other or cancel one another as far as electrical conductivity is concerned. Hence a material containing equal numbers of donors and acceptors is, "electrically speaking," an intrinsic semiconductor, but hardly "pure."

In the above discussion all the donor atoms were considered to be "ionized"; that is, their extra electrons were considered to be contributing one electron per atom to conduction. At very low temperatures (below 200°K for Si) there is not sufficient thermal energy available to ionize these donors; some remain attached to their nuclei and hence, being immobile, do not contribute to electric conductivity. A very similar argument applies to acceptors. At very low temperatures some holes which normally contribute to conduction remain attached to their acceptor nuclei and hence do not contribute positive carriers to conduction. The acceptor atoms are no longer "ionized."

It is possible to calculate the number of ionized donors or acceptors in a semiconductor in thermal equilibrium if the Fermi energy level is known. The number of ionized donors is equivalent to the number of empty donor energy states. The probability of occupation of any energy state is defined in terms of the Fermi energy. If there are N_d donor atoms or states per cubic meter present in a crystal occupying an energy state E_d, the density of empty ionized states is given by

$$N_{\text{ionized donors}} = N_d[1 - f(E_d)]$$

$$= N_d\left(1 - \frac{1}{\frac{1}{2}e^{(E_d - E_F)/kT} + 1}\right). \tag{5.24}$$

FIGURE 5.8

Calculation of the density of
electrons in n-type silicon at
room temperature. There are
very few holes in the
valence band and nearly
all the donors are ionized.
a) Density-of-states function
at the bottom of the
conduction band and top of
the valence band.
b) The Fermi function with
$f(E)$ greatly magnified at the
edge of the conduction band
and shown in the circle.
c) Density of electrons and
holes.

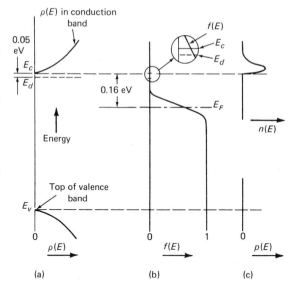

(a)   (b)   (c)

(Here the Fermi statistics have to be modified. The factor $\frac{1}{2}$ is due to donor
electron spin considerations.) It follows from this equation that if $E_d - E_F >
kT$, then nearly all the donors are ionized. On the other hand, if $E_F - E_d > kT$,
then practically all donor sites have electrons trapped or immobilized on them.
This is illustrated in Fig. 5.8. This type of reasoning applies to acceptor or
other energy levels as well. However, one must remember that an ionized
acceptor is an acceptor level with an electron trapped on it and that the Fermi
function refers to the probability of occupation of the state by electrons.

D. Calculation of the Fermi Energy

The question often arises as to the position of the Fermi energy level in a given
material. To calculate the Fermi level in a uniform semiconductor in thermal
equilibrium, the number of donor and acceptor atoms, the energy gap, the
temperature, and the density-of-state functions must be known. That is, the
state of the system must be defined. Note that the Fermi level reflects all these
factors, and any change in these quantities is reflected by a shift in the Fermi
energy. Note too that the Fermi energy is defined only under thermal equilibrium conditions although sometimes a quasi-Fermi level is defined in nonequilibrium situations. This is done to enable one to express the number of
electrons and holes present in a simple manner, and will be discussed later in
connection with semiconductor devices.

For calculation of the Fermi energy the condition of charge neutrality can be
applied to a homogeneous semiconductor. That is, it is nearly always assumed
that no net charge can exist locally anywhere in a uniformly doped semiconductor containing the same number of impurities throughout its volume. This

result is explained in terms of the Faraday "ice pail" experiment. That is, if any net charge occurs in the interior of a conductor, it will disperse itself to the surface in an extremely short time of the order of the dielectric relaxation time of the material.[7] Charge neutrality is expressed as

$$p + (N_d - n_d) = n + (N_a - p_a), \qquad (5.25)$$

where p and n refer to hole density in the valence band and electron density in the conduction band respectively, N_d and N_a are the density of impurity donor and acceptor atoms respectively, and n_d and p_a refer to the density of electrons trapped on donor sites and holes trapped on acceptor sites respectively. Note that the quantities in parentheses on the left- and right-hand sides of Eq. (5.25) represent the concentration of ionized donors and acceptors respectively, the ionized donor being positively charged and the ionized acceptor having a negative charge.

When expressions for all the quantities in this equation are introduced from Eqs. (5.24), (5.11), and (5.17), the Fermi energy is determined. Note that the solution for the Fermi energy generally involves a complicated transcendental equation containing exponentials. Graphical techniques have been devised to solve for the Fermi energy.[8] However, it is often possible to obtain an algebraic solution using very good approximations. For example, in a moderately doped donor impurity semiconductor at room temperature the number of conduction electrons may be assumed to be equal to the number of ionized donors if $E_g \sim 1$ eV.

Equation (5.25) indicates that at 300°K the Fermi level is somewhere above the middle of the energy gap for a donor-doped semiconductor, since then many donor atoms are ionized; it is correspondingly below the middle of the gap for acceptor-doped semiconductors. The more donor impurities added, the closer the Fermi energy approaches the edge of the conduction band. The Fermi energy level approaches the edge of the valence band as more acceptor impurities are added.

EXAMPLE 5.2

Given the n-type semiconductor silicon at 300°K with an energy gap of 1.15 eV. The material contains only donor type impurities, all of which are ionized. The donor density is 1.0×10^{22} m^{-3}. Assuming that the electron effective mass in this material is the same as the free electron mass, calculate the Fermi energy.

SOLUTION   Beginning with the charge-neutrality condition given in Eq. (5.25), since no acceptors are present and all donors are ionized, $n_d = 0$ and $N_a - p_a = 0$. This equation then reduces to

$$p + N_d = n.$$

---

[7] Of the order of 10^{-12} sec for semiconductors.
[8] See, for example, W. Shockley, *Electrons and Holes in Semiconductors*, Section 16.3, Van Nostrand Rheinhold, New York (1950).

Since the np product for silicon is $(1.5 \times 10^{16})^2$ m^{-6} and the electrons in conduction must be at least 10^{22} m^{-3}, the hole concentration must be less than $(1.5 \times 10^{16})^2/10^{22}$ m^{-3}. Hence $p \ll n$, N_d. So to a good approximation

$$N_d = n,$$

or essentially all the conduction electrons come from ionized donors. Using Eqs. (5.11) and (5.20a), we have

$$N_d = 2\left(\frac{2\pi m_n^* kT}{h^2}\right)^{3/2} e^{-(E_c - E_F)/kT}$$

$$= 4.83 \times 10^{21} \left(\frac{m_n^*}{m}\right)^{3/2} T^{3/2} e^{-(E_c - E_F)/kT}.$$

Hence

$$1.0 \times 10^{22} = (4.83 \times 10^{21})(1)(300)^{3/2} e^{-(E_c - E_F)/0.026},$$

where $E_c - E_F$ is in electron-volts. Taking the natural logarithm of this equation, and some algebra, yields

$$E_c - E_F = 0.20 \text{ eV}.$$

Hence the Fermi energy is about 0.20 eV below the conduction-band edge.

E. Temperature Dependence of the Fermi Level

At high temperature, when many electrons are thermally excited into the conduction band, thus exceeding the electron concentration from donor atoms, the Fermi level tends toward the center of the gap. This occurs because thermally excited electrons leave behind an equal number of conducting holes, and the

FIGURE 5.9

Calculated variation of the Fermi energy with temperature. This is for silicon, energy gap assumed constant at 1.15 eV, with impurity concentration as a parameter. A donor-level ionization energy of 0.05 eV and an acceptor level of 0.06 eV is assumed. The energy at the top of the valence band E_v is here taken as zero.

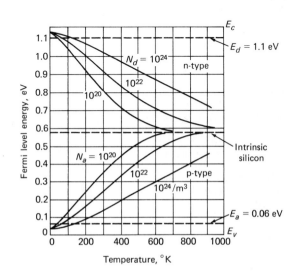

material is sometimes said to become "intrinsic." Reducing the temperature
causes the Fermi level to move toward the conduction-band edge for n-type,
and toward the valence-band edge for p-type semiconductors. The variation
with temperature of the Fermi level of acceptor- and donor-doped crystals of
Si is given in Fig. 5.9.

F. Variation of Electrical Conductivity of Semiconductors with Temperature

The electrical conductivity of a donor semiconductor crystal at any temperature
can be determined if the Fermi energy, the energy gap, the electron mobility,
and the excess of donor over acceptor impurities are known. At room temperature for both n-type silicon and germanium nearly all donors are ionized.
Then the density of electrons in the conduction band is essentially the density
of donor atoms in the crystal, and hence the electrical conductivity is to a good
approximation

$$\sigma_n = nq\mu_n = N_d q\mu_n,   \tag{5.26}$$

FIGURE 5.10

Conductivity versus reciprocal absolute temperature
for the arsenic-doped
germanium samples shown in
Fig. 5.11a and b. The
dashed line represents the
intrinsic conductivity. The
numbers on the curves
identify the samples as
follows:
No. 55, $N = N_d - N_a$
$\quad = 1.0 \times 10^{19}$ m^{-3}
$\quad (N_a < 2 \times 10^{18}$ m$^{-3})$;
No. 53, $N = 9.4 \times 10^{19}$ m^{-3}
$\quad (N_a < 2 \times 10^{18}$ m$^{-3})$;
No. 64, $N = 1.7 \times 10^{21}$ m^{-3}
$\quad (N_a < 2 \times 10^{19}$ m$^{-3})$;
No. 54, $N = 7.5 \times 10^{21}$ m^{-3}
$\quad (N_a < 1.5 \times 10^{20}$ m$^{-3})$;
No. 61, $N = 5.5 \times 10^{22}$ m^{-3}
$\quad (N_a < 5 \times 10^{21}$ m$^{-3})$;
No. 58, $N \sim 10^{23}$ m^{-3}.
(These data, and those of
Fig. 5.11a and b, were
reported by E. M. Conwell,
Proc. IRE, Vol. 40, p. 1327,
1952.)

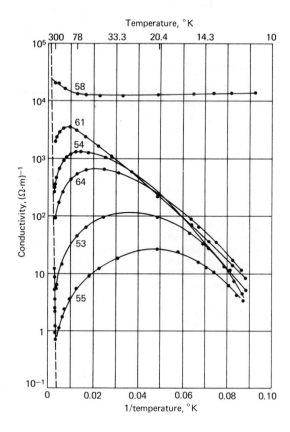

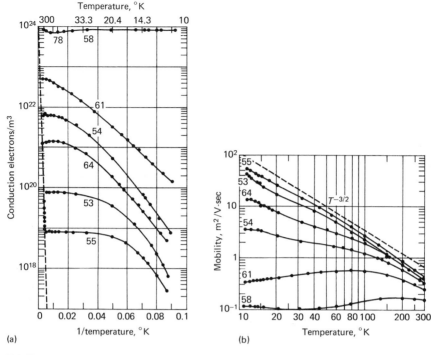

FIGURE 5.11

a) Concentration of charge carriers versus reciprocal absolute temperature for the arsenic-doped germanium samples shown in Fig. 5.10. The dashed line represents the concentration of intrinsic carriers $n_i(T)$.

b) Mobility versus temperature for a set of arsenic-doped germanium crystals. Note that for pure germanium the mobility decreases with increasing temperature as the three-halves power of the absolute temperature.

where μ_n is the electron mobility, and N_d is the number of donor atoms per cubic meter. For acceptor-doped material the electrical conductivity is correspondingly

$$\sigma_p = pq\mu_p = N_a q\mu_p, \qquad (5.27)$$

where μ_p is the hole mobility and N_a is the number of acceptor atoms per cubic meter. As the temperature is raised above room temperature the mobility of carriers decreases owing to increased scatter by the thermally vibrating lattice ions. This causes the electrical conductivity to decrease initially. As the temperature is further increased, the number of thermally excited electron-hole pairs increases exponentially and the material's conductivity rapidly increases and tends toward "intrinsic." Curves indicating the variation in the conductivity of n-type germanium with temperature are given in Fig. 5.10 for samples ranging in donor concentration from 10^{19} m^{-3} to 10^{24} m^{-3}. These curves can be broken down into the variation of electron carrier density and mobility with temperature. Such graphs for the same samples are shown in Fig. 5.11a and b respectively.

5.4 CARRIER TRANSPORT IN SEMICONDUCTORS

In Section 4.5 the subject of electronic transport in metals was discussed. An expression was derived connecting the current density and the electric field in a metallic conductor (see Eq. 4.22). This relationship is known as Ohm's law and relates to the *drift* of electrical carriers under the influence of an electric field. The picture here is that there is a net movement of electrons in the direction opposite to the electric field; this movement is superimposed on the equilibrium random motion of the carriers which they possess by virtue of their position in the energy band and the temperature.

In a semiconductor such as silicon at room temperature, there are electrons available for movement at the bottom of the conduction band, and holes available for motion at the top of the valence band. These latter carriers behave as positive charges and the manner in which they participate in electrical conduction has already been described. Hence, in general, the net current in a semiconductor is given by the sum of the electron and hole currents, $J_n + J_p$, so that in one dimension Ohm's law becomes

$$J_x = (J_n)_x + (J_p)_x = \sigma_n \mathcal{E}_x + \sigma_p \mathcal{E}_x = nq\mu_n \mathcal{E}_x + pq\mu_p \mathcal{E}_x. \qquad (5.28)$$

Here n and p refer to the density of electrons and holes respectively; μ_n and μ_p are the electron and hole mobilities respectively. In addition to this conduction current the transport of carriers in a semiconductor can also occur by a process, not observed in metals, known as *diffusion*.

A. The Diffusion Current Density

The motion of carriers by diffusion occurs when there is a nonuniform distribution of these particles. This does not occur to any appreciable extent in metals since for the most part conduction in these materials is due to one type of carrier, and any nonuniformity would be quickly dispersed. In a semiconductor there can be considerable spatial differences in the distribution of electrons; of course, space-charge neutrality requires that these charges must be balanced locally by comparable numbers of holes.

Consider a distribution of electrons in distance as shown in Fig. 5.12. If we take an incremental region Δx about x_1, there are more carriers to the left of Δx than to the right. Hence there exists a tendency, not due to Coulomb repulsion of these charges, for a net flux of carriers to move to the right to minimize this nonuniformity. This results from the random motion of the electrons, some moving to the right and others to the left. Statistically, it is evident that more electrons in any given time interval will cross to the right of

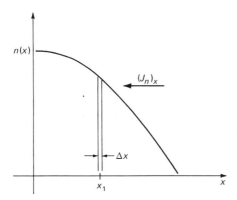

FIGURE 5.12

Nonuniform distribution of electron carrier density $n(x)$, resulting in an electron diffusion current density $(J_n)_x$.

Δx than to the left of this region. The effect is to even out the concentrational nonuniformity, similar to the dispersion of an ink drop in a tank of water, and is called diffusion. No electric field is required for this motion which results from the net to and fro motion of the carriers.

When the concentration of carriers varies with distance in a semiconductor this constitutes a concentration gradient. This concentration gradient in one dimension is measured by the change in carrier concentration Δn in a distance Δx, in the limit as $\Delta x \to 0$. This concentration gradient is denoted by dn/dx in the case of electrons and by dp/dx in the case of holes. A reasonable assumption is that the magnitude of this diffusive flux of electrons at x_1 is proportional to the extent of the nonuniformity which is expressed by the electron gradient there. That is,

$$\text{Diffusion electron flux} \propto \left. \frac{dn}{dx} \right|_{x=x_1} \tag{5.29}$$

expressed in electrons per quartic meter. The diffusion electric current density for electrons may be expressed by multiplying the flux by the electron charge. This can then be written in the form of an equation as

$$(J_n)_x = q D_n \frac{dn}{dx}, \tag{5.30}$$

where D_n is a proportionality constant called the *diffusion constant* for electrons, with the dimension square meters per second. A corresponding expression for holes in a semiconductor is

$$(J_p)_x = -q D_p \frac{dp}{dx}, \tag{5.31}$$

where D_p is the diffusion constant for holes. Note that the minus sign is necessary if the electron charge q is considered to have a pure numerical value, for if the concentration of holes decreases in the direction of the $+x$-axis, dp/dx is negative, and J_p represents a positive current density which travels in the $+x$-direction. For electrons, a decreasing carrier concentration in the $+x$-direction means dn/dx is negative and hence Eq. (5.30) predicts a current density in the $-x$-direction. For the same type of gradient the hole and electron currents are in opposite directions owing to the opposite signs of their electrical charges.

B. Current Flow with Drift and Diffusion

In the general case of current flow in a semiconductor, movement of charge occurs due both to the presence of an electric field (drift) and to concentration gradients (diffusion). The mathematical treatment of problems involving such a case is normally difficult. Fortunately, however, a good number of practical problems exist in which one or the other transport process predominates, and we shall only concern ourselves with these cases. This reduces the complexity enormously. However, for completeness the total current flow in a semiconductor in one dimension is expressed as

$$J_x = J_n + J_p = q\left(n\mu_n\mathcal{E}_x + D_n\frac{dn}{dx}\right) + q\left(p\mu_p\mathcal{E}_x - D_p\frac{dp}{dx}\right). \tag{5.32}$$

C. The Einstein Relation

The two material parameters which express the ability of carriers to drift or diffuse in a semiconductor are the mobility μ and the diffusion constant D. The concept of mobility, previously developed, derives from the net carrier motion resulting from random collisions of these charges with the lattice atoms under the action of an applied potential gradient or electric field. Similarly, the diffusion process can be described in terms of a net motion of carriers superimposed on their random thermal motion, under the effect of a concentration gradient and involving collisions with the lattice. (This is analogous to the manner in which heat flows in a material under the influence of a thermal gradient.)

Since drift and diffusion result from similar statistical mechanisms, it can be shown that the parameters μ and D for a semiconductor material are not independent. They are related by an equation known as the Einstein relation. This relation is given for electrons and holes respectively by

$$D_n = \frac{kT}{q}\mu_n \quad \text{and} \quad D_p = \frac{kT}{q}\mu_p. \tag{5.33a, b}$$

Here k is the gas constant, q the electronic charge, and T the absolute temperature. Since the units of the diffusion constant are square meters per second and the mobility is in square meters per volt-second, kT/q must be expressed in volts and equals 0.026 V at room temperature. Some measured values for μ are given in Table 4.5.

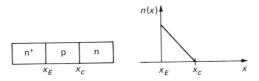

EXAMPLE 5.3

The p-type base region of an n-p-n bipolar silicon transistor of the type shown in Fig. 1.3 has a width of 2.0×10^{-6} m and is doped with 1.0×10^{21} acceptors/m^3. Electrons are injected into this region from the emitter at x_E, producing a uniform gradient of electrons there. The electron concentration drops to zero at the collector at x_c. If 2.0×10^{20} electrons/m^3 are present at the emitter edge of the base region (x_E), calculate the diffusion current density of electrons through this base region under steady-state conditions. What electric field must be present in this base region to yield an electron drift current density just equal to the diffusion current density just calculated? Determine the voltage drop across the base width corresponding to this field.

SOLUTION From Problem 5.2 (end of chapter), $\mu_n = 0.14$ m^2/V-sec, and by the Einstein relation (Eq. 5.33a), $D_n = 0.026$ V^{-1} $(0.14$ m^2/V-sec) or $D_n = 0.0036$ m^2/sec. From Eq. (5.30),

$$(J_n)_{\text{diffusion}} = qD_n \frac{dn}{dx} = 1.6 \times 10^{-19} \text{ C}(0.0036 \text{ m}^2/\text{sec}) \frac{2.0 \times 10^{20}}{2.0 \times 10^{-6}} \text{ m}^{-4}$$

$$= 5.8 \times 10^4 \text{ A/m}^2.$$

From Eq. (5.28),

$$(J_n)_{\text{drift}} = qn\mu_n \mathcal{E}$$

$$5.8 \times 10^4 \text{ A/m}^2 = 1.6 \times 10^{-19} \text{ C}(1.0 \times 10^{21}\text{m}^{-3})(0.14 \text{ m}^2/\text{V-sec})(\mathcal{E})\text{V/m}$$

$$\mathcal{E} = 2.6 \times 10^3 \text{ V/m}.$$

The voltage drop is $V = \mathcal{E}W = 2.6 \times 10^3(2.0 \times 10^{-6}) = 5.2 \times 10^{-3}$ V.

5.5 GENERATION AND RECOMBINATION OF MINORITY CARRIERS IN SEMICONDUCTORS

The discussion of homogeneous semiconductors thus far has been confined to thermal equilibrium conditions. However, semiconductor devices in general operate under nonequilibrium conditions. For example, a bar of n-type germanium acting as a photodetecting device is not in equilibrium in the presence of light. When the bar is illuminated, *excess charge carriers* (above the equilibrium numbers) are produced in the material and the electrical conductivity of the bar increases. This can be demonstrated by the experiment

FIGURE 5.13

Apparatus for observation of
the photoconductivity of a
semiconductor and excess
carrier recombination. The
effect of the light is to
increase the conductivity of
the semiconductor bar,
increasing the circuit
current, and hence causing
an increased voltage output
as viewed across the resistor
R on the oscilloscope.

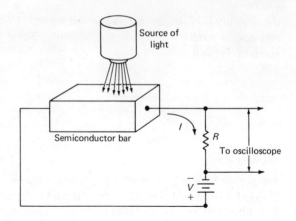

illustrated in Fig. 5.13. Energy has been absorbed from the light by the germanium in producing these excess carriers. Hence the semiconductor is no longer in thermal equilibrium.

The extra carriers are produced by the light energy $h\nu$ breaking electrons free from their covalent tetrahedral bonds in the valence band and raising them into the conduction band. A conducting hole is left behind for every excess electron carrier so produced. This is referred to as electron-hole *pair production* or generation and is illustrated in Fig. 5.14a. The electrical neutrality of the bar is, of course, maintained. Generally, a donor impurity semiconductor has many orders of magnitude more electrons than holes. When weak light introduces electron-hole pairs, the number of electrons is hardly increased above the equilibrium number already present, but the number of holes normally increases significantly. Hence this process is usually referred to as *minority carrier hole injection*, since this positively charged carrier is greatly outnumbered by the electrons or *majority* carriers present in thermal equilibrium. In transistors minority carrier introduction or injection takes place by the supply of electric energy instead of by light energy, via a p-n junction. Nevertheless, this type of carrier injection is basically the same, and charge neutrality is normally maintained (i.e., for every hole injected an electron must be introduced).

A. Minority Carrier Recombination

Electron-hole pairs are only produced by light of high enough frequency ν such that $h\nu > E_g$, where E_g is the semiconductor forbidden energy gap. When the light is shut off, the material must return to thermal equilibrium and the excess carriers must disappear. This occurs by electron-hole *recombination*. The recombination can take place when one of the conducting electrons is bound again on a normal covalent tetrahedral site. This is referred to as *band-to-band* recombination since the electron drops from the conduction band back into the valence band in the energy band picture. This is illustrated in Fig. 5.14b and is generally coupled with the emission of photons with energy equal to that of the semiconductor energy gap. It has been found, however, that this band-to-band recombination process is not favored in silicon and germanium, whereas it is in gallium arsenide, for example. The reason for this will be discussed later.

In Ge and Si, recombination is generally found to take place via a recombination *center* introduced by some impurity atom such as copper or gold. Physically, these impurity atoms tend to immobilize a hole, which immediately recombines with a conduction electron. This speeds up the process, and the recombination center "catalyzes" the electron-hole "reaction." This is illustrated on the energy band diagram in Fig. 5.14c. Atoms such as copper and gold are impurities which tend to introduce energy levels at about the center of the energy gap. The recombination process may be described as follows: When a hole jumps up to the recombination energy level, an electron promptly drops into this level, and annihilation takes place. A completely analogous way of describing the same process is: An electron from the conduction band drops back into a hole in the valence band using a recombination level as a "stepping stone" (see Fig. 5.14c).

The main point is that the one hole and one electron disappear and the time taken for this recombination process is called the minority carrier *lifetime*. In silicon and germanium the lifetime of carriers recombining through an impurity recombination center can range from less than 1 nsec to 1 msec. The recombination time is shorter in samples with more impurity centers and depends on the facility with which a center catalyzes the reaction, which is referred to as the *capture cross section*.

B. Light Emission from Semiconductors

Since energy is required to raise an electron from the valence band to the conduction band producing an electron-hole pair, energy must be released on recombination. If the electron-hole pair is produced by energy supplied by an electric field and most of the recombination energy is released via light or

FIGURE 5.14

a) Generation of an electron-hole pair.
b) Direct recombination of an electron and a hole with the emission of a photon.
c) Electron recombining with a hole via a recombination center.

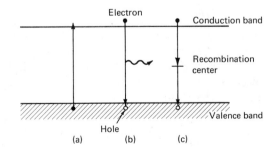

photons, this forms the basis of a *light-emitting diode* (LED) as well as an *injection p-n junction laser*. The former produces incoherent light while the latter yields coherent light as will be described later. In Si and Ge most of the recombination energy is absorbed in heating up the crystal, whereas in gallium arsenide photons are emitted. The reason that GaAs emits photons and hence is intrinsically a better material for fabricating semiconductor lasers will be discussed soon. However, let us first mathematically formulate the recombination process. This will be of importance later in the discussions of transistors, integrated circuits, and lasers.

C. Minority Carrier Lifetime

It has been shown that under thermal-equilibrium conditions the product of electrons and holes in any semiconductor must be a constant at any temperature. The value of this constant will depend on semiconductor parameters such as the energy gap (see Eq. 5.18). When carriers are injected via the energy of an electric field or light, equilibrium is disturbed and carriers in excess of the equilibrium np product are introduced. For reasons of electrical neutrality, the numbers of electrons and holes so introduced must be equal; we shall also assume that their number is small compared to the equilibrium number so that quasi-thermal equilibrium occurs. This possibility of introducing excess carriers, yet maintaining electrical neutrality, distinguishes semiconductors from metals and results from the two-carrier nature of the semiconductor. This accounts for the fact that transistors utilizing two electric carriers — *bipolar* transistors — are fabricated from semiconductor materials and not metals.

It should be understood that in thermal equilibrium generation and recombination of carriers are constantly taking place, but on the average for every electron excited into conduction one recombines or drops back into the valence band. If the equilibrium generation rate is G carriers/m^3-sec and the recombination rate is R, then this fact may be expressed as $R - G = 0$. In the case of band-to-band recombination, a reasonable assumption is that the recombination rate is linearly proportional to the number of electrons and holes present. Hence at equilibrium we can write

$$rn_0 p_0 - G = 0, \tag{5.34}$$

where r is a constant of proportionality, and n_0 and p_0 refer to the equilibrium density of electrons and holes respectively. If now the crystal is slightly disturbed from equilibrium by a few excess holes Δp and an equal number of electrons Δn, the net rate of excess hole-carrier recombination (or loss) at any time will be given by

$$-\frac{d(\Delta p)}{dt} = r(n_0 + \Delta n)(p_0 + \Delta p) - rn_0 p_0. \tag{5.35}$$

This assumes r is unchanged since the density of excess carriers introduced is small. Then Eq. (5.35) reduces to

$$\frac{d(\Delta p)}{dt} = -r(n_0 + p_0)\,\Delta p, \tag{5.36}$$

if we neglect second-order terms. The quantity $1/r(p_0 + n_0)$ is usually defined as the minority carrier lifetime τ, and Eq. (5.36) is then written as[9]

$$\frac{d(\Delta p)}{dt} = -\frac{\Delta p}{\tau}. \tag{5.37}$$

Integrating this equation yields

$$\Delta p = A e^{-t/\tau}. \tag{5.38}$$

To evaluate the integration constant A, let the initial injected density of holes be $(\Delta p)_0$ at time $t = 0$, and let the source of injection be removed at that time. Then

$$\Delta p = (\Delta p)_0\, e^{-t/\tau}, \tag{5.39}$$

which indicates that the excess holes (and electrons) disappear exponentially with time so that after a time τ only a fraction $1/e$ are left. Note that the speed of annihilation of excess holes varies inversely as τ, which in Si and Ge depends on the density of recombination centers and the capture cross section of these centers. In the fabrication of transistors and integrated circuits impurities like Cu and Au are selectively introduced into the semiconductor material by high-temperature processing to establish the lifetime at some value predetermined by design.

D. The Direct Transition

Let us return to the question of the important difference between the recombination process in Si and Ge as compared to GaAs. The answer must be sought in terms of the band theory of solids. The solution of the Schrödinger equation for a crystal has already been shown to yield a relationship between the energy E of an electron in the solid and its wave vector k. In the simple case discussed, this relationship was indicated by the graph of Fig. 3.10, which shows that there exist certain forbidden bands of energy.

A theorem of solid-state physics states that each segment of the E-versus-k curve may be translated in the k-direction by $\pm n\pi/a$ without loss of generality (n being an integer); this follows from the periodicity of the Schrödinger solution, which results from the periodicity of the potential due to the lattice ions in the crystal. In this way all segments of the E-versus-k curve of Fig. 3.10 may be included between $k = +\pi/a$ and $-\pi/a$, in a *reduced-zone* representation.

---

[9] τ here is not to be confused with the dielectric relaxation time.

FIGURE 5.15

a) Photon-induced direct transition of an electron from the valence band to the conduction band. Here the lowest energy of the conduction band occurs at the same value of wave vector k as the highest energy of the valence band. The energy gap E_g corresponds to the separation of this minimum and maximum, in this reduced zone representation. Note that in this illustration the hole effective mass is much greater than the electron effective mass.
b) Indirect transition of an electron from the valence band to the conduction band. Here the lowest energy of the conduction band occurs at a larger value of k, say k_1, than the highest energy of the valence band. This transition requires the participation of both a photon (vertical) and a phonon (horizontal) to provide the energy and momentum, respectively, for the transition. The energy gap still corresponds to the separation of the top of the valence band from the bottom of the conduction band as shown.

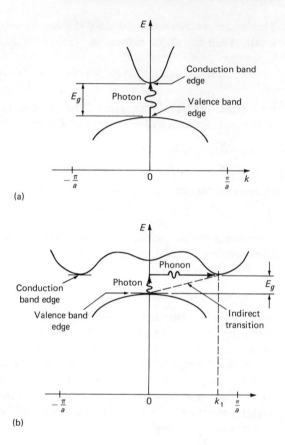

(a)

(b)

The shape of the E-versus-k curves now becomes symmetric about $k = 0$; portions of these curves representing two adjoining energy bands are shown in Fig. 5.15a. In this way the E-versus-k structure near $k = 0$ for adjoining bands may be conveniently studied. For example, for a semiconductor material the lower band may be the valence band and the upper band the conduction band. The uppermost parabola represents the energy states of electrons in the conduction band. The inverted parabola corresponds to electrons in the valence band; the negative curvature near the top of the band is indicative of the negative effective mass of these electrons (see Section 4.5B) leading to the concept of holes (see Section 5.2A).[10] The range of energies between the top of the valence

[10] Note that the magnitude of the curvature is different for electrons and holes since, in general, their effective masses are different.

band at $k = 0$ and the bottom of the conduction band at $k = 0$ constitutes the forbidden energy gap.

As an example of the usefulness of this diagram, consider a photon of light striking the semiconductor represented by the energy band diagram of Fig. 5.15a.[11] The excitation of an electron into the conduction band occurs most probably for electrons with zero momentum (that is, $k = 0$). Note that the transition takes place directly upward, since this requires minimum energy expenditure as shown in Fig. 5.15a; it is referred to as a *direct transition*. That is, the photon hardly alters the momentum of the electron on the scale of k shown. This can be proved by a simple calculation which is the subject of one of the exercises at the end of this chapter.

E. The Indirect Transition

The E-versus-k solutions obtained from the Schrödinger equation do not always result in curves such as that in Fig. 5.15a but may instead yield a behavior as indicated in Fig. 5.15b. Although the maximum energy in the valence band here corresponds to $k = 0$, the minimum electron energy in the conduction band occurs at a value of k greater than zero. Now a direct upward transition is not favored since the minimum energy for excitation of an electron into the conduction band does not occur at $k = 0$. The energetically favored transition, in fact, is from $k = 0$ in the valence band to $k = k_1$ in the conduction band. Hence this process does not take place at constant momentum. The additional momentum cannot come from the photon as previously suggested; hence it must be supplied otherwise. It is contributed by collisions with the atoms of the crystal vibrating in their lattice sites. The quantization of the lattice vibrational energy gives rise to the concept of the *phonon*.

This type of *indirect transition* is indicated in Fig. 5.15b. Here again the minimum-energy principle is illustrated, as well as the law of conservation of momentum. The participation of a phonon as well as a photon is necessary for this transition. Note that the energy gap for the semiconductor represented by this energy band diagram corresponds as always to the minimum separation of the valence and conduction bands, which in this case doesn't occur at $k = 0$.

F. Band Structure and Semiconductor Devices

The two different types of semiconductor band structures, sometimes referred to as "direct" and "indirect", as illustrated in Fig. 5.15a and b, correspond to gallium arsenide and silicon respectively. Although the crystal structures of these semiconductors are basically diamond cubic, subtle differences in the bonding forces separate these materials into dramatically different electrical categories. For example, excellent semiconductor lasers can be constructed

[11] This diagram is valid only for electron motion along one crystalline direction.

from GaAs but not Si. Because the direct excitation is favored energetically, the inverse reaction, direct recombination, is also highly probable. This causes very poor current gain in GaAs transistors since the injected carriers recombine (are lost) in transit between emitter and collector (see Section 1.2). However, the directness of this recombination results in efficient energy emission in terms of photons and hence in light-emitting or laser action.

The indirect transition typical of silicon is inefficient since momentum must be exchanged with the crystal-lattice atoms. Hence most of the recombination energy goes into the vibrational energy of the crystal atoms, which "heats up" the crystal solid. Little energy is left for the production of photons and hence Si is a poor light emitter. However, because of the inefficient recombination, this process is slow and the lifetime correspondingly long, accounting for the general use of Si for transistors and integrated circuits.

These concepts will be useful later in the detailed description of semiconductor devices.

5.6 SEMICONDUCTOR SURFACES AND SEMICONDUCTOR-DEVICE STABILITY

The discussion of the physics of semiconductor materials thus far has been confined to a treatment of the properties of the crystal bulk or volume. The electrical characteristics of all the semiconductor devices to be discussed in succeeding chapters are, however, affected to a greater or lesser extent by the properties of the *surface* of the semiconductor crystal. For example, success or failure in the quality fabrication of insulated-gate field-effect transistors (Chapter 10) and charge-coupled devices (Chapter 11) is almost totally determined by the control of the surface characteristics of the semiconductor material, since all the electronic action in these devices takes place within a few micrometers of the surface. Current may be conducted across the p-n junctions in the diode and bipolar transistor (Chapter 1) at the surface of the semiconductor crystal, where the p-n junction intersects the surface, instead of through the semiconductor bulk. The control of surface properties requires an understanding of the surface of the semiconductor crystal and its interaction with the surrounding ambient. The intent of this section is to gain insight into the electrical behavior of semiconductor surfaces.

The outstanding stability and reliability of solid-state devices results from the fact that the structure of the crystal, as well as metal contacts to the crystal, is time invariant at moderate temperatures and in somewhat hostile environments. In normal use there are few short-term burnout mechanisms for semiconductor devices, and any aging problems are usually very long term. Radiation of an atomic, nuclear, optical, or thermal variety can penetrate the bulk of the solid-state material and cause permanent damage therein. But this usually requires comparatively high-energy radiation. However, the outer surface of the solid is particularly vulnerable to low-energy interaction with its environment. Since the bulk electrical properties of semiconductor materials are sensitive to minute quantities of impurities (donors and acceptors), it would be expected that a small amount of contaminant on the semiconductor surface will significantly affect the electrical properties of a device fabricated from this material. This is the case, and it accounts for the fact that all semiconductor devices are encapsulated in some way to protect them from the ambient. The primary task is to stabilize the semiconductor surface so that the inherent reliability of solid-state semiconductor devices can be maintained. For this purpose these devices are sometimes sealed in vacuum or, more often, coated with electrically insulating materials such as pure silicones or epoxies. Silicon devices specifically are most often protected by the material's natural oxide, which forms on the silicon crystal surface on exposure to pure oxygen at high temperatures, or which may be deposited onto its surface. The excellent electric insulating properties of the uncontaminated form of this material and its ease of application make it particularly desirable to coat silicon transistors and integrated circuits using the planar technology illustrated in Fig. 1.2; all the p-n junctions intersecting the surface are silicon-dioxide coated, greatly enhancing the electrical stability of these devices.

P-N junctions which terminate at the crystal surface may exhibit spurious electric conduction due to current caused by the mobility across the junction of ion contaminants on the surface. Even fixed charges on the surface can attract mobile holes and electrons from the semiconductor bulk to the surface, where they may undergo electric conduction due to a voltage applied across the p-n junction. Without some form of protection these effects could be cumulative, increasing with time or else varying in time, causing unpredictable changes in device electrical behavior. Time-invariant electrical characteristics are a requisite for the modern semiconductor devices used in vast numbers in complex electronic systems. Let us now turn our attention to the physics of semiconductor surfaces in order to better understand the influence of surface phenomena on the devices to be discussed in subsequent chapters.

A. Semiconductor Surface Physics

Consider a semiconductor material such as silicon freshly cleaved or cut in high vacuum to expose an ultraclean surface. Although the crystal structure described in Chapter 2 was generally considered to be a perfectly regular arrangement of atoms, this can no longer be the case at the cleaved surface of the crystal. From a symmetry viewpoint one would expect that two of the four covalent bonds which hold the silicon diamond-type lattice together would be broken, and hence unsaturated, at the crystal surface, in contact with the vacuum (see Fig. 2.10b). These broken covalent bonds are lacking two electrons per atom, and hence about 10^{19} atoms/m^2 are in a position to attract negative charges which might alight on the crystal surface giving rise to possible energy states in the forbidden energy gap. This represents an enormous quantity of acceptor electronic states which, if they existed in practice, would dominate the behavior of most semiconductor devices.

Fortunately, any silicon surface when exposed to air will quickly form an oxide layer about 40 Å thick. Apparently the silicon surface atoms become bound to the oxygen atoms in the air, forming a silica (SiO_2) polyhedron structure. This satisfies the silicon "dangling" bonds at the surface and reduces these surface states under clean conditions to $10^{15}/m^2$ or less. These are the so called *fast surface* or *interface* states which are responsible for electron and hole recombination and generation effects at the silicon surface similar to those already described in Section 5.5 for the semiconductor bulk. The time constants are correspondingly within the nanosecond to microsecond range. This is in contrast to the much slower states, in the range of seconds to days or even months, observed in some heavily oxidized silicon surfaces. The surface densities of these slow states are comparable to the fast states and seem to be located in the oxide, within 20 Å of the silicon-oxide interface. Both apparently act as donor states, are positively charged when ionized, and hence induce a tendency toward an n-type skin just below the silicon surface, owing to the electrons in the semiconductor bulk being attracted to the silicon surface. The slow surface states have been attributed to excess silicon ions in the oxide near the silicon interface which have not yet reacted with the negative oxygen ions which diffuse through the silicon dioxide at high temperature to form a stable SiO_2 tetrahedral structure.[12]

Figure 5.16a shows an equilibrium energy band diagram for the interface between a p-type silicon crystal and vacuum, assuming a perfect silicon crystal structure out to the surface and ignoring the effects of broken surface bonds. This may be referred to as the *flat-band* surface situation; it is taken as a reference, but obviously doesn't occur normally. Figure 5.16b shows the equilibrium energy band structure for the same p-type silicon crystal coated with a 2000 Å thick oxide layer, including the effects of fast and slow interface state charges. The electric field lines from these positive charges terminate on some electrons collected at the silicon surface as well as on some negatively charged ionized acceptors fixed in the silicon crystal close to the surface. Hence the silicon surface just under the oxide tends toward n-type, as indicated by the closer proximity of the edge of the conduction band to the Fermi level. This *band bending* near the silicon surface represents a potential gradient and hence an electric field in the silicon, due to the electric dipole at the interface. The extent of this band bending (and the tendency of the original p-type surface to become less p-type) depends on the chemical treatment of the silicon surface prior to oxidation as well as the care taken during the oxidation process.[13]

[12] A. S. Grove, *Physics and Technology of Semiconductor Devices*, Chapter 12, John Wiley and Sons, New York (1967).
[13] Indeed this treatment is a carefully guarded secret of individual semiconductor-device manufacturers.

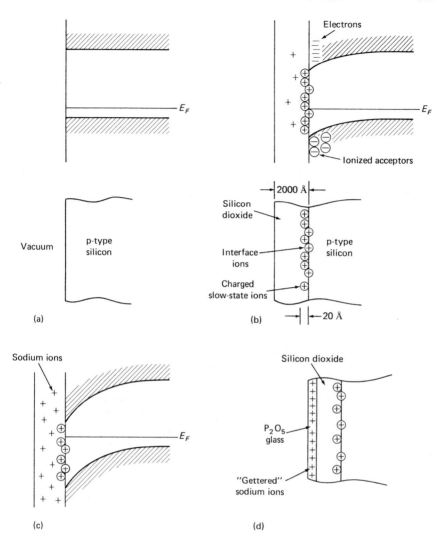

FIGURE 5.16

a) Interface between p-type silicon and vacuum ignoring broken surface bonds, with an energy band diagram indicating flat-band conditions.

b) Equilibrium energy band diagram for p-type silicon showing fast and slow states at the interface and in the oxide. Also shown are electrons collected in the silicon at the interface as well as ionized acceptors.

c) Equilibrium energy band diagram for p-type silicon with its surface inverted to n-type by positive ions in the silicon dioxide as well as at the oxide-silicon interface.

d) Sodium ions gettered in the phosphosilicate glass formed on the silicon dioxide surface.

The crystal-plane orientation of the silicon surface also affects the density of surface states. Cleaving the crystal along (1 1 1) planes gives rise to the rupturing of the greatest number of bonds per square meter, which accounts for the observation of the largest density of surface states in this situation. The lowest density of surface states is observed when the oxide is grown onto the (1 0 0) surface planes, which include a minimum number of broken bonds per square meter.

B. Charges in Silicon Dioxide

Surface effects in semiconductors may also be due to ionic charges in the bulk of the silicon dioxide layer which in turn induce charges in the silicon, near the silicon-oxide interface. For example, positively charged sodium and other alkali-metal ions are known to have a high solubility as well as high mobility in SiO_2 or quartz. These positive charges embedded in the oxide will tend to drive the silicon near the interface even more toward n-type, causing additional band bending. This is shown in Fig. 5.16c. Note that the p-type silicon crystal now has a surface which has been *inverted* to n-type; this is indicated by the fact that the Fermi level is closer to the conduction-band edge than the valence-band edge near the surface. (Positively charged oxygen vacancies have been reported to be another source of mobile charge in the oxide.) The high mobility of these ions in silicon dioxide accounts for their motion even at temperatures of 150°C and lower, particularly when an external voltage is applied across the oxide, giving rise to an electric field therein. The motion of these charges in the silicon dioxide causes corresponding changes in the electrical nature of the silicon surface just under the oxide. It has already been indicated how the surface layer on a p-type silicon crystal may be inverted to n-type by charges in the coating oxide. The extent of this conversion depends not only on the number of such contaminant charges but also on their proximity to the silicon-oxide interface (the nearer the interface, the larger the effect). The normal fields present in device operation can cause the motion of these charges and hence time-variable changes in device characteristics that are surface sensitive. Methods are needed to reduce this contamination to significantly less than 1 ppm.[14]

Techniques have been developed to minimize the effect of the sodium ion, which is a primary contaminant and difficult, if not impossible, to eliminate during semiconductor-device processing. The final processing of an n-p-n planar transistor often involves exposure of the oxidized silicon to phosphorus pentoxide (P_2O_5) to form the n^+-type emitter region (see Fig. 1.2). This reacts with the upper surface of the SiO_2 layer already on the silicon to form a phosphosilicate "glass." Sodium is much more soluble in this phosphorus glass than the pure silicon dioxide and hence tends to collect or be "gettered"

---

[14] E. H. Nicollian, "Surface Passivation of Semiconductors," *J. Vac. Sci. Technol.*, p. 539, Sept./Oct., 1971. This paper provides a good brief summary of surface effects in silicon.

by this layer, far from the silicon-oxide interface, minimizing the electrical effect in the silicon under the oxide. This is shown in Fig. 5.16d. Another effect of this P_2O_5 treatment is for this superoxidant to supply excess oxygen to fill the oxygen vacancies in the SiO_2 structure which can also act as mobile positive charges or else enhance the sodium rate of drift.

Another way to inhibit the drift of sodium ions toward the silicon-oxide interface is to provide an intervening layer of silicon nitride, Si_3N_4. This material may be deposited onto an oxidized silicon wafer by reacting silicon tetrachloride, $SiCl_4$, and ammonia, NH_3, at a temperature of about 800°C to form this compound. The mobility of sodium or alkali-metal ions in silicon nitride is known to be significantly slower than in silicon dioxide, owing to the denser nature of the nitride; hence it can help "seal off" a device so coated from sodium contamination from the environment.

Finally it has been observed that hole traps may be introduced into a silicon dioxide layer by high-energy electron or ultraviolet radiation. Again these immobilized positive charges can induce an n-type layer at the silicon surface under the oxide. In fact, the effect of all these oxide-silicon surface phenomena is invariably to drive the silicon toward n-type. The limitation of semiconductor-device behavior due to surface effects will be mentioned in later chapters.

PROBLEMS

5.1 a) Show that for an intrinsic semiconductor the Fermi energy is

$$E_F = E_v + \frac{E_g}{2} + \frac{3}{4}kT \ln \frac{m_p^*}{m_n^*}.$$

b) If $m_p^*/m_n^* = 6$, as in GaAs, which has an energy gap of 1.43 eV, find the deviation of the Fermi energy from the center of the energy gap at 300°K.

c) Repeat part b for a temperature of 600°K.

5.2 The effective mass, as determined by cyclotron resonance measurements, for electrons and holes in Ge, Si, and GaAs is given in the table below.

a) Determine the number of carriers in intrinsic Si at 300°K.

b) Repeat part a at 600°K.

	Energy gap (300°K), eV	Electron effective mass/m	Hole effective mass/m	Mobility at 300°K, m²/V-sec	
				Electrons	Holes
Ge	0.66	0.55	0.37	0.38	0.18
Si	1.15	0.40	0.50	0.14	0.048
GaAs	1.43	0.08	0.50	0.85	0.04

5.3 a) Determine the conductivity of intrinsic silicon at 300°K from the energy gap value.

b) Repeat part a for germanium.

c) The energy gap is a function of temperature. Will it increase or decrease as the temperature is reduced from room temperature? (*Hint*: Use the Krönig-Penney model.)

5.4 In analogy with Eq. (5.24), write an expression for the density of ionized acceptors (acceptor atoms which have captured an electron).

5.5 Refer to Fig. 5.9 for the variation with temperature of the Fermi energy level in silicon doped with 10^{22} donors/m^3.

a) Determine the number of ionized donors at 300°K and at 30°K.

b) Using the concept of charge neutrality, prove that the Fermi energy increases at 300°K as the number of donors is increased from 10^{22} to $10^{24}/m^3$.

c) Prove that adding acceptors to this crystal will decrease the Fermi energy.

d) If the crystal contains exactly 10^{22} donors/m^3 and 10^{22} acceptors/m^3, what will be the conductivity of the crystal?

5.6 What is the average distance between dopant atoms (measured in parent-crystal interatomic spacings) for a donor level of $10^{22}/m^3$ in silicon?

5.7 Calculate the position of the Fermi level at 300°K for

a) silicon containing 10^{23} boron atoms/m^3,

b) germanium containing 10^{23} arsenic atoms/m^3 plus 5×10^{22} atoms of indium.

c) Repeat part a at 600°K.

5.8 Refer to Fig. 5.11. Explain why the conductivity of germanium at first increases as the temperature is raised from 0°K, then begins to decrease, and finally increases, as indicated in Fig. 5.10.

5.9 Using the Einstein relation, determine the carrier diffusion constant for electrons and holes in pure

a) germanium at 300°K,

b) germanium at 30°K.

5.10 a) Determine the maximum value of the energy gap which a semiconductor, used as a photoconductor, can have if it is to be sensitive to yellow light ($\lambda = 6000 \times 10^{-10}$ m).

b) A photodetector whose area is 5.0×10^{-6} m^2 is irradiated with yellow light whose intensity is 20 W/m^2. Assuming each photon generates one electron-hole pair, calculate the number of pairs generated per second.

5.11 a) From the known energy gap of the semiconductor gallium arsenide, calculate the primary wavelength of photons emitted from this crystal as a result of electron-hole recombination.

b) Is this light visible?

c) Will a silicon photodetector be sensitive to the radiation from a GaAs laser? Why?

5.12 a) Determine the magnitude of the wave vector k for yellow light ($\lambda = 6 \times 10^{-7}$ m).

b) Calculate the width of the first Brillouin zone as indicated in Fig. 3.10 and partly duplicated in Fig. 5.15a for a material with a lattice constant $a = 5.0 \times 10^{-10}$ m.

c) Using the results of parts a and b, show that the momentum (or wave vector) of a yellow photon is negligibly small compared to the momentum of nearly all the electrons in the first Brillouin zone.

5.13 Explain with the aid of a simple calculation why the crystal gallium phosphide ($E_g = 2.25$ eV) is transparent to red light ($\lambda = 7 \times 10^{-7}$ m), while the crystal silicon is opaque to this light. (*Hint*: When a photon incident on a crystal has sufficient energy to raise an electron from the valence band to the conduction band, its energy will be absorbed by the crystal.)

6 Introduction to Semiconductor Diodes

6.1 THE P-N JUNCTION DIODE IN EQUILIBRIUM

In Chapter 5 the properties of semiconductor materials containing a uniform distribution of impurity atoms were considered in some detail. There are a number of electronic devices which employ uniformly doped semiconductors such as germanium and silicon. Some of these devices, such as thermistors, photoconductive cells, and strain gauges, are linear devices.[1] However, the devices most extensively in use today contain at least one p-n junction and exhibit a nonlinear variation of current with voltage.

The general construction and I-V characteristic of a typical semiconductor p-n junction diode have already been described in Chapter 1. There is extensive general use of this type of device in electronic circuitry. For example, semiconductor diodes are employed as voltage limiters and references, current rectifiers, signal demodulators, etc. Some circuit applications of junction diodes are discussed later. Perhaps the most important application of the p-n junction is as the emitter of the bipolar transistor. Hence the following diode considerations will have application in the description of transistors.

This chapter presents a derivation of the dc and low-frequency current-voltage characteristic of the p-n junction diode by considering charge-carrier flow through the junction. The behavior of the p-n junction photodiode and the metal-semiconductor diode will also be described. In the next chapter, the higher-frequency properties of the junction diode will be treated, as well as the transient switching behavior.

---

[1] In linear devices the electric current through the device is directly proportional to the applied voltage; i.e., Ohm's law applies.

FIGURE 6.1

a) P-N junction diode grown
in the form of a single-
crystal bar.
b) Planar p-n junction
formed by selectively diffusing a p-type impurity into an
n-type semiconductor crystal.
The darkened region
represents the insulating
layer protecting the junction
from the environment.

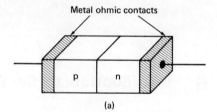

(a)

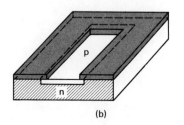

(b)

A. A Physical Description of the p-n Junction

A schematic diagram of a p-n junction diode grown in the form of a semiconductor single-crystal bar is shown in Fig. 6.1a. The more common diode
geometry is similar to that shown in Fig. 6.1b. This configuration can be
formed by selectively diffusing a p-type impurity (such as boron) into the
n-type semiconductor crystal by exposure to the impurity at a very high temperature. Similarly, a junction could be formed by diffusing an n-type impurity
(such as arsenic) into a p-type semiconductor.[2] In either case, the structure
consists of a single crystal of semiconductor material with a p-n junction,
shown schematically in Fig. 6.2a. On one side of the junction are found predominantly p-type impurities, whereas the other side has mainly n-type doping
atoms. In this analysis it will be assumed that the doping is uniform on each
side of the p-n junction as indicated in Fig. 6.2b.

The electrical characteristics of this diode depend mainly on the properties of
the junction.[3] Figuratively, one can consider the formation of the p-n junction
as bringing together two uniformly doped bars of semiconductor material, one
p-type and the other n-type. The p-side contains holes as the majority carrier,
coming mainly from negatively charged acceptors, and very few minority
carrier electrons.[4] The n-side contains electrons derived mainly from the

---

[2] Impurity diffusion technology will be described in Chapter 11 in connection with the
technology of integrated circuits.

[3] It will be shown that voltage drops in the uniform semiconductor material away from the
junction and across the metal contacts are normally small at moderate current levels. Also
surface effects can influence the diode characteristics, as discussed in Section 5.6 and later in
this chapter.

[4] The greater the number of acceptors and hence holes, the smaller the number of electrons,
since the product is constant at any particular temperature (see Eq. 5.18).

FIGURE 6.2

a) Schematic of a p-n junction diode.
b) Uniform distribution of acceptor and donor impurities on each side of an abrupt p-n junction.
c) Space-charge region under thermal equilibrium conditions showing ionized acceptor and donor charges in a region otherwise depleted of mobile carriers.
d) The variation of electric field with distance in the depletion layer, reaching a maximum negative value at the junction.
e) The variation of electric potential with distance in the diode, showing that all the potential drop is across the space-charge region. ΔV_0 is the barrier height or contact potential.

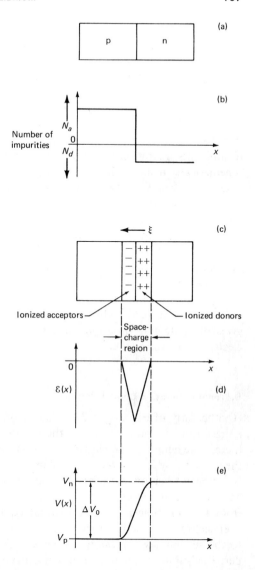

positively charged donors as the majority carriers and very few minority holes. Before contact, each side is electrically neutral owing to the balance of the mobile carriers and fixed ion charges. However, as these two sections are brought together, one can imagine some charge transfer taking place across the junctions; for the sharp carrier-concentration gradient cannot be sustained in the same sense as was discussed in Section 5.4A, and the mobile charges begin to diffuse, causing a smoothing out of the drastically nonuniform distribution of mobile carriers in the neighborhood of the junction. That is, the holes tend to move toward the n-region where there are few holes, leaving behind negatively charged acceptor ions, and the electrons tend to flow toward the p-side where there are few electrons, leaving behind positively charged donor ions.[5] The redistribution tendency is opposed by an electric field which is set up in the neighborhood of the metallurgical p-n junction. This retarding field results from the fixed electric charges due to the positively charged donor ions and the negatively charged acceptor ions in the proximity of, and on opposite sides of, the junction, left behind by the mobile carriers.[6] This very field, caused by the migration of the mobile carriers across the junction, tends to limit this migration and establish an equilibrium situation. This is shown schematically in Fig. 6.2c.

One can visualize the creation of an electric field at the p-n junction as resulting from the transfer of electrons from the electrically neutral donor atoms in the n-region to the electrically neutral acceptor atoms in the p-region and taking place in close proximity to the p-n junction. This redistribution of charge occurs only in a space-charge region within a few micrometers of the metallurgical junction which is depleted of mobile carriers and contains only fixed charges. The exact distribution of electric field near the p-n junction can be obtained by solution of the Poisson equation applied to this region; it is graphed in Fig. 6.2d. This formulation will be presented later. Figure 6.2e gives the potential variation through the diode obtained by integrating the electric field over distance.

B. The p-n Junction Contact Potential

Another way of expressing the effect of the electric field which is built in by ionized donors and acceptors at the junction is in terms of a potential barrier there. The height of the barrier in volts is called the contact potential and is of the order of $\frac{1}{2}$ V for silicon and germanium p-n junctions. This built-in potential is somewhat reminiscent of the electrochemical potential at a junction between an electrode and electrolyte in a battery. However, no current can be drawn from the p-n junction under thermal equilibrium conditions. There is no chemical reaction and so power derived from such a device would be contrary to the second law of thermodynamics—the device would then comprise a perpetual-motion machine. Hence if a metal wire is connected between the two ends of the p-n junction, contact potentials must be set up at the metal-semiconductor end connections to balance the contact potential appearing at the p-n junction. Now the net voltage in the loop is zero so that no current can flow, as required on theoretical grounds.

---

[5] If these concentration gradients occurred in a uniformly doped bar of semiconductor crystal, the carrier gradients would be reduced quickly to zero in a period equal to the dielectric relaxation time of the semiconductor material, this being of the order of 10^{-12} sec.
[6] It can be shown that a very abrupt transition from p-type to n-type (in less than a few micrometers) must occur to support such a field.

The junction contact potential can be estimated indirectly by extrapolation of junction-capacitance measurements as will be described later. This is possible because the capacitance effectively measures the charge distribution near the junction, which, in turn, determines the barrier potential. The contact potential will now be analytically calculated for a semiconductor p-n junction in thermal equilibrium at a temperature T, in terms of the number of acceptors and donors on each side of the p-n junction.

The analysis is based on the concept that in equilibrium not only is the total current zero, but the electron and hole currents must *separately* be zero. (This derives from a very fundamental and useful physical concept which is called the principle of *detailed balance*.) The hole current in one dimension generally can be written (see Eq. 5.32) as

$$J_p = q\left(p\mu_p \mathcal{E}_x - D_p \frac{dp}{dx}\right) \tag{6.1a}$$

and the electron current as

$$J_n = q\left(n\mu_n \mathcal{E}_x + D_n \frac{dn}{dx}\right). \tag{6.1b}$$

Here $\mathcal{E}_x$ represents the field at any place x, and dn/dx and dp/dx are the carrier gradients there which cause diffusive current flow. Setting each of these equations equal to zero yields

$$-p\mu_p \frac{dV}{dx} = D_p \frac{dp}{dx} \tag{6.2a}$$

and

$$n\mu_n \frac{dV}{dx} = D_n \frac{dn}{dx}, \tag{6.2b}$$

where V is the electric potential which is related to the electric field by definition as

$$\mathcal{E} \equiv -\frac{dV}{dx}. \tag{6.3}$$

Our attention is directed mainly at this point to the space-charge region where the electric field is not zero. Now integrating Eq. (6.2a) yields

$$\ln \frac{p}{p_c} = \frac{-\mu_p}{D_p}(V - V_c) = -\frac{q}{kT}(V - V_c), \tag{6.4}$$

where p_c and V_c are arbitrary hole concentration and voltage reference values, and the Einstein relationship (Eq. 5.33) between μ_p and D_p is used. Let us call V_p the potential on the p-side of the junction, far to the left and remote from the junction, as shown in Fig. 6.2e. The potential on the n-side, far to the right and remote from the junction, is taken as V_n. Equation (6.4) now gives

$$\ln \frac{p_{p0}}{p_c} = -\frac{q}{kT}(V_p - V_c), \tag{6.5a}$$

where p_{p0} represents the equilibrium density of majority holes in the p-region remote from the junction. Similarly for the p_{n0} minority holes on the n-side, remote from the junction,

$$\ln \frac{p_{n0}}{p_c} = -\frac{q}{kT}(V_n - V_c). \tag{6.5b}$$

Subtracting Eq. (6.5b) from (6.5a) yields

$$\ln \frac{p_{p0}}{p_{n0}} = -\frac{q}{kT}(V_p - V_n). \tag{6.6}$$

Then

$$\Delta V_0 = \frac{kT}{q} \ln \frac{p_{p0}}{p_{n0}} \quad \text{or} \quad \frac{p_{p0}}{p_{n0}} = e^{q\Delta V_0/kT}, \tag{6.7}$$

where ΔV_0 is the potential difference (or contact potential) between the two sides of the p-n junction, remote from the junction, under thermal equilibrium conditions.

In Section 5.3F it was pointed out that at room temperature in germanium and silicon all donors and acceptors are ionized so that essentially $p_{p0} = N_a$ and $n_{n0} = N_d$. Also Eqs. (5.18) and (5.19) yield $n_{n0} p_{n0} = n_i^2$. Hence Eq. (6.7) can be written as

$$\Delta V_0 = \frac{kT}{q} \ln \frac{N_a N_d}{n_i^2}. \tag{6.8}$$

This expression relates the semiconductor p-n junction built-in voltage to the impurity level on each side of the junction. An analogous treatment for electrons would begin with Eq. (6.2b). The identical result would be obtained, indicating the strong interdependence between electron and hole flow in a semiconductor.

The same value of the contact potential for a p-n junction may be obtained in quite another manner. This is accomplished with the aid of the energy band diagram for electrons and holes in semiconductors using the concept of the Fermi energy. In Fig. 6.3a the energy band diagrams for separate p-type and n-type semiconductor crystal segments are shown. Note the position of the Fermi level E_{Fp} near the valence-band edge for the p-type semiconductor and E_{Fn} near the conduction-band edge for the n-type semiconductor. Now consider that the two segments are brought toward each other, finally being brought into contact in thermal equilibrium. Once this has occurred, the Fermi energies on the p- and n-sides must agree. This, in fact, defines thermal

FIGURE 6.3

a) Energy band diagrams for separate p-type and n-type crystal segments.
b) Energy band diagram for a p-n junction diode under thermal equilibrium conditions in which the Fermi energy is constant throughout the device.

E_∞ is the reference energy of an electron infinitely distant from the semiconductor, out of its field of influence

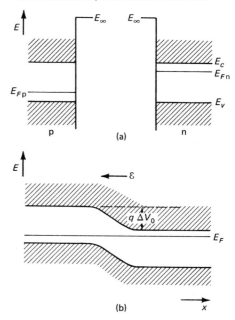

equilibrium for a crystal, namely that the Fermi energy is the same throughout the sample.[7] This requirement causes a potential barrier to be formed between the p- and n-sides of this junction as shown in Fig. 6.3b. That is, the n-side must be depressed in energy relative to the p-side to permit the respective Fermi levels to line up. (Or conversely the p-side is raised relative to the n-side.) Since this energy diagram is for electrons, and the electron energy E is given by $E = -qV$, decreasing energy means a numerically greater potential V. Hence the electric potential on the n-side of the p-n junction is higher than that on the p-side. This, of course, is in agreement with the picture obtained by consideration of carrier flow as shown in Fig. 6.2e. The potential barrier indicates the existence of an electric field in the depletion region directed from the n- to the p-side of the junction which prevents the large number of electrons in the n-material from diffusing to the p-region. It also keeps the large quantity of holes on the p-side from entering the n-region of the crystal.

It should be pointed out that although the p-region contains predominantly holes, it also includes some minority electrons. The electric field of the p-n junction tends to transport any such electrons that diffuse into the space-charge region across this region and over to the n-side. Although the number

---

[7] This is analogous to two objects, each at a different temperature, being brought into contact and coming to a uniform temperature after some time has elapsed as thermal equilibrium is established.

of such electrons is small relative to the other charges present in materials such as extrinsic germanium at room temperature, this number becomes quite substantial as the temperature is raised. (A corresponding argument applies for the minority holes on the n-side of the junction being transported to the p-side by the field present at the junction.) The electron carrier flow must, of course, balance to zero in thermal equilibrium. That is, any minority electron flow from the p-region to the n-side of the junction must be balanced by a comparable flow of the more energetic majority electrons on the n-side mounting the potential barrier and traveling to the p-side. A corresponding argument applies to hole flow. However, it will become clear later that when an external reverse voltage is applied this minority carrier flow through the junction is the source of current flow through a reverse-biased germanium p-n diode.

EXAMPLE 6.1

Determine the change in barrier height of a p-n junction diode at 300°K when the doping on the n-side is changed by a factor of 1000 and the doping on the p-side remains unchanged.

SOLUTION From Eq. (6.8),

$$(\Delta V_0)_1 = \frac{kT}{q} \ln \frac{(N_a)_1(N_d)_1}{n_i^2}$$

and

$$(\Delta V_0)_2 = \frac{kT}{q} \ln \frac{(N_a)_2(N_d)_2}{n_i^2},$$

where the subscript 1 refers to the lightly doped case, and 2 to the heavily doped case. Subtracting the first equation from the second gives

$$(\Delta V_0)_2 - (\Delta V_0)_1 = \frac{kT}{q} \ln \frac{(N_a)_2(N_d)_2}{(N_a)_1(N_d)_1}$$

$$= 0.026 \text{ V (ln } 1000) = 0.18 \text{ V}.$$

Hence a thousandfold change in doping alters the barrier height by only 180 mV.

C. The Space-Charge Region in Equilibrium

A quantitative calculation of the electric field distribution in the space-charge region will now be given. In a practical p-n junction diode one side of the junction is usually much more heavily doped with impurities than the other side. The main reason for this has to do with the fabrication process for semiconductor p-n junction diodes. If the junction is the emitter junction of a

FIGURE 6.4

a) Charge density due to impurities versus distance within a p^+-n junction depletion region in equilibrium. N_a represents the density of ionized acceptors, and N_d that of ionized donors.
b) Electric field versus distance in the depletion region, indicating the maximum negative value of field at the metallurgical junction.

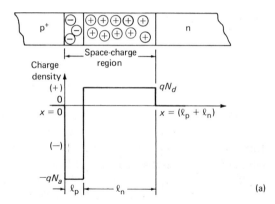

(a)

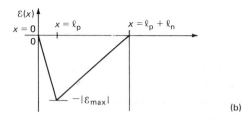

(b)

bipolar transistor, there are also important electrical reasons for requiring an unsymmetrically doped p-n junction.

For our calculation here, assume that the p-region of the junction contains many more impurity atoms than the n-side. This junction will now be referred to as a p^+-n junction. Again we will assume that there is a uniform distribution of impurities on each side of the junction. The total charge density on each side is shown schematically in Fig. 6.4a. Note that the negative charges on the p-side and the positive charges on the n-side originate from ionized acceptors and donors respectively. Note too that the region containing the positive charges is considerably wider than the negatively charged region. This results from the assumption that the density of impurity atoms is much greater on the p-side than the n-side and that the total number of positive ion charges must equal the number of negative ion charges. The latter results from the fact that the field flux lines which originate on the positive charges in the n-region must all terminate on the negative charges of the p-region, since the electric field is everywhere zero in the uniformly doped material outside this *space-charge* region. (Charge neutrality in semiconductors dictates that there can be no space-charge region in a crystal unless a relatively sharp impurity gradient occurs in a region of the order of a few micrometers, as at a p-n junction.) Equating the number of charges on each side of the junction of Fig. 6.4a yields

$$N_a l_p A = N_d l_n A, \qquad (6.9)$$

so that the ratio of the space-charge width on the n-side to that on the p-side is

$$\frac{l_n}{l_p} = \frac{N_a}{N_d}. \tag{6.10}$$

In this discussion it has been tacitly assumed that there are no electrons or holes in the space-charge region and that the only charges there are ionized donors or acceptors. This is a good approximation since the high electric field present in the space-charge region tends to sweep out mobile electrons and holes, and this volume is sometimes referred to as the *depletion* region. At the outer edges of the space-charge layer the field is small and some mobile carriers are present. However, this causes only a small correction to the space-charge calculation, and so the effect of mobile charges will be neglected here and only the fixed charges considered, in a so-called "depletion-layer" approximation.[8]

Gauss's law applied to the p-side of the p-n junction depletion region yields

$$\epsilon_0 \epsilon_r \frac{d\mathcal{E}}{dx} = -qN_a. \tag{6.11}$$

Here $\mathcal{E}$ is the electric field intensity, N_a the acceptor concentration, ϵ_0 the permittivity of free space, and ϵ_r the semiconductor dielectric constant. Integrating this equation gives

$$\mathcal{E} = -\frac{qN_a}{\epsilon_0 \epsilon_r} x + B. \tag{6.12}$$

The electric field vanishes at the edge of the space-charge layer, that is, $\mathcal{E} = 0$ at $x = 0$. Using this boundary condition requires that $B = 0$. So the electric field decreases linearly with x and reaches a maximum negative value at $x = l_n$. This is illustrated in Fig. 6.4b. Hence

$$|\mathcal{E}_{max}| = \frac{qN_a l_p}{\epsilon_0 \epsilon_r}. \tag{6.13a}$$

The edge of the space-charge region of the n-side of the junction is at $x = l_p + l_n$. Integrating Gauss's equation for the n-side gives

$$|\mathcal{E}_{max}| = \frac{qN_d l_n}{\epsilon_0 \epsilon_r}. \tag{6.13b}$$

The total potential drop across the space-charge layer is the sum of the drops across the p- and n-sides and is obtained by integrating the electric field across the depletion region. That is, the equilibrium barrier height is

$$|\Delta V_0| = \int_0^{l_p + l_n} \mathcal{E}(x)\, dx, \tag{6.14}$$

---

[8] See P. E. Gray et al., *Physical Electronics and Circuit Models of Transistors*, John Wiley & Sons, New York (1964), p. 20, for a discussion of the accuracy of this approximation.

which represents the area under the field curve of Fig. 6.4b. Integrating gives the barrier height,

$$|\Delta V_0| = \frac{q}{2\epsilon_0 \epsilon_r} (N_a l_p^2 + N_d l_n^2). \tag{6.15}$$

With Eqs. (6.13a) and (6.13b) this becomes

$$|\Delta V_0| = |\mathcal{E}_{max}| \frac{l_p + l_n}{2}. \tag{6.16}$$

Note that this equation may be deduced by inspection of Fig. 6.4b, by determining the area under the curve geometrically. This can be rewritten, using Eq. (6.10), as

$$|\Delta V_0| = \frac{|\mathcal{E}_{max}| l_n}{2} \left(1 + \frac{N_d}{N_a}\right). \tag{6.17}$$

Combined with a previously derived expression for the equilibrium barrier height given in Eq. (6.8), Eqs. (6.13) and (6.17) now permit calculation of the space-charge widths in terms of device design parameters as

$$l_p = \left[\underbrace{\left(\frac{kT}{q} \ln (N_a N_d / n_i^2)\right)}_{\Delta V_0} \left(\frac{2\epsilon_0 \epsilon_r}{q N_a} \frac{N_d}{N_d + N_a}\right) \right]^{1/2} \tag{6.18a}$$

and

$$l_n = \left[\underbrace{\left(\frac{kT}{q} \ln (N_a N_d / n_i^2)\right)}_{\Delta V_0} \left(\frac{2\epsilon_0 \epsilon_r}{q N_d} \frac{N_a}{N_a + N_d}\right) \right]^{1/2}. \tag{6.18b}$$

For a p^+-n junction where $N_a \gg N_d$,

$$l_n \simeq \left(\Delta V_0 \frac{2\epsilon_0 \epsilon_r}{q N_d}\right)^{1/2} = \left(\frac{2\epsilon_0 \epsilon_r kT}{q^2 N_d} \ln \frac{N_a N_d}{n_i^2}\right)^{1/2} \tag{6.19}$$

and $l_p \ll l_n$. Hence the depletion-layer width is approximately inversely proportional to the square root of the impurity level in the most lightly doped side of the p-n junction and penetrates mainly into this latter region.

Precisely the same procedure may be followed in cases where the transition from p- to n-type is not abrupt as assumed here. Gauss's law can be integrated easily if, for example, an inverse linear variation in impurity concentration is assumed as one proceeds from the p- to the n-region. Then the space-charge width under equilibrium conditions varies inversely as the *cube root* of the most lightly impurity-doped region.[9]

[9] A. B. Phillips, *Transistor Engineering*, p. 114, McGraw-Hill Book Company, New York (1962).

EXAMPLE 6.2

Determine the space-charge width of the n-region of a silicon p^+-n junction in thermal equilibrium at $300°K$ if the doping on the n-side is 1.0×10^{20} donors/m^3 and on the p-side is 5.0×10^{25} acceptors/m^3. The relative dielectric constant of silicon is 12.

SOLUTION   From Eq. (6.10),

$$\frac{l_n}{l_p} = \frac{N_a}{N_d} = \frac{5.0 \times 10^{25}}{10^{20}} = 5.0 \times 10^5.$$

Therefore, the space-charge width in the p-region is negligibly narrow compared to that in the n-region. Hence by Eq. (6.19)

$$l_n = \left[\left(\frac{2\epsilon_0 \epsilon_r kT}{q^2 N_d}\right) \ln \frac{N_a N_d}{n_i^2}\right]^{1/2}$$

$$= \left[\underbrace{\frac{kT}{q} \ln \left(\frac{N_a N_d}{n_i^2}\right)}_{\Delta V_0} \frac{2\epsilon_0 \epsilon_r}{q N_d}\right]^{1/2}$$

$$= \left\{\left[(0.026 \text{ V}) \ln \frac{5.0 \times 10^{25}(1.0 \times 10^{20})}{(1.5 \times 10^{16})^2}\right]\right.$$

$$\left. \times \left[\frac{2(8.85 \times 10^{-12} \text{ F/m})(12)}{(1.6 \times 10^{-19} \text{ C})(10^{20})\text{m}^{-3}}\right]\right\}^{1/2}$$

$$= [(0.8)(13.3 \times 10^{-12})]^{1/2} = 3.3 \times 10^{-6} \text{ m}.$$

This derivation has assumed no externally applied voltage. An inverse voltage bias applied to the junction (p-side negative relative to n-side) creates an additional electric field in the space-charge region, in the same direction as the built-in field. Hence, according to Eq. (6.18), the space-charge width will tend to increase until $\int \mathcal{E} \, dx$ over this space-charge width is equal to the sum of the built-in voltage and the absolute value of the applied voltage. For this to occur, additional ionized acceptors must be added to the p-side of the depletion layer and an equal number of ionized donors to the n-side. A schematic representation of this is shown in Fig. 6.5.

This method of calculating the space-charge width is of interest in the design of a charge-coupled device[10] as well as the varactor diode, which is a variable-capacitor-type semiconductor device. The capacitor character of the junction diode is indicated by the fact that an incremental applied voltage results in an incremental change in charge on each side of the depletion layer, in analogy with the charging of a parallel-plate capacitor. It will be shown quantitatively in the next chapter how the capacitance of the junction can be adjusted by applying an external voltage, thereby altering the space-charge width.

[10] This device produces a simple type of shift register and will be described in Chapter 11.

FIGURE 6.5

a) Schematic diagram showing the additional fixed charge (above equilibrium) of ionized donors and acceptors (crosshatched areas) introduced into the p^+-n junction space-charge region under reverse bias conditions. Space-charge widening results.
b) Graph showing the electric-field increase in the space-charge region under reverse bias conditions. The crosshatched area represents the applied reverse voltage.

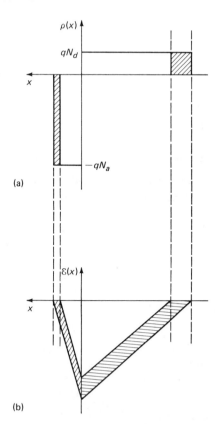

6.2 STEADY-STATE CURRENT FLOW IN A P-N JUNCTION DIODE AT LOW FREQUENCIES

The previous discussion was a description of a semiconductor p-n junction under equilibrium conditions with no external voltage applied. Consider now the case of a p-n junction under applied-voltage conditions and hence current flow. If the p-side of the junction is brought to a higher or more positive potential (V_{pA}) than the n-side (V_{nA}), this is referred to as "forward" bias. "Reverse" or "inverse" bias refers to the n-side being raised to a higher potential than the p-side. It will next be shown analytically that the current I through the p-n junction and the voltage drop $V (= V_{pA} - V_{nA})$ across the junction are related generally by a diode equation of the form

$$I = I_0(e^{qV/kT} - 1). \tag{6.20}$$

That is, when the junction is forward biased (V is made positive), the current increases rapidly for small changes in voltage. In reverse bias the current is much smaller and becomes essentially equal to I_0 as the voltage is increased. Physically this is explained in terms of reduction of barrier height due to forward bias, causing appreciable current to flow; in the reverse direction, the increased barrier height limits carrier flow across the junction. This is illustrated in Fig. 6.6. Note that an increase in the potential in a positive sense depresses the energy levels in the electron energy band diagram. Again this results from the negative charge on the electron. The Fermi energy is not defined under nonequilibrium conditions, but levels E_{Fp} and E_{Fn} are called the quasi-Fermi levels.

A. Steady-State Minority Carrier Flow

Consider a simple one-dimensional model of the p^+-n junction as shown in Fig. 6.7a. A forward voltage bias causes the reduction of barrier height and permits many holes from the p-region to enter the n-region as minority carriers. This process is known as *minority carrier injection*. Correspondingly, electrons from the n-side cross the lowered barrier and enter the p-side as minority carriers. This is shown in Fig. 6.7b. Although the current through the junction consists of the sum of both the electron and hole components, the hole flow dominates, owing to the preponderance of holes on the heavily doped p-side compared to electrons on the n-side of the p^+-n junction. In addition, although the total hole current in the n-region consists of both a drift component due to

FIGURE 6.6

a) Energy diagram for a p^+-n junction under reverse bias. The barrier height is increased over equilibrium conditions.
b) Energy diagram for a p^+-n junction under forward bias. The barrier height is decreased below equilibrium conditions.

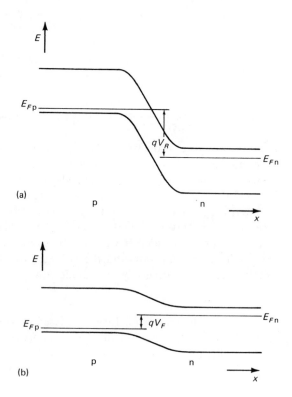

FIGURE 6.7

a) Energy diagram for p⁺-n junction diode under forward bias showing injection of electrons and holes. x is taken as zero at the n-side edge of the space-charge region. The n-region is assumed to extend to infinity in the x-direction (long-diode approximation).

b) Injection of excess holes (Δp_n) into the n-region and electrons (Δn_p) into the p⁺-region of a p⁺-n junction diode. p_{no} and n_{po} correspond respectively to the equilibrium concentrations of holes in the n-type and electrons in the p-type regions. The exponential falloff of excess holes and electrons with distance is due to minority carrier recombination.

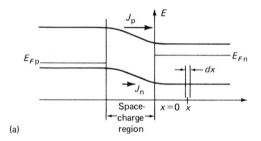

(a)

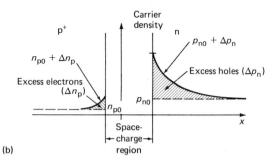

(b)

the electric field and a diffusion component (see Eq. 5.32), the latter current dominates completely in the n-region near the junction under low-level injection conditions. (Low-level injection refers to the case where the density of injected minority holes is smaller by about an order of magnitude than the density of majority electrons in the n-region at equilibrium.)

Dominance of diffusion current occurs in this case since the applied voltage appears across the space-charge region and creates an electric field primarily in this region, and very little penetrates into the n-region to produce a drift component of the current there.[11] This is illustrated by Fig. 6.7a since the slope of the energy band edge is nearly zero ($dE/dx = 0$) everywhere except in the space-charge region. Since $E = -qV$, then $dE/dx = -q\,dV/dx = 0$ and the electric field is nearly zero outside the space-charge region.

Now consider a very small volume of width dx and unit area in the n-region, located a distance x from the edge of the space-charge region as shown in Fig. 6.7a. Under steady-state conditions the current of carriers entering the region dx differs from that leaving the region by the amount of minority hole loss by recombination in that volume. This statement is an expression of the continuity of carrier flow and can be expressed mathematically as

$$-\frac{1}{q}\frac{dJ_p}{dx} = \frac{\Delta p_n}{\tau_p}, \tag{6.21}$$

[11] The demonstration that at low injection levels the current in the n-region near the junction is almost exclusively diffusion current is given in Appendix A.

where $(1/q)(dJ_p/dx)$ is the change of hole flux with respect to x in dx, and $\Delta p_n/\tau_p$ represents the rate of recombination of excess injected hole density Δp_n in the n-region, within dx, where the hole lifetime is τ_p (see Eq. 5.37). Since only hole diffusion current is considered, this is given by

$$J_p = -qD_p\frac{dp_n}{dx}. \qquad (6.22)$$

Combining Eqs. (6.21) and (6.22) gives the steady-state *continuity* equation for hole density in the n-region:

$$D_p\frac{d^2p_n}{dx^2} = \frac{\Delta p_n}{\tau_p}. \qquad (6.23a)$$

In terms of injected carriers above the equilibrium number, the excess hole density continuity equation becomes, since $p_n = p_{n0} + \Delta p_n$,

$$D_p\frac{d^2(\Delta p_n)}{dx^2} = \frac{\Delta p_n}{\tau_p}. \qquad (6.23b)$$

The solution of this equation may be written as

$$\Delta p_n(x) = Be^{-x/L_p} + Ce^{x/L_p}, \qquad (6.24)$$

where $\sqrt{D_p\tau_p}$ is defined as the *minority carrier diffusion length* L_p. Just as τ_p represents the *time* taken for minority holes to decrease in number by a factor of $1/e$ by recombination, L_p represents the *distance* that holes can diffuse on the average before their number decreases by a factor of $1/e$ due to recombination. Since recombination occurs all along the path of motion of the hole carriers in the n-region, Δp_n must approach zero for large positive values of x. (We assume in this derivation that the diode thickness is much greater than L_p.) This boundary condition requires that C in Eq. (6.24) be identically zero, for otherwise Δp_n would grow with x.

B. The p-n Junction Current-Voltage Relation

Making use of Eq. (6.22) for the hole diffusion current density and introducing Eq. (6.24) gives, for this current density at x,

$$J_p(x) = \frac{qD_p\,\Delta p_n(x)}{L_p} = \frac{qD_p\,Be^{-x/L_p}}{L_p}. \qquad (6.25)$$

This expression holds for any distance x from the edge of the space-charge region. Hence it must also be true at the n-side edge of this region where $x = 0$. There Eq. (6.25) may be written as

$$J_p(0) = \frac{qD_p\,\Delta p_n(0)}{L_p}, \qquad (6.26a)$$

FIGURE 6.8

a) Plot of electron density n and hole density p in a long p^+-n junction diode in forward bias.
b) Plot of the electron current density J_n and the hole current density J_p in a long p^+-n junction diode versus distance under forward bias.

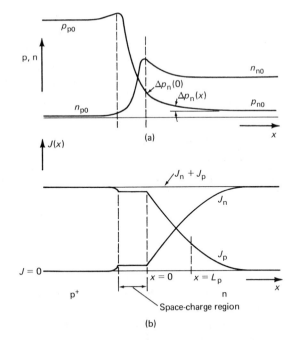

where $\Delta p_n(0)$ represents the excess minority hole density present at the n-side edge of the space-charge region under forward voltage bias conditions. By combining Eqs. (6.25) and (6.26a), the hole current density at any value of x in the n-region can be written as

$$J_p(x) = \frac{q D_p \Delta p_n(0)}{L_p} e^{-x/L_p}. \qquad (6.26b)$$

This implies that the hole current falls off with x, so that it reaches $1/e$ of its value at $x = 0$ in a diffusion length L_p (see Fig. 6.8b). However, under the steady-state conditions assumed, the current density must be the same at all values of distance x and time t. Hence the loss of hole current with distance due to hole recombination must be made up by the generation of electron current just to compensate this effect. This is shown diagrammatically in Fig. 6.8b. Note that at a distance x, far (several L_p) from the junction, the current is nearly entirely electron drift current, as one would expect in an n-type semiconductor lacking injected holes.

It is now necessary to determine a relationship between $\Delta p_n(0)$ and the voltage V applied across the space-charge layer, to derive the p-n junction current-voltage characteristic sought. It is a reasonable assumption that, because of the large hole density in the p-region under normal forward voltage bias conditions, the junction is intrinsically capable of supplying very much more current than it actually supplies. This means that the balance of drift and diffusion currents in the space-charge region, which is absolute under equilibrium conditions, is not much disturbed by the applied voltage. In fact, the applied bias raises the diffusion component of current only slightly above the drift component. This is demonstrated in Example 6.3. Then in analogy with Eq. (6.7), since the equilibrium is hardly disturbed, the density of holes on the n-side of the depletion region and holes on the p-side edge are related by

$$\frac{(p_p)_{\text{p-side edge}}}{(p_n)_{\text{n-side edge}}} = e^{q(\Delta V_0 - V)/kT}, \tag{6.27}$$

where ΔV_0 is the equilibrium built-in barrier potential, and V represents the forward voltage (p-side positive with respect to n-side) applied across the space-charge layer. Combined with (6.7) to eliminate ΔV_0, Eq. (6.27) becomes

$$\frac{(p_n)_{\text{n-side edge}}}{(p_p)_{\text{p-side edge}}} = \frac{p_{n0}}{p_{p0}} e^{qV/kT}. \tag{6.28}$$

For low-level injection conditions the injected minority hole density is assumed much smaller than the density of the equilibrium number of majority electrons present in the n-region; similarly, the injected minority electron density is considerably less than the hole density present in the heavily doped p-region. Since the condition of charge neutrality requires that electrons and holes are injected into the p-region in equal numbers, this latter statement requires that $(p_p)_{\text{p-side edge}} \simeq p_{p0}$. Then Eq. (6.28) reduces to

$$p_n(0) = p_{n0} e^{qV/kT}. \tag{6.29}$$

This is the extremely useful *junction law*. Multiplying both sides by $n_n(0)$ and noting that $n_n(0) \simeq n_{n0}$, we obtain $p_n(0)n_n(0) = p_{n0} n_{n0} e^{qV/kT} = n_i^2 e^{qV/kT}$, which is the nonequilibrium version of $p_n n_n = n_i^2$ (see Eq. 5.18).

When the voltage applied to the junction is zero, Eq. (6.29) shows that the minority hole density at the n-side edge of the space-charge region reduces to the equilibrium density, as expected. It also predicts that the density of holes injected into the n-region increases exponentially with applied voltage. That is, the density of holes from the heavily hole-populated p-side of the junction, which can surmount the junction potential barrier and enter the n-side, depends exponentially on the amount by which the barrier is reduced. This is analogous to the classical problem, for example, of gas atoms in free space mounting a potential barrier. This was treated in the nineteenth century by Boltzmann, and hence Eq. (6.29) is sometimes referred to as the Boltzmann relation and is sometimes taken as a basic assumption, even at moderate injection levels.

We can now write, for the excess holes just appearing at the n-side edge of the space-charge region,

$$\Delta p_n(0) = p_n(0) - p_{n0} = p_{n0}(e^{qV/kT} - 1). \tag{6.30}$$

Finally, the current density due to holes crossing the p-n junction is obtained by introducing Eq. (6.30) into (6.26b) as

$$J_p(x) = \frac{q D_p p_{n0}}{L_p} (e^{qV/kT} - 1) e^{-x/L_p}. \tag{6.31a}$$

The expression represents only the hole diffusion current density. In general, the hole drift current density must be added to this to get the total hole current density. The drift and diffusion components of electron current density must also be added. However, for analysis of the p^+-n junction considered here, it has already been argued that the hole diffusion component dominates near the junction at $x = 0$. Hence in this case Eq. (6.31a) at $x = 0$ approximately represents the relation between the total junction current density and applied voltage:

$$I = \frac{q A D_p p_{n0}}{L_p} (e^{qV/kT} - 1) = \frac{q p_{n0} L_p A}{\tau_p} (e^{qV/kT} - 1), \tag{6.31b}$$

where A is the p-n junction cross-sectional area; this is the form predicted in Eq. (6.20). The coefficient $q p_{n0} L_p A / \tau_p$ in Eq. (6.31b) represents the *saturation* or *leakage* current through the junction when a few volts of negative or reverse potential are applied to the device, for then $e^{qV/kT} \to 0$ and $I \to q p_{n0} L_p A / \tau_p$. Physically this means that the inverse leakage current in such a p^+-n junction is made up of the flow across the junction of the equilibrium number of minority holes within a diffusion length of the junction $(p_{n0})(L_p A)$, which are hence "collected" during their lifetime τ_p. These carriers flow "downhill"[12] from the n-side to the p-side of the potential barrier (see Fig. 6.7a). Minority electrons from the p-side also can flow downhill through the barrier to the n-side. However, due to the high acceptor concentration on the p-side, the number of majority holes is very large and so the number of minority electrons is small there, due to constancy of the np product. Therefore, in a p^+-n junction, although the leakage current consists of both diffusing holes and electrons, the hole component dominates. In the more general case of a p-n junction which has comparable doping on both the p- and n-sides, the diode current-voltage equation becomes

$$I = qA \left(\frac{D_p p_{n0}}{L_p} + \frac{D_n n_{p0}}{L_n} \right) (e^{qV/kT} - 1) \simeq qA \left(\frac{D_p}{L_p N_d} + \frac{D_n}{L_n N_a} \right)$$
$$\times n_i^2 (e^{qV/kT} - 1), \tag{6.32}$$

where A is the area of the p-n junction. Note that p_{n0}, n_{p0}, and n_i are highly temperature dependent as given by Eqs. (5.11), (5.17), and (5.19). However, D_p, D_n, L_p, and L_n are normally only mildly temperature dependent. Hence the temperature variation of the junction saturation current is mainly dependent

---

[12] Note that "downhill" for holes is "uphill" for electrons as pictured in Fig. 6.7a.

on the change of the semiconductor minority carrier densities and n_i with temperature. This leakage current is contributed mainly by the diffusion of minority carriers from the least-doped side of the p-n junction.

EXAMPLE 6.3

Assuming the silicon p^+-n junction diode design in Example 6.2, show that the balance of drift and diffusion currents in the space-charge region, which is absolute under equilibrium conditions, is not much disturbed by the application of a normal forward bias voltage, say 0.60 V at 300°K.

SOLUTION   First let us roughly approximate the hole diffusion current in the space-charge region (which is equal and opposite to the hole drift current) under equilibrium conditions:

$$(J_{\text{diffusion}})_p = -qD_p \frac{dp}{dx}.$$

From Example 6.2 the density of holes at the left-hand edge of the space-charge region is 5.0×10^{25} m^{-3}. Since the np product in equilibrium at 300°K is $(1.5 \times 10^{16})^2$, the hole density at the right-hand edge of the space-charge region is $(1.5 \times 10^{16})^2/10^{20}$ m^{-3} or 2.25×10^{12} holes/m^3. Since the depletion-layer width from Example 6.2 is 3.3×10^{-6} m,

$$-\left(\frac{dp}{dx}\right)_{\text{space-charge}} \simeq \frac{5.0 \times 10^{25} - 2.25 \times 10^{12}}{3.3 \times 10^{-6}} \text{ m}^{-4} = 1.5 \times 10^{31} \text{ m}^{-4}$$

and

$$(J_{\text{diffusion}})_p \simeq 1.6 \times 10^{-19} \text{ C}(1.3 \times 10^{-3} \text{ m}^2/\text{sec})(1.5 \times 10^{31} \text{ m}^{-4})$$
$$= 3.1 \times 10^9 \text{ A/m}^2.$$

So $(J_{\text{drift}})_p \simeq 3.1 \times 10^9$ A/m^2, also. Let us now calculate the hole current at the right-hand edge of the depletion layer, assuming in advance that the equilibrium drift and diffusion current balance are little upset by the application of a favored bias of 0.60 V. Using Eq. (6.31), assuming that $L_p = 1.0 \times 10^{-4}$ m, we have

$$J_p = \frac{qD_p p_{n0}}{L_p}\left(e^{qV/kT} - 1\right)$$

$$= \frac{1.6 \times 10^{-19} \text{ C} \,(1.3 \times 10^{-3} \text{ m}^2/\text{sec})(2.25 \times 10^{12} \text{ m}^{-3})}{1.0 \times 10^{-4} \text{ m}}$$

$$\times (e^{0.60/0.026} - 1)$$

$$= 4.7 \times 10^4 \text{ A/m}^2.$$

This represents only $[4.7 \times 10^4/(3.1 \times 10^9)]$ 100 or 1.5×10^{-3} percent, a minute upset of the balance of equilibrium drift and diffusion currents, and the proposition is proved.

C. Space-Charge Generated Current

In the above calculation it was assumed that essentially all the current was derived from the diffusion of minority carriers through the junction. This is a valid assumption for germanium and semiconductor materials with smaller energy gaps. In the case of semiconductors with wider energy gaps, like silicon and gallium arsenide, there is still another important component to the current. This component derives from carriers that are generated and recombine in the space-charge region, and was neglected in the diffusion-current treatment. In the space-charge region in equilibrium, hole-electron pairs are continuously being created thermally and recombining in a time called the lifetime. In forward bias minority current carriers from the p-side of the junction can recombine in the space-charge region before reaching the n-side, and this recombination current must be supplied. Also the reverse biased junction space-charge region has a large electric field present causing mobile electrons and holes generated within it to be swept out before recombining, constituting a current. A formulation of this problem of space-charge generated junction current[13] yields approximately the following current-voltage characteristic:

$$I \simeq \frac{qAWn_i}{\tau} (e^{qV/2kT} - 1), \tag{6.33}$$

where W is the space-charge width wherein the minority carrier lifetime is τ, A is the junction area, and n_i is the density of intrinsic carriers in the semiconductor. Note the factor-of-two difference in the exponential term in this formulation compared to the expression which considers only diffusion current (see Eq. 6.32). Note too that the coefficient of the exponential term represents the space-charge generated junction leakage current under reverse bias. The form of this coefficient can be made plausible by identifying qn_i/τ as the charge density generated per unit time in the space-charge volume AW. The charge density n_i is involved since the space-charge region is depleted of mobile carriers and hence effectively is "intrinsic." Since the space-charge width increases with increased reverse potential bias, the space-charge generated leakage current rises with increased inverse voltage. This is in contrast to the diffusion component of leakage current, which is voltage independent above a small voltage and hence is termed "saturation" current.

In general, both the diffusion and space-charge generated components of the junction leakage current should be considered. However, by comparing the magnitude of the currents given by Eqs. (6.33) and (6.32) it can be ascertained

[13] C. T. Sah, R. N. Noyce, and W. Shockley, "Carrier Generation and Recombination in p-n Junctions and p-n Junction Characteristics," *Proc. IRE*, Vol. 45, p. 1228 (1957).

which is dominant in any particular case. Note that the diffusion current dominates over space-charge generated current for narrow-energy-gap semiconductors since the former varies as n_i^2 whereas the latter varies as n_i. Hence in germanium only diffusion current normally needs to be considered. However, in silicon (and wider-energy-gap materials) space-charge generated current dominates at low forward currents and for reverse bias. At relatively large forward currents in silicon junctions, diffusion controls again since the diffusion component increases rapidly with voltage, whereas the space-charge generated component is mainly unchanged.

D. Surface Leakage Current

The reverse current of a real p^+-n junction diode includes a surface leakage component in addition to the bulk leakage terms due to diffusion and space-charge generation derived above. This may be of the ohmic type, simply caused by the electric conduction properties of spurious impurity ions which appear on the diode surface, to a greater or lesser extent depending on the manufacturer's cleaning techniques. Surface ionic conduction is particularly enhanced in the presence of moisture or under high humidity conditions. This type of leakage current can be reduced many orders of magnitude by depositing a protective insulating layer on the p^+-n junction surface. Silicon planar junctions are commonly protected by depositing a thin *passivating* silicon dioxide layer on the junction surface as shown in Fig. 1.2. This is standard practice in the fabrication of integrated circuits and will be discussed in Chapter 11. (Some limitations of oxide passivation have been explained in Section 5.6.) P-N junction surfaces are alternatively cleaned by chemically etching away contaminated and disturbed surface layers of silicon with a combination of hydrofluoric and nitric acids; a thorough washing in ultrapure water and drying then follows. Junctions so prepared are uniformly flat in contrast to the curvature at the edges of the planar junctions of Fig. 1.2 and are referred to as "mesa" type junctions. When p-n junctions with blocking voltage capabilities of several hundred volts or more are required, mesa junctions are fabricated; for a given applied voltage a portion of the space-charge region of a planar junction reaches a higher electric field than a comparably doped mesa junction since the highest field is developed at the point of greatest junction curvature. Hence planar junctions have limited blocking voltage capability (see Section 6.3c).

Still another type of junction leakage is referred to as junction "channeling." This refers to the conduction of current in the surface of the semiconductor crystal adjoining the junction due to mobile charges induced in the semiconductor by fixed charges on the crystal surface, or even in the oxide "protecting" the surface.[14] This produces a leakage current essentially independent of

[14] For a full discussion of this phenomenon, see A. Grove, *The Physics and Technology of Semiconductor Devices*, John Wiley & Sons, New York (1965), p. 298ff. See also Section 5.6 of this text

voltage, in contrast to ohmic ionic leakage where the current and voltage are linearly related. An n^+-p planar junction consists of a small n^+-region embedded in a p-type semiconductor substrate. As has already been discussed in Section 5.6, the oxide and oxide-surface interface charges are always positive so that an n-type surface skin is induced which can invert the p-type surface covered by the oxide. This effectively increases the n-p junction area, thereby increasing the leakage current. Since this induced n-layer is thin, a current-limiting action similar to that observed in a field-effect transistor (see Chapter 10) occurs so that the leakage current is constant, although higher, above a certain voltage and remains fixed as the voltage is raised. This will obviously not occur in the case of a p^+-n junction; instead the surface charges will tend to reduce the blocking voltage of this type of junction (see Section 6.3C, Voltage breakdown).

6.3 OTHER ELECTRICAL PROPERTIES OF P-N JUNCTION DIODES

The successful circuit use of the p-n junction diode requires that some additional properties of the junction and electrical contacts to the device be understood. For example, successful circuit applications of the diode require a knowledge of the temperature behavior of the device. The electrical characteristics of semiconductor devices in general are quite sensitive to temperature (see Eq. 6.32). Self-heating caused by the current flowing through the junction diode provides a high-current limitation for the device. A high-voltage limitation results from excessive junction leakage current due to current multiplication in the reverse-biased space-charge region. High-frequency or high-speed performance is limited by the time necessary to charge the space-charge capacitance of the junction through the bulk series resistance of the semiconductor body and ohmic contacts. High-speed switching performance can also be limited by injected minority carrier transit time as well as storage effects in the semiconductor bulk. These diode limitations will be discussed in the remainder of this chapter and in the following chapter.

A. Temperature Behavior of the p-n Junction

The electrical characteristic of the semiconductor p-n junction is quite sensitive to temperature. This becomes apparent when the current-voltage relationship for the junction is examined. In addition to the explicit exponential variation with temperature of the junction current as indicated in Eq. (6.20), the leakage or saturation current is also temperature sensitive. To illustrate this we can write the saturation current for a germanium p-n junction from Eq. (6.32) as

$$I_0 = qA \left(\frac{D_p}{L_p N_d} + \frac{D_n}{L_n N_a} \right) n_i^2. \tag{6.34}$$

Here n_i^2 represents the square of the density of intrinsic carriers, which can be written as (see Eq. 5.19)

$$n_i^2 = KT^3 e^{-E_g/kT}, \tag{6.35}$$

where K is a constant independent of temperature. This drastic temperature dependence of n_i^2 completely dominates the leakage current indicated in Eq. (6.34) over a wide temperature range, since the minority carrier diffusion constant and lifetime vary rather weakly with temperature by comparison. The approximate fractional change in the saturation current in germanium diodes with temperature is then

$$\frac{1}{I_0}\frac{\partial I_0}{\partial T} = \frac{3}{T} + \frac{E_g}{kT^2} \tag{6.36}$$

or about 0.1 per °K at room temperature.

For moderate forward bias voltages such that $\exp q(V/kT) \gg 1$, the fractional change in junction current for a fixed applied voltage is obtained by differentiating Eq. (6.20):

$$\frac{1}{I}\frac{\partial I}{(\partial T)_V} = \frac{1}{I_0}\frac{\partial I_0}{\partial T} - \frac{qV}{kT^2} \tag{6.37}$$

$$= \frac{3}{T} + \frac{E_g - qV}{kT^2}.$$

This temperature variation is somewhat less than that of the saturation current, as given by Eq. (6.36).

Consideration of the temperature variation of the diode voltage drop for a fixed bias current yields, for current values $I \gg I_0$,

$$\left(\frac{\partial V}{\partial T}\right)_I = \frac{V}{T} - \frac{kT}{q}\frac{1}{I_0}\frac{\partial I_0}{\partial T} = \frac{V - E_g/q}{T} - \frac{3k}{q}. \tag{6.38}$$

This averages about -2 mV/°K at room temperature for germanium junctions.

In the case of silicon the temperature dependence of the junction leakage current as given by Eq. (6.33) is dominated by the temperature variation of n_i. Hence the fractional change in leakage current of a reverse-biased silicon junction is given by

$$\frac{1}{I_0}\left(\frac{\partial I_0}{\partial T}\right)_V = \frac{1}{2}\left(\frac{3}{T} + \frac{E_g}{kT^2}\right) \tag{6.39}$$

or about 0.08 per °K at room temperature. For moderate forward-bias voltages such that $\exp (qV/kT) \gg 1$, the fractional change in silicon diode current for a given applied voltage is

$$\frac{1}{I}\left(\frac{\partial I}{\partial T}\right)_V = \frac{1}{I_0}\frac{\partial T_0}{\partial T} - \frac{qV}{2kT^2} = \frac{1}{2}\left(\frac{3}{T} + \frac{E_g - qV}{kT^2}\right). \tag{6.40}$$

This temperature variation is somewhat less than that of the leakage current given by Eq. (6.39). The temperature variation of the diode forward voltage drop at a fixed bias current, for current values of $I \gg I_0$, is given by

$$\left(\frac{\partial V}{\partial T}\right)_I = \frac{V - E_g/q}{T} - \frac{3k}{q}. \tag{6.41}$$

This coefficient averages about -2.5 mV/°K at room temperature for silicon diodes.

The junction temperature coefficients which have been discussed are of particular importance when the junction is the emitter of a transistor. It is generally desired that transistor circuits be unaffected by temperature fluctuations. Temperature variations are often compensated for in temperature-sensitive circuit applications by balancing the emitter voltage against the voltage drop across a similar diode in thermal proximity to the emitter junction. This is referred to as a *differential* arrangement. It is in general use in integrated-circuit amplifiers.

B. Bulk and Contact Resistance in Diodes

Thus far the discussion has been restricted to the electrical properties of the p-n junction of the semiconductor diode. The applied voltage is assumed to appear completely across the junction space-charge layer. However, in a practical diode structure there is bulk semiconductor material on both the p- and n-sides of the junction. This provides resistance in series with the junction and in general there is an ohmic voltage drop in these regions. In addition, there may be potential drops at metal-semiconductor contacts at the ends of the diode which provide electrical connections to an external circuit. In a practical silicon or germanium diode, these potential drops are negligible except at high current densities. A demonstration of this is given in Fig. 6.9, which shows a semilog plot of the current in a typical silicon diode versus applied voltage. From Eqs. (6.32) and (6.33) it may be seen that when a diode is biased with only a few kT/q units of voltage in the forward direction, the diode equation becomes strictly exponential, that is, $e^{qV/kT} \gg 1$. Hence a semilog plot of I versus V should yield a straight line. This is often found to be true over a range of eight or more orders of magnitude in current, which verifies the diode equation and confirms that nearly all the applied voltage appears across the p-n junction. At some high current value, though, the voltage drop across the diode series resistance is no longer negligible, and deviation from strictly logarithmic behavior is observed. This is seen in Fig. 6.9; the extent of the deviation from exponential behavior is a measure of the diode series resistance as shown in the figure.

The value of diode current at which some of the applied voltage begins to appear across the diode series resistance can be estimated by comparing the effective junction resistance with the diode series resistance, at increasing

FIGURE 6.9

Graph of the logarithm of forward current versus forward voltage for a typical silicon p-n junction diode. Deviation from the simple exponential law occurs at high current where a contribution from the potential drop across the n-type semiconductor bulk and the metal-semiconductor end contacts, $I(R_s + R_c)$, is observed. Deviation at low voltage is due to charge generation in the space-charge region, which supplies an additional component to the diffusion current. Surface leakage current is also more apparent at low voltages.

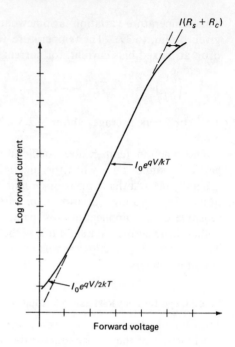

values of current. The effective junction resistance is current-dependent since the diode is basically a nonlinear device. The junction resistance at any current, usually called the *dynamic* or *incremental* resistance, is defined as the derivative dV/dI and is obtained from Eq. (6.20) as

$$\frac{dV}{dI} = \frac{kT}{qI}. \tag{6.42}$$

As an example, the dynamic resistance of a semiconductor p-n junction at 1 mA is 26 Ω.[15] At 26 mA this resistance reduces to 1 Ω. Now in a p^+-n junction the bulk series resistance on the p-side of the junction is usually negligible, owing to the high acceptor content and high electrical conductivity. The n-side resistance may not be negligible. Consider a germanium junction of the form illustrated in Fig. 6.1a. Assume the diode cross-sectional area is 10^{-6} m^2 and the 0.01 Ω-m resistivity n-region is 2×10^{-4} m thick. The bulk series resistance of the diode can be calculated by Ohm's law to be 2 Ω.[16] This is nearly negligible compared to the 26 Ω junction resistance at 1 mA current but certainly not at the 26 mA level of diode current. It should be pointed out, however, that the bulk series resistance of the diode at 26 mA of current would typically be much less than 2 Ω owing to the increased number of electrical carriers on the n-side

[15] Note that the dynamic resistance is independent of junction area.
[16] $R = \rho l/A$, where ρ is the resistivity, l the n-region thickness, and A the cross-sectional area.

due to hole injection. This increases the conductivity of that region and is called conductivity modulation.

Account must also be taken of any potential drops at the metal-semiconductor contacts at the ends of the diode. Significant contact potentials may be obtained at metal-semiconductor junctions; this will be discussed later under the heading of Schottky barriers. In practical devices these end-contact drops are made negligible by heavily doping the ends of the semiconductor junction crystal with acceptor impurities on the p-side and donors on the n-side to the extent that they are nearly metallic. Then the metal-semiconductor contacts become nearly metal-metal contacts and yield very small contact potentials. Since the current in a p-type semiconductor which is carried almost entirely by holes must convert completely to electron current characteristic of most metals at the p-semiconductor-metal contact, this must be a region of high hole carrier recombination. There are plenty of electrons in the metal available for recombination with these holes. In fact, an ohmic contact is sometimes defined as a region of infinitely large recombination.

Now the total voltage drop across a germanium p^+-n junction diode can be written as

$$V = \frac{kT}{q} \ln\left(\frac{I}{I_0} + 1\right) + I(R_s + R_c), \tag{6.43}$$

by making use of Eq. (6.20). Here R_s represents the modulated bulk series resistance of the diode structure, and R_c the metal-semiconductor contact resistance. The latter is primarily of interest in the case of diode rectifiers which operate at high current levels.

C. Breakdown in the p-n Junction

Some volume and surface sources of leakage current in a reverse-biased junction diode have already been identified. The bulk components due to minority carriers diffusing through and generated within the space-charge layer have been described. An additional source of inverse leakage current in a p-n diode is *impact* ionization of the semiconductor atoms in the space-charge layer by high-energy carriers moving through this region under the effects of the high electric field there. This is called current multiplication or *avalanching*. Another source of leakage current is due to carriers penetrating the junction potential barrier by quantum mechanical tunneling. This is known as *Zener* tunneling after the physicist who predicted the effect. Both these sources of inverse leakage current are extremely voltage sensitive and hence occur suddenly when a sufficiently high electric field is reached in the depletion layer. Above the value of applied voltage corresponding to this critical field the current increases drastically for a small increase in voltage. This rather well-defined voltage is called the junction *breakdown* voltage. The complete current-voltage characteristic of the junction diode including avalanching is

FIGURE 6.10

Complete current-voltage
characteristic of a typical
junction p-n diode showing
avalanching at high reverse
negative voltage. A complete
set of specification sheets for
a typical commercial silicon
p-n junction diode is
included in Appendix C.

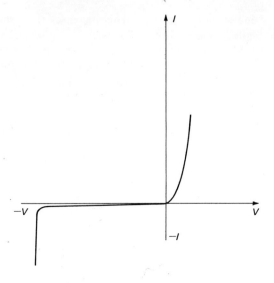

shown in Fig. 6.10. This breakdown sometimes limits the voltage-blocking ability of the p-n junction but also often has important circuit uses in voltage limitation or regulation. The excessive leakage current in the diode in the avalanche region must be limited or else thermal runaway can occur, owing to the high voltage and high current, which can destroy the device.

Avalanche breakdown   The phenomenon of avalanching results when electrons or holes, on diffusing into the junction space-charge region, acquire sufficient energy from the electric field there to knock bound valence electrons out of the lattice atoms in that region. This high electric field usually results from a large inverse voltage applied to the diode. The junction depletion layer must be wide enough so that the mobile carriers can gain sufficient energy from the field to cause ionization. If one electron or hole produces on the average less than one additional carrier, then the junction leakage current is hardly increased. If, however, on the average one additional carrier is produced and these extra carriers each produce one additional carrier, etc., then avalanching results. This process is, in many ways, identical to the one which causes inert gases to ionize in a fluorescent light tube. The mathematics for describing this effect is already known from the treatment of gaseous breakdown phenomena. When this formulation is applied to the semiconductor breakdown case, the current multiplication factor M can be explicitly derived. A simple empirical expression for M is

$$M = \frac{1}{1 - (V/V_B)^n},$$
(6.44)

where V_B is the junction breakdown voltage, V is the applied potential, and n is a numerical factor which depends on the semiconductor crystal utilized in the

FIGURE 6.11

Graph of critical field for junction breakdown versus carrier density for n-type silicon indicating both the avalanche and Zener regions.

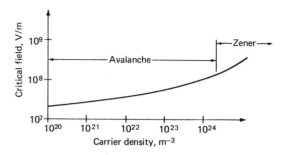

fabrication of the diode ($n = 3$ to 6 for silicon). This factor multiplies the normal junction leakage current denoted by I_0 in Eq. (6.20).

Note that $M \to \infty$ as $V \to V_B$, and avalanche breakdown occurs. In silicon p^+-n junctions the breakdown voltage is approximately proportional to the resistivity of the lightly doped n-region. This result may be derived by combining Eqs. (6.13b) and (6.17) to give

$$\mathcal{E}_{max} = \left[\frac{2qN_d(|\Delta V_0| + |V|)}{\epsilon_0 \epsilon_r}\right]^{1/2}. \tag{6.45}$$

Note that the fields produced by the barrier voltage and the applied voltage are in the same direction and their effects are additive. Now for an extrinsic n-region, the resistivity $\rho_n = 1/N_d q \mu_n$, so that Eq. (6.45) becomes

$$|\mathcal{E}_{max}| = \left|\frac{2(\Delta V_0 + |V|)}{\rho_n \mu_n \epsilon_0 \epsilon_r}\right|^{1/2}. \tag{6.46}$$

Since normally at breakdown $|V|_B \gg \Delta V_0$,

$$|V|_B = \frac{\rho_n \mu_n \epsilon_0 \epsilon_r |\mathcal{E}_{max}|_B^2}{2}. \tag{6.47}$$

In high-resistivity (greater than 0.001 Ω-m) silicon, the breakdown field $|\mathcal{E}_{max}|_B$ is appreciably constant[17] at about 3×10^7 V/m. Hence the breakdown voltage increases in proportion to resistivity. For example, a 100 V breakdown silicon diode requires a resistivity of a little over 0.01 Ω-m. For lower resistivity, the breakdown field increases somewhat. A graph of the breakdown field versus resistivity for n-type Si is given in Fig. 6.11.

Zener breakdown Excessive leakage current may also result when the electric field in the semiconductor becomes so high in value that there is a finite probability that electrons in their covalent bonds can be directly excited into conduction. The quantum mechanical calculation of this probability was performed by Clarence Zener. No carrier acceleration or collision is required

---

[17] S. L. Miller, "Ionization Rates for Holes and Electrons in Silicon," *Phys. Rev.*, Vol. 105, p. 1246 (1957).

for this breakdown mechanism. Hence it is only observed in p^+-n^+ junctions where the space-charge region is so narrow due to the heavy doping that the carriers do not have a sufficiently long path to gain enough energy for impact ionization. Under these conditions the depletion region is so thin that tunneling of the type discussed in Section 3.6B takes place. In fact even heavier doping results in the tunnel diode device as described in Section 1.3A.

A Zener breakdown field of about 10^8 V/m is observed in silicon junctions when the doping level is about 10^{24} donors/m^3. This is illustrated in Fig. 6.11 and corresponds to a breakdown voltage of about 6 V. Hence silicon diode breakdown occurs by the avalanche mechanism above approximately 6 V and by the Zener mechanism below 6 V. Because of the different mechanisms involved, it is interesting to note that the breakdown voltage of diodes which avalanche at greater than 6 V increases somewhat with temperature, whereas those which undergo Zener breakdown below 6 V have breakdown voltages which decrease with increasing temperature. Hence the breakdown voltage has essentially a zero temperature coefficient of variation for approximately 6 V breakdown diodes. Diodes of this type are often used as a temperature-stable voltage reference.

Voltage breakdown due to surface effects   In addition to the bulk mechanisms for p-n junction breakdown just described, there are surface effects which often limit the reverse blocking voltage of a junction diode. Contamination and imperfections on the semiconductor surface near where the p-n junction intersects the surface can result in premature voltage breakdown, i.e., at a voltage lower than that predicted by the bulk considerations discussed in the two sections immediately preceding this one. Hence mesa-type junctions must be heavily etched to remove the damaged semiconductor surface layer and then thoroughly washed and dried to maintain surface cleanliness. Surfaces so prepared are then usually coated with a pure, highly insulating silicone coating to stabilize the surface over a long period of time. Breakdown voltages in excess of 1000 V are achieved in mesa p-n junctions fabricated from silicon crystals.

For an oxide-protected planar p^+-n junction the positive charge normally present in the oxide and at the silicon-oxide interface tends to induce additional electron charges at the surface of the lightly doped n-substrate. This effectively causes a p^+-n^+ junction to occur at the silicon surface. Previous arguments have indicated that the breakdown voltage of a junction with heavy doping on both sides is lower than that of a junction with low doping on at least one side. Hence reduced breakdown voltage can occur at the semiconductor surface of such a device, to the extent of the density of charges in the oxide.

For an oxide-protected planar n^+-p junction the positive oxide charge will tend to induce an n-layer on the lightly doped p-substrate. As discussed previously, this will cause the n-p junction to extend over the whole silicon upper surface, causing higher leakage current and even reduced blocking voltage due to a surface defect somewhere on the upper surface. This tendency can be reduced by introducing a p-type annular "guard ring" around the n^+-p junction to limit its extension. Sometimes this induced n-layer can tend to reduce the sharp curvature at the edge of a planar junction, which will in fact raise the junction blocking voltage.

EXAMPLE 6.4

Determine the thickness in micrometers of the space-charge region of a p^+-n^+ junction, reverse biased by 6.0 V, near the onset of Zener breakdown. Also find the peak field in the space-charge region under this condition. The p^+-region is doped with 5.0×10^{25} acceptors/m^3 and the n^+-region contains 2.0×10^{24} donor impurities/m^3.

SOLUTION   The calculation is identical to that of Example 6.2 with the exception that an external voltage is applied in addition to the built-in voltage. Calculation of the built-in voltage by Eq. (6.8) yields $\Delta V_0 = 1.1$ V. The space-charge width is then given by Eq. (6.19) as

$$
\begin{aligned}
l_n &= \left[(|\Delta V_0| + |V|) \frac{2\epsilon_0 \epsilon_r}{q N_d} \right]^{1/2} \\
&= \left[(1.1 + 6.0) \frac{2(8.85 \times 10^{-12})(12)}{1.6 \times 10^{-19}(2.0 \times 10^{24})} \right]^{1/2} \\
&= 6.9 \times 10^{-8} \text{ m} \quad \text{or} \quad 0.069 \ \mu\text{m}.
\end{aligned}
$$

Equation (6.16) gives the peak electric field as

$$
\begin{aligned}
\mathcal{E}_{max} &= \frac{2(|\Delta V_0| + |V|)}{l_n} = \frac{2(1.1 + 6.0) \text{ V}}{6.9 \times 10^{-8} \text{ m}} \\
&= 2.1 \times 10^8 \text{ V/m}.
\end{aligned}
$$

6.4 THE P-N JUNCTION PHOTODIODE

The semiconductor p-n junction as a light emitter has already been discussed in Sections 1.3C and 5.5F. The p-n junction as a light *detector* or *photodiode* will now be discussed. If a reverse-biased semiconductor p-n junction is illuminated with photons having energy in excess of the forbidden energy gap, hole-electron pairs may be generated in the vicinity of the junction.[18] Those generated in the junction space-charge region will be swept through the junction by the electric field there, constituting an extra source of reverse current. In addition, hole-electron pairs generated in the p- and n-regions on either side of the junction region can diffuse to the space-charge region and then be collected there by the electric field. Because of the direction of the electric field only minority holes

---

[18] See Section 5.5 for a discussion of carrier generation by light.

FIGURE 6.12

Schematic diagram of an
illuminated p-n junction
photodiode. Only photons
with energy greater than the
semiconductor energy gap
produce electron-hole pairs.
Only those produced within
a diffusion length of the
junction space-charge
region are collected.

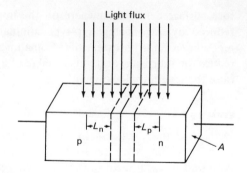

generated on the n-side and minority electrons generated on the p-side will be
swept through the junction to constitute a current which can be detected at the
terminals of the diode. The magnitude of this current will be shown to be
proportional to the number of incident photons per unit time.

If no external bias voltage is impressed on the diode, the effect of illumination
on a p-n junction is to produce an open-circuit potential or *photovoltage*. This
type of photovoltaic cell is the basic element of the solar cell used to energize
modern-day space satellites. The source of this potential may be understood
by considering the effect of the built-in field in the space-charge region. This
field is in a direction such that generated holes are swept toward the p-side,
and electrons toward the n-side of the junction. Under open-circuit conditions,
the positively charged holes will pile up near the p-terminal and the electrons
near the n-terminal, constituting a batterylike photovoltage. In contrast to
the p-n junction in the dark, power can be delivered to a resistive load connected
across this illuminated p-n diode.[19] The energy, of course, is supplied by the
incident photons. Practically speaking, solar cells exhibiting a conversion
efficiency of solar energy to electrical energy of 14 percent can be fabricated
from the semiconductor silicon; they are extremely lightweight direct energy
converters and hence are used extensively in space projects.

A. The Photodiode *I-V* Characteristic

Consider light impinging on a simple p-n junction bar as shown in Fig. 6.12.
Since only the holes generated on the n-side within a hole diffusion length L_p
of the space-charge region can be collected by the junction, giving rise to a
photocurrent, this hole current can be estimated as

$$I_p = qAL_pG. \qquad (6.48)$$

Here A is the junction cross-sectional area, and G is the hole-electron generation
rate per unit volume which is proportional to the incident light flux. A corresponding expression can be written for the electrons generated on the p-side.

---

[19] See Section 6.1B, where it is indicated that the open-circuit voltage of a p-n junction in the
dark must be zero and it cannot deliver power.

Physically, this expression can be interpreted as follows: G is the rate of production of holes per unit volume, and the product AL_p represents the volume in which generated holes are produced and from which these carriers can diffuse to the space-charge region before recombination. Hence when the magnitude of the hole charge q is taken into account, Eq. (6.48) can be interpreted as the total hole charge produced by the light that is collected in the space-charge region per unit time, which when swept through the junction constitutes the photocurrent due to the illumination. Since each photon with energy in excess of the semiconductor energy gap can produce one electron-hole pair, the photocurrent is proportional to the number of these incident photons per unit time. The total photodiode current I can then be written as

$$I = I_0(e^{qV/kT} - 1) - (qAL_p G + qAL_n G), \tag{6.49}$$

where the first term on the right refers to the ordinary p-n junction dark current when a voltage V appears across the device, and the second term on the right corresponds to the photocurrent due to hole generation on the n-side and electron generation on the p-side of the p-n junction.

6.5 THE METAL-SEMICONDUCTOR (SCHOTTKY) DIODE

The metal-semiconductor contacts to the p-n junction diode are normally made as low in resistance and as ohmic[20] as possible. This is to ensure that the electrical characteristics of the device are solely those of the carefully designed junction. In general, though, the metal-semiconductor contact is a rectifying junction. This is due to the presence of a potential barrier located at a contact between a metal and the carefully prepared surface of a semiconductor[21] which has a substantially different work function than the metal. Here the more general definition of the work function is implied: that is, the energy necessary to remove an electron from a level of energy equal to the Fermi energy, to a distance where the material no longer exerts any force on the electron.

The energy band diagram at thermal equilibrium for the contact between a metal and an n-type semiconductor crystal is shown in Fig. 6.13b. In Fig. 6.13a, the conduction band of the metal and the semiconductor conduction and valence bands are pictured for the case of the materials completely separate from each other. For the situation illustrated, bringing the materials into intimate contact in thermal equilibrium means that the energy levels of the semiconductor must be depressed relative to those of the metal so that their respective Fermi levels can line up. This means that the semiconductor is now

---

[20] Here "ohmic" refers to the electrical symmetry of the contact; the resistance is the same, independent of the direction of current flow.
[21] Since the electrical characteristics of this diode depend very sensitively on the metal-semiconductor interface, the surface states discussed in Section 5.6 must be reduced to a very low density to preclude their having a significant influence on the properties of this device.

FIGURE 6.13

a) Energy band diagram of a metal (left) and an n-type semiconductor crystal (right) when separate from each other. The metal work function is the energy necessary to carry an electron in the metal far from the influence of the metal to an energy E_∞.
b) Energy diagram for a metal in intimate contact with an n-type semiconductor under thermal equilibrium conditions. Note that "band bending" in the semiconductor provides a field which keeps additional electrons in the conduction band of the semiconductor from spilling over into the metal.

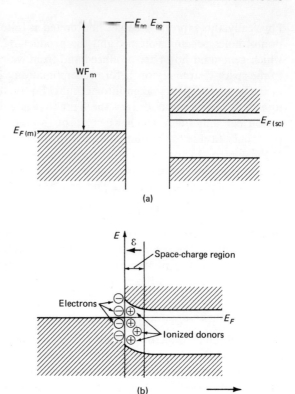

at a higher potential than the metal in this electron energy diagram. This can be considered to occur via a transfer of negative charge from the semiconductor to the metal on contact. Before contact there are electrons in the conduction band of the semiconductor which are in energy states above those of the electrons in the conduction band of the metal. Hence, on contact, the electrons from the semiconductor "spill over" into the metal, leaving behind fixed positive donor charges. This sets up an electric field which tends to discourage additional electrons from leaving the semiconductor, and an equilibrium potential barrier results as shown in Fig. 6.13b. The electric field lines originate on the ionized donors in the semiconductor bulk and terminate on electrons in the metal attracted to the metal-semiconductor interface.

Note that the space-charge region occurs only in the semiconductor here, and not in the metal. The reason is that the semiconductor has a greater ability than the metal to support space charge, owing to the much smaller density of charges in the semiconductor. This creates a so-called "bending of the bands" in the semiconductor near the metal contact. It should be clear that the height of the potential barrier depends on the difference in work function between the metal and semiconductor. A similar consideration applies to a metal and p-type semiconductor contact. This analysis is the subject of a problem at the end of this chapter.

The asymmetry of current flow through this metal-semiconductor or Schottky junction can be understood by considering the application of forward and reverse potential biases to the diode. Consider that the semiconductor is biased positively with respect to the metal (reverse bias). The electrons in the metal are urged toward the barrier by the field but few have sufficient energy to overcome the barrier potential. Similarly, the holes in the n-type semiconductor are moved toward the barrier by the applied field but there are few of these minority holes available. Hence these carriers constitute a small leakage current through the barrier. Also, electrons in the semiconductor are repelled from the junction region, and this depletion causes the space-charge region of ionized donors to widen. This is illustrated in Fig. 6.14a and may be compared to reverse bias in a p-n junction which is in the low current direction and which also produces space-charge widening.

When the metal is now biased positively with respect to the semiconductor (forward bias), the potential barrier is reduced and electrons in the conduction band of the semiconductor can flow more easily into the metal. This is shown

FIGURE 6.14

a) Metal-semiconductor junction energy diagram under reverse bias conditions. V_R is the reverse voltage applied.
b) Metal-semiconductor junction energy diagram under forward bias conditions. V_F is the forward voltage applied.

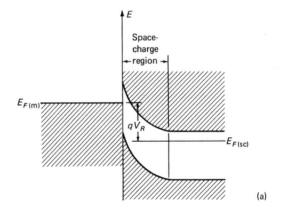

(a)

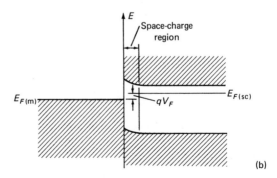

(b)

in Fig. 6.14b. The electrons crossing the potential barrier to enter the metal have energies much greater than the Fermi energy in the metal. These are quite energetic and are called *hot* electrons. Note that the majority carrier electrons carry the current in this case, whereas the minority injected carriers carry the main part of the forward current in the p-n junction. This fact expresses perhaps the most important practical difference between a Schottky-barrier diode and a p-n junction diode. As will be shown in the next chapter, the speed with which a p-n diode can be switched from the forward (conducting) mode to the reverse (blocking) mode may be limited by the minority carrier lifetime, which can be of the order of 10^{-6} sec. Since the Schottky diode conducts by majority carriers, the charge redistribution or relaxation time is about 10^{-12} sec and so only the charging time of the space-charge capacitance limits switching speed.

The current-voltage characteristic of the Schottky diode can be approximated by an expression similar to Eq. (6.20) for the p-n junction, namely, $I = I_0 \times (e^{qV/kT} - 1)$. However, I_0 for the Schottky barrier is usually substantially greater than the leakage current for Ge and Si p-n junctions. Hence substantial current can be obtained at moderately low voltages[22] compared with the p-n junction. This aspect of the device, coupled with its high-speed or high-frequency properties, makes it particularly suitable for a high-frequency low-level detector such as the radio-frequency stage of a radio or TV receiver. It is also used to great advantage in increasing the switching speed of transistor-transistor logic (TTL) digital computer circuits, as will be described in Chapter 11.

PROBLEMS

6.1 A germanium p-n junction has 1.0×10^{24} p-type impurities/m^3 uniformly distributed on the p-side and 9.4×10^{19} n-type impurities/m^3 on the n-side.

a) Calculate the junction contact potential under thermal equilibrium conditions at 300°K.
b) If the impurity concentration on the n-side is increased to 1.7×10^{21} m^{-3}, what is the new contact potential?
c) Determine the width of the depletion layer on the n-side for part a. Repeat for the p-side. The relative dielectric constant for germanium is 16.
d) Calculate the maximum electric field in the depletion region for part a.
e) Repeat part a at 450°K.
f) Repeat part a for a similarly doped silicon p$^+$-n junction.

6.2 Beginning with Eq. (6.2b), derive Eq. (5.8).

6.3 Show that Eq. (6.15) follows from Eq. (6.14).

---

[22] A few tenths of a volt.

6.4 The cross-sectional area of a silicon p^+-n junction is $1.0 \times 10^{-6} \, m^2$. The n-region is 200 μm wide and is doped with 5.0×10^{20} donor atoms/m^2. The p^+-region is 100 μm wide and is doped with 5.0×10^{25} acceptors/m^3. The minority hole lifetime in the n-region is 0.10 μsec. The minority electron lifetime in the p-region is 0.005 μsec. The diode current is 1.0 mA.

a) Determine the injected hole density at the n-side edge of the space-charge region at 300°K.
b) Roughly estimate how far these holes penetrate the n-region.
c) Determine the ratio of the density of minority holes injected to the density of majority electrons which are present in the n-region.
d) Calculate the total amount of excess charge in the n-region due to injected holes.

6.5 Show that Eq. (6.24) is a solution of Eq. (6.23).

6.6 For a p-n junction with very wide p- and n-regions, starting with expressions of the form of Eq. (6.26b), show that the ratio of hole to electron minority carrier current injected through the junction is given by $\sigma_p L_n / \sigma_n L_p$. Here σ_p and σ_n are the electrical conductivities on the p-side and n-side respectively, and L_p and L_n are the minority carrier diffusion lengths on the n- and p-side respectively.

6.7 Calculate the junction leakage current for the silicon p^+-n junction of Problem 6.4, due to:

a) Minority carriers in the p- and n-regions diffusing through the junction.
b) Current generated in the space-charge region when the junction is reverse biased with 10 V. (Take the average lifetime in the depletion region as 0.10 μsec and assume it is relatively temperature insensitive.)
c) Derive an expression for the ratio of the diffusion component of the saturation current to the space-charge limited component. Is this ratio larger for germanium or silicon?
d) Determine the ratio of the diffusion component of the saturation current at 300°K to that at 450°K for the germanium p-n junction of the same design as the silicon p^+-n junction of Problem 6.4.
e) By what factor does the space-charge generated leakage current of the silicon p-n junction of part b change as the reverse voltage is increased from 1.0 to 100 V?

6.8 Two germanium p-n junction diodes are connected in series with a reverse voltage of 100 V applied to the combination through a resistance of 1000 Ω. Diode A has a reverse saturation current of 0.10 μA while diode B has a saturation current of 0.01 μA at 300°K. Assume the diodes have negligible series and contact resistances.

a) Estimate the voltage supported by diode A and diode B. What is the voltage drop across the 1000 Ω resistor?

b) If the applied potential of 100 V is now reversed in direction, determine the current through the 1000 Ω resistance and the voltage drop across this resistor.

c) Calculate the ratio of the power dissipated in diode B in reverse bias to that in the forward-biased state.

6.9 Consider the two diodes of Problem 6.8 connected in parallel. If 50 V is applied to this parallel combination in the forward direction through a 1000 Ω resistor, calculate the current in each diode.

6.10 a) Discuss the conversion of electron current to hole current when the current in a copper wire enters a bar of p-type silicon through an ohmic contact.

b) How will this differ if the contact is partially rectifying?

6.11 a) Calculate the approximate breakdown voltage for the p-n junction diode of Problem 6.4.

b) What is the width of the space-charge region at the voltage at which the junction avalanches?

c) Repeat part a if the doping on the n-side of the junction is increased to 1.0×10^{22} donors/m^3.

6.12 Plot the current multiplication factor M versus reverse voltage for the p-n junction diode described in Problem 6.4. Assume the factor n in Eq. (6.44) is equal to 3.

6.13 A Schottky barrier is formed by depositing a thin film of copper onto a surface of clean n-type 1 Ω-m germanium. If the work function of copper is 4.5 eV and that of the germanium is 4.0 eV, calculate the height of the potential barrier preventing the electrons in the conduction band of the semiconductor from entering the metal under equilibrium conditions.

6.14 Draw the energy band diagram for a metal–p-type semiconductor contact in thermal equilibrium, starting with Fig. 6.13a.

6.15 The p-n junction diode described in Problem 6.4 is employed as a photodetector. Illumination causes a hole-electron generation rate of 3.0×10^{27} pairs/m^3-sec. Determine the photocurrent due to the incident light.

7 Diode Applications and Frequency Performance

7.1 THE P-N JUNCTION DIODE AS A CIRCUIT ELEMENT

The junction diode is an extremely useful circuit component utilized frequently in power supplies, amplifier circuits, and computer circuits. Its usefulness results primarily from the fact that its impedance to the flow of electric current can differ by many orders of magnitude, depending on the direction of the current flow through it. In addition, the sharp voltage-breakdown characteristic provides for voltage-reference uses. The next few sections describe a circuit model for the p-n junction diode and some elementary circuit applications of the device.

A. Circuit Modeling of the p-n Junction Diode

An ideal diode would have zero resistance in the forward direction and infinite resistance in the reverse direction. The *I-V* characteristic of such a diode is given in Fig. 7.1a. This type of behavior is often symbolized by the arrowlike symbol shown in Fig. 7.1b. The inverted triangle points in the direction of "easy" current flow and constitutes the anode of the diode. The horizontal bar represents the cathode of the diode. This model is called a *piecewise linear* diode model because of the straight-line segments of the idealized *I-V* characteristic. It is useful for analyzing electronic circuits containing this element if the applied voltage is significantly greater than the junction barrier voltage, and the circuit resistances are much in excess of the diode series resistance. Should this not be the case the piecewise linear model for the junction diode given in Fig. 7.2b can be used; its *I-V* characteristic is shown in Fig. 7.2a. Here V_0 is related to the built-in junction voltage[1] and R_s represents some diode internal series resistance.

---

[1] The exponential nature of the diode law calls for a continuous variation of current with voltage. However, in a practical sense, the current increases rapidly when the voltage exceeds V_0.

FIGURE 7.1

a) Ideal piecewise linear diode I-V characteristic. Piecewise linear refers to the straight-line segments of this nonlinear characteristic. The model ignores the influence of built-in barrier voltage, series resistance, and reverse diode leakage and is described by $V = 0$ for $I > 0$ and $I = 0$ for $V < 0$.
b) Circuit symbol for *ideal* piecewise linear diode.

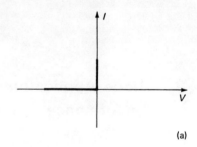

(a)

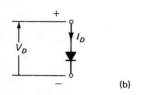

(b)

FIGURE 7.2

a) Ideal piecewise linear diode I-V characteristic including the effect of built-in voltage plus series resistance.
b) Ideal piecewise linear circuit model including a barrier voltage V_0 and series resistance R_s.

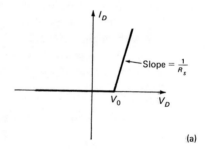

(a)

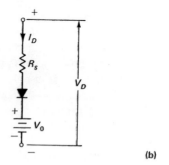

(b)

    In these models it is assumed that the voltage applied to the diode never exceeds the diode breakdown voltage. In circuits where the voltage-breakdown property of the diode is of primary importance, the breakdown diode may be represented by the piecewise linear model of Fig. 7.3b, whose I-V characteristic

FIGURE 7.3

a) Ideal piecewise linear diode I-V characteristic including the effect of the breakdown voltage and current multiplication plus resistance in the breakdown region.
b) Ideal piecewise linear circuit model including breakdown voltage V_z plus resistance in the breakdown region R_z.

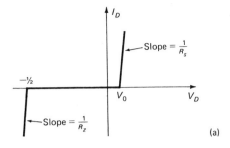

(a)

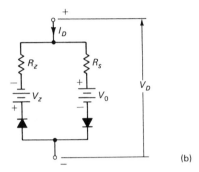

(b)

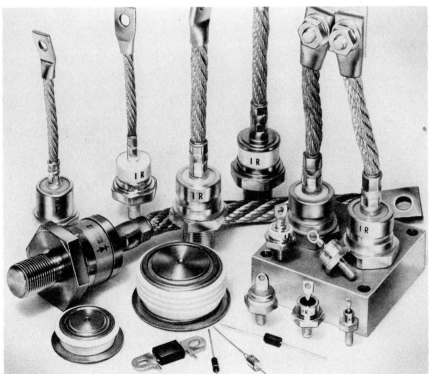

Selection of low- to high-power silicon diode rectifiers. (Courtesy of International Rectifier.)

195

is shown in Fig. 7.3a. Diodes designed primarily with a specific breakdown characteristic are commonly called *Zener* diodes, in spite of the fact that the voltage breakdown is most often by the avalanche mechanism.

B. Power Supply Circuits

The usefulness of the junction diode in converting ac power to dc will now be demonstrated. Generally speaking, in ground-based equipment the available power is ac, obtained from rotating ac generators. These machines yield a nearly sinusoidal voltage output at a frequency of 60 or 50 Hz. Since the transistors and integrated circuits in most electronic equipment require a dc voltage supply, use is made of the rectifying property of the diode to transform the ac to dc.

In the circuit analysis which follows, the total value of a variable is expressed as a lower-case letter with an upper-case subscript: v_{CE}; the ac value of a variable is indicated with a lower-case letter and subscript: v_{ce}; and the dc value of a variable is expressed as an upper-case letter with an upper-case subscript: V_{CE}.

FIGURE 7.4

Half-wave rectifier circuit where v_g is the voltage supplied by a generator having internal resistance R_G, R_L is the load resistance, and D is a symbol representing an actual semiconductor diode. Note that the open triangle of D represents a *real* diode, whereas the filled triangle of Fig. 7.3 symbolizes an *ideal* diode.

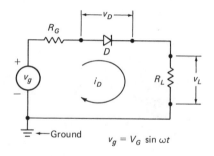

$$v_g = V_G \sin \omega t$$

FIGURE 7.5

a) Voltage waveform supplied by the voltage generator. b) Voltage waveform observable across the load resistor. This varying voltage is equivalent to the time-averaged dc voltage indicated as "dc output."

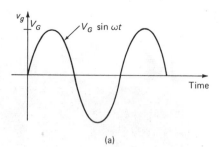

(a)

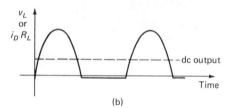

(b)

Half-wave rectification   The simplest circuit for converting an ac voltage into a dc supply is shown in Fig. 7.4. An ac voltage, $V_G \sin \omega t$, is supplied by a sinusoidal voltage generator (of internal resistance R_G) and rectified by the semiconductor diode D, resulting in a dc current through a resistive load R_L. During the positive half-cycle of the applied voltage, the diode is forward biased, and considerable current flows through the load resistance. During the negative half-cycle, the diode is in reverse bias and hence little current flows in the circuit. The voltage waveforms supplied by the voltage source and observable across the load resistor are shown in Fig. 7.5. Basically, when the supply voltage is positive with respect to ground, nearly all the voltage is across the load resistor; nearly all the voltage is across the diode rectifier (and little across the load) when the supply voltage becomes negative relative to ground. The loop equation for this circuit is given by

$$v_g = v_D + i_D(R_G + R_L), \tag{7.1}$$

where the current i_D is determined by the diode law as

$$i_D = I_0(e^{qv_D/kT} - 1). \tag{7.2}$$

Any series resistance in the diode may be added to the source resistance R_G.

Equations (7.1) and (7.2) are two equations in two unknowns i_D and v_D; all other quantities normally are known. The analytical solution for these unknown quantities is difficult, involving the solution of a transcendental equation. Often estimation techniques may be employed to yield a more or less approximate solution. However, it is more instructive to use a graphical method, frequently useful in solving much more difficult problems. This technique involves the utilization of a *load line*. This line represents the linear relationship between i_D and v_D, which can be obtained by rearranging Eq. (7.1) as

$$i_D = \frac{v_g}{R_G + R_L} - \left(\frac{1}{R_G + R_L}\right) v_D. \tag{7.3}$$

This is the equation of a straight line, on a graph of i_D versus v_D, with a negative slope $1/(R_G + R_L)$, and i_D intercept of $v_g/(R_G + R_L)$, and a v_D intercept of v_g. If Eq. (7.2) is now plotted on the same graph, the intersection of this diode curve with the load line represents the simultaneous solution of Eqs. (7.1) and (7.2), when v_g is positive relative to ground. This is shown in Fig. 7.6. The intersection of the two curves at point O defines i_D and v_D at the circuit *operating point*. Also shown as dashed lines on this graph are load lines for larger and smaller values of load resistance R_L than the original solid load line. Note that the voltage drop v_D across the diode is fairly constant over a broad range of values of load resistance. This diode drop is normally of the order of 0.7 V for silicon p-n junction diodes and about 0.25 V for germanium diodes. The difference results from the difference in energy gaps and hence the built-in voltages of these semiconductor materials (see Eq. 6.32).

Note that when the supply voltage goes negative, the diode leakage current I_0 determines the circuit current since v_D is now negative and the diode now assumes a very high impedance value. Now nearly all the applied voltage appears across the diode, and a small diode current, of the order of nanoamps to milliamps (depending on the area of the device), flows in a negative direction in the circuit.

A point-by-point plot of the current in this circuit versus time during the positive source-voltage half-cycle may be obtained by updating the load line in Fig. 7.6 for different values of v_g corresponding to its sinusoidal time variation.

FIGURE 7.6

Typical resistive load-line graph showing how the circuit operating point O is obtained graphically. Load lines having half or double the original circuit resistance are indicated by the dashed lines. The new operating points are indicated respectively by O' and O''.

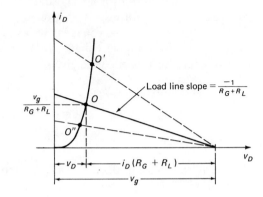

FIGURE 7.7

a) Load lines for different values of v_g corresponding to the sinusoidal variation of the generator voltage with time. b) Variation of current with time in a half-wave rectifier circuit corresponding to sinusoidal generator voltage.

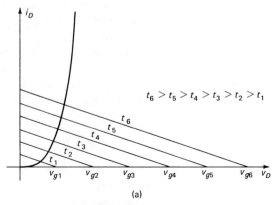

(a)

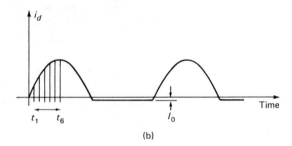

(b)

FIGURE 7.8

a) Full-wave rectifier circuit.
Note that only one-half the
transformer voltage is effective in yielding output voltage across the load.
b) Full-wave rectified current
waveform.

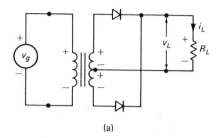

(a)

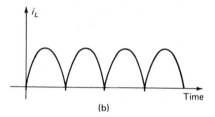

(b)

This is shown in Fig. 7.7a. The variation of current with time is shown in
Fig. 7.7b. Note that the current flow now is essentially of a unidirectional
nature and hence the diode is called a *current rectifier*.

Full-wave rectification   Half-wave rectification is inefficient in the sense
that substantial current is delivered to the load only one-half of the time. This
situation may be remedied by using two diode rectifiers in a full-wave rectification circuit as shown in Fig. 7.8a. Transformer coupling permits the upper
diode to conduct current when the supply voltage is positive with respect to
ground; the lower diode now is reverse biased and is normally conducting
negligible current. When the supply voltage goes negative relative to ground,
the lower diode conducts appreciable current while the upper diode is passive.
The rectified current delivered to the load R_L versus time is plotted in Fig. 7.8b.

Bridge rectifier   Note that for a given voltage delivered to the load the full-
wave rectifier circuit requires approximately twice as much supply voltage as
the half-wave circuit. This situation can be remedied by the use of a total
of four diodes in a bridge-rectifier circuit arrangement. This is shown in
Fig. 7.9. Now during one half-cycle, diodes D_1 and D_4 will conduct appreciable
current, with diodes D_2 and D_3 in the high-impedance or *blocking* condition.
During the next half-cycle, diodes D_2 and D_3 will conduct, with D_1 and D_4 in
the blocking state. In this case the voltage across the load resistance will be two
diode drops lower than the supply voltage; in the circuits discussed previously
the difference is only one diode drop. However, this is often of little consequence
since the diode drops are normally small compared to the supply voltage.
     Since the usual objective in building a rectifier circuit is to deliver a dc supply

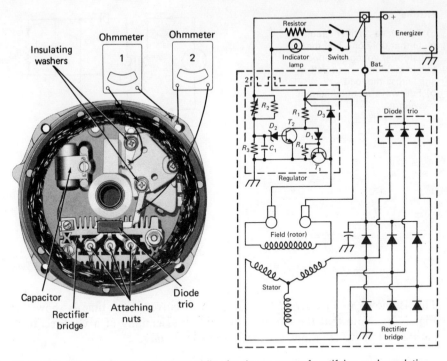

Typical circuit package used in automobiles for the purpose of rectifying and regulating ac current supplied by an alternator. Note the use of a rectifier bridge containing six diodes. A circuit diagram is at the right. (Photo courtesy of Delco-Remy, Division of General Motors.)

FIGURE 7.9

Bridge-rectifier circuit. Note that here the entire transformer voltage is effective in producing output voltage. The output waveform is similar to that obtained in the full-wave circuit of Fig. 7.8.

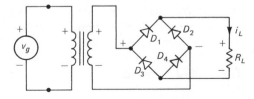

to the load, these rectification schemes are deficient in the sense that although a unidirectional current is obtained, it is pulsating in nature. The pulsation of the rectified current may be reduced by the use of a shunt-capacitor filter. The bridge-rectifier circuit with a shunt-capacitor filter is shown in Fig. 7.10. The capacitor connected across the load resistor stores energy from the voltage source when the diode rectifiers are conducting maximum current and delivers some of this energy to the load as the supply voltage tends to fall off, reducing the current through the diode rectifier. If the filter capacitor is large enough in value so that the discharge time constant through the load, $R_L C$, is long com-

FIGURE 7.10

Bridge-rectifier circuit with a shunt filter capacitor to reduce output-voltage fluctuations. The capacitor stores charge when the voltage is high and returns this charge to the load as the voltage falls off.

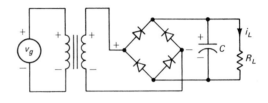

FIGURE 7.11

Load voltage versus time for a bridge-rectifier circuit with a shunt filter capacitor, showing the reduction of voltage fluctuation (solid line) compared with the unfiltered full-wave voltage waveform (dashed line). The dc voltage output, averaged in time, is also indicated.

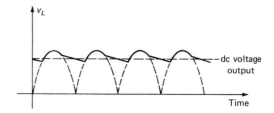

pared with the period of the supply voltage, the potential across the load resistance will be maintained at nearly the peak supply voltage. That is, as the sinusoidal supply voltage passes its peak and the supply current drops off, the capacitor discharges some of its stored energy, which tends to maintain the current. This is shown diagrammatically in Fig. 7.11. Note that although the load-voltage fluctuation has been significantly reduced, there is still some residual voltage fluctuation, called *ripple*. More critical dc power supplies require larger filter capacitors or more efficient forms of filtering.

Zener-diode-regulated supply A simple Zener-diode-regulated full-wave power supply is shown in Fig. 7.12. Here the output voltage across the load resistor is fixed at essentially the value of the breakdown voltage of the Zener diode employed. When the filter capacitor becomes charged to a potential in excess of the Zener breakdown voltage, creating a tendency to force additional current through the load, this excess current is bypassed to ground through the avalanching Zener diode, maintaining the breakdown voltage across the load and holding the load current constant.

C. The Diode Demodulator

An amplitude-modulated radio signal basically consists of a moderately high-frequency sinusoidal carrier wave whose amplitude varies in time in response to an impressed audio (low-frequency) signal. The audio information is contained in the envelope of this wave as shown in Fig. 7.13a. To extract the information from the wave it is necessary to generate a signal which varies as the envelope of the modulated signal. This process is known as *demodulation* or *detection*. The circuit for accomplishing this is shown in Fig. 7.13b. In this circuit the

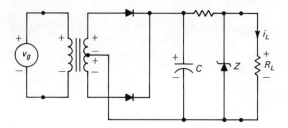

FIGURE 7.12

Zener-diode-regulated full-wave voltage supply. The output voltage is fixed essentially at the breakdown voltage of the Zener diode indicated by the special diode symbol marked Z. Any tendency to exceed the Zener breakdown voltage causes the Zener diode to conduct and hence shunt off any current that might be delivered to the load from this fluctuation. The voltage waveform obtained across the load has less ripple than that obtained in the circuit for Fig. 7.11.

diode serves the purpose of rectifying the modulated wave so that the output to the load resistance is unidirectional. In addition, the capacitor across the load serves the same basic function as the power-supply filter capacitor. That is, the capacitor is initially charged nearly to a particular peak value of the modulated voltage, such as A in Fig. 7.13a. Now as the signal voltage drops toward B, the diode tends toward reverse bias and cuts off, forcing the capacitor to discharge through the load resistance R_L. If the discharge time constant $R_L C$ is long compared with the period of the alternating signal, the peak capacitor voltage is maintained across the load resistor until the next cycle peak, such as E in the figure, when the capacitor is charged to this peak value. In this way the demodulated voltage, which is essentially the envelope of the modulated wave, appears across the load resistor as shown in Fig. 7.13c. Thus the audio information is removed from the received wave and is ready for amplification.

D. Diode Gates in Digital-Computer Circuits

Diodes are important components in the electronic performance of logic functions in the digital computer. Here the semiconductor diode performs the function of an electronic switch as does the transistor switch described in Section 1.2C. It is ON when a forward voltage in excess of the barrier voltage is applied, biasing it into the low-impedance condition; it is OFF in reverse bias, or in the high-impedance state.

Information is transmitted through a digital computer in the form of electric pulses. Should 1 V pulses be incident at either A or B in Fig. 7.14a, then 1 V pulses will be observed at the output marked D. If silicon diodes are used, the voltage output at D with both A and B at ground potential (taken as zero) will

FIGURE 7.13

a) Amplitude-modulated
high-frequency carrier wave.
b) Diode demodulator or
envelope detector.
c) Demodulated wave (solid
line) obtained by passing
modulated signal (dashed
line) through the diode
demodulator circuit.

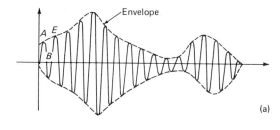

(a)

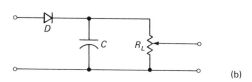

(b)

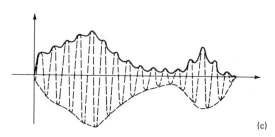

(c)

be about -0.7 V below ground potential, and both diodes will be conducting
in the forward direction. If a 1 V input pulse is applied to A (and B is maintained at ground potential), the diode connected to A will still be forward
biased, maintaining the normal voltage drop of 0.7 V. Hence the output
potential at D will rise to $+0.3$ V relative to ground, reflecting an output
voltage rise of 1 V for an input pulse signal of 1 V. The same argument may be
repeated for a 1 V pulse applied to B with A maintained at ground potential, and a
1 V output pulse will occur at B. Of course, the same 1 V output will be observed
if either A or B or both A and B have pulses applied. This is the OR function.

The logic aspects of this digital circuit may be conveniently expressed in terms
of the *truth table* of Fig. 7.14a. A rise in voltage when a diode is turned ON is
referred to by the logic symbol **1**. A lowered voltage when the diode is OFF
corresponds to the logic symbol **0**. Hence a **1** at A and a **0** at B in this OR
circuit yields a **1** at D, as indicated in the second line of the truth table. The
other entries in the truth table may easily be confirmed.

In contrast, consider the diode logic circuit of Fig. 7.14b, which requires that
pulses be applied to *both* F and G for an output pulse to be observed at H. In
this case the requirement is that both pulses be applied simultaneously.

Assume that either the F or G input is at ground potential. Then the output
at H will be at approximately one diode drop, 0.7 V, positive with respect to
ground. A positive-going 2 V pulse applied at F will tend to reverse bias or cut

FIGURE 7.14

a) Diode OR circuit.
b) Diode AND circuit.

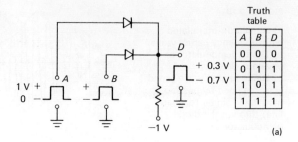

A	B	D
0	0	0
0	1	1
1	0	1
1	1	1

Truth table

(a)

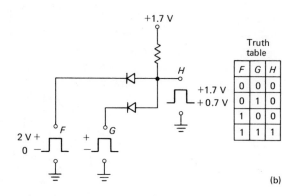

F	G	H
0	0	0
0	1	0
1	0	0
1	1	1

Truth table

(b)

off the diode connected to F since its cathode is at a higher potential than its anode. That is, the potential at F will rise to 2 V, with the potential at point H still held at 0.7 V by the forward-conducting diode connected to G. Now consider that a 2 V pulse is applied simultaneously to F and G. Since this cuts off both diodes connected to points F and G, only an extremely small leakage current can flow in these diodes, and correspondingly in the load resistor; hence with essentially no IR drop across the load resistor, point H rises nearly to the battery potential $+1.7$ V. Hence only with an input pulse applied to both F and G does the output potential at H rise from 0.7 V to $+1.7$ or 1 V. This constitutes an AND function. Its logic table is shown in Fig. 7.14b.

Here a positive input pulse results in a positive-going output pulse. If the positive input signal resulted in a negative-going output pulse, this would constitute a NAND function. A corresponding definition for the OR-type circuit would define a NOR function.

Note that the nonlinear character of the current-voltage characteristic of the semiconductor diode makes possible the performance of these logic functions.

E. Diode Photodetectors and Solar Cells

In Section 6.4A the I-V characteristic of the photodiode was discussed. The equivalent circuit of this diode can be derived using Eq. (6.49), and it is shown in Fig. 7.15a. The effect of illumination on the diode is indicated by the current source I_L in parallel with the ordinary dark or light-free p-n junction diode characteristic. The I-V characteristics of the photodiode in the dark and in the presence of light are shown in Fig. 7.15b. The photodetector aspects of the device are indicated by noting that the change in reverse bias current of the device is proportional to the incident light flux (see Eq. 6.49). The utility of the photodiode as a solar cell or electric generator is indicated by noting that the open-circuited illuminated device develops a photovoltage across the terminals marked V_{OC} in Fig. 7.15b. The magnitude of this voltage may be calculated by setting $I = 0$ in Eq. (6.49). This yields

$$V_{OC} = \frac{kT}{q} \ln\left[\frac{qAG(L_p + L_n)}{I_0} + 1\right]. \tag{7.4}$$

Practically, the value of this quantity is somewhat less than one-half the energy gap or about 0.5 V for silicon solar cells. The manner in which this voltage is developed using the energy band concept is diagrammed in Fig. 7.15c.

If the device is short-circuited in the presence of light, a short-circuit current I_{SC} is observed. If a load resistor is then placed across the terminals of an illuminated solar cell, a voltage somewhat less than V_{OC} and current somewhat less than I_{SC} will deliver power to this load. If the load is matched to the solar

FIGURE 7.15

a) Equivalent circuit of a p-n junction photovoltaic cell showing the polarity of photovoltage produced by illumination.
b) I-V characteristic of a p-n junction photocell in the dark and under illumination. The open-circuit voltage V_{OC} and the short-circuit current I_{SC} under illumination are indicated.
c) Energy band diagram for a photodiode open-circuited and illuminated. The shift of the Fermi level on the n-side relative to the p-side, far from the junction, represents the open-circuit voltages.

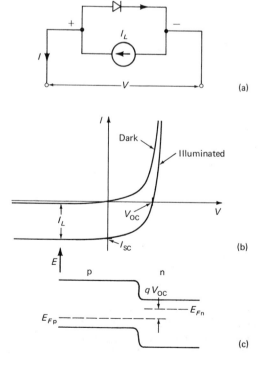

TIROS I (Television Infrared
Observation Satellite) showing its two hemicylindrical
solar arrays used for converting solar energy into electrical
energy. (Photo courtesy of
National Aeronautics and
Space Administration.) The
drawing is a closeup of an
individual solar cell.

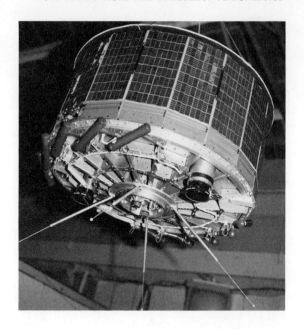

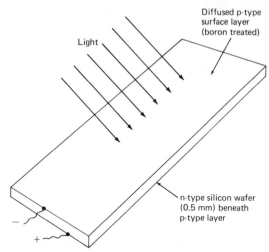

generator, maximum power will be transmitted. A typical 2 cm^2 silicon solar
cell can deliver about 10 mW in noonday sun and will convert solar to electric energy at about 12 percent efficiency. Series and parallel connection of
thousands of these devices yields a power supply of the type used to energize
space vehicles, typically capable of 28 V and a few kilowatts.

EXAMPLE 7.1

The leakage current I_0 of a 2.0 cm² silicon p⁺-n junction solar cell at 300°K is 1.0 nA. The short-circuit current of this device exposed to noonday sun is 20 mA, and the electron-hole pair generation rate in the silicon is 3.0×10^{24} m⁻³-sec⁻¹.

a) What is the lifetime of minority holes in the n-region of this device? (*Hint:* Assume the electron lifetime in the p-region is very small because of the high impurity level there.)
b) What resistance value must be connected across the cell in order that 10 mA of load current are delivered to this load?
c) Calculate the power delivered to the load of part b.

SOLUTION a) From Eq. (6.49),

$$I_{SC} = I_L = qAG(L_p + L_n),$$

so that

$$L_p = \frac{I_{SC}}{qAG},$$

where it is assumed that $L_n \ll L_p$. Hence,

$$L_p = \frac{20 \times 10^{-3} \text{ A}}{1.6 \times 10^{-19} \text{ C} \,(2.0 \times 10^{-4} \text{ m}^2)(3.0 \times 10^{24} \text{ m}^{-3})}$$

$$= 2.1 \times 10^{-4} \text{ m},$$

and

$$\tau_p = \frac{L_p^2}{D_p} = \frac{(2.0 \times 10^{-4})^2 \text{ m}^2}{1.25 \times 10^{-3} \text{ m}^2/\text{sec}} = 32 \times 10^{-6} \text{ sec} \quad \text{or} \quad 32 \text{ } \mu\text{sec}.$$

b) Equation (6.49) is

$$|I| = I_L - I_0(e^{qV/kT} - 1).$$

Solving for V gives

$$V = \frac{kT}{q} \ln \left(\frac{I_L - |I|}{I_0} + 1 \right)$$

$$= (0.026 \text{ V}) \ln \left[\frac{(20 \times 10^{-3} - 10 \times 10^{-3}) \text{ A}}{1.0 \times 10^{-9} \text{ A}} + 1 \right]$$

$$= 0.42 \text{ V},$$

$$R = \frac{V}{I} = \frac{0.42}{10 \times 10^{-3}} = 4.2 \times 10^{-3} \text{ } \Omega.$$

c) Power $= IV = 10 \times 10^{-3} \text{ A} \,(0.42 \text{ V}) = 4.2 \times 10^{-3} \text{ W}.$

7.2 SMALL-SIGNAL HIGH-SPEED HIGH-FREQUENCY DIODE PERFORMANCE

Chapter 6 was confined to a discussion of the p-n junction electrical characteristic under dc or low-frequency quasi-static conditions. One of the important applications of the semiconductor diode is its use as a high-speed switch in computers. Hence it is of interest to investigate the physical characteristics of the device which limit high-speed performance so that very fast diodes can be designed. The small-signal high-frequency limitation of the diode used as a varactor diode or radio-frequency detector is also of interest. Hence the p-n junction diode frequency cutoff will also be discussed.

The time-dependent response of a p^+-n junction diode operating in the forward direction, of the type illustrated in Fig. 6.7, is governed partly by the flow of minority carriers in the lightly doped n-region of the diode. This charge flow may be investigated by solving the time-dependent minority carrier flow equation in the n-region:

$$D_p \frac{\partial^2 \Delta p_n}{\partial x^2} = \frac{\Delta p_n}{\tau_p} + \frac{\partial \Delta p_n}{\partial t}. \tag{7.5}$$

Equation (7.5) is derived in the same manner as Eq. (6.23b) except that in the latter steady-state case $\partial \Delta p_n / \partial t = 0$. The low-level hole current density is then obtained from $J_p = -qD_p(\partial \Delta p_n(0)/\partial x)$, where $\Delta p_n(0) = p_{n0}(e^{qv/kT} - 1)$ in analogy with Eqs. (6.26a) and (6.30). Note that the time-dependent voltage is indicated here by the lower-case v, and that quasi-equilibrium conditions are assumed.

When a time-varying voltage is applied to the junction, the space-charge width and hence charge content of that region also vary. This variation of charge with time constitutes a displacement current through the junction, indicating the capacitivelike behavior of the junction space-charge region. The speed of response of the p^+-n diode to a time-varying voltage depends both on the time necessary for charge redistribution in the space-charge region and the time necessary for charge redistribution of injected carriers in the n-region in series with the junction. The p-n junction space-charge capacitance will first be considered.

A. Junction Space-Charge Capacitance

In Section 6.1C the p-n junction space-charge region in equilibrium was considered for an abrupt junction. The increase in the space-charge width with reverse applied voltage is illustrated in Fig. 6.5. This increase causes more fixed charge in the form of ionized donors and acceptors not neutralized by electrons and holes to appear within the depletion region. Forward bias reduces the number of unneutralized ions in this region. The incremental charge increase dq_{sc} of ionized donors or acceptors, taken per incremental voltage change dv, is called the depletion-layer capacitance and is given by

$$C_j = -\frac{dq_{sc}}{dv}. \tag{7.6a}$$

Here the negative sign results from the fact that the reverse voltage must increase in a negative sense for a charge increase. The additional current passed by this junction when the inverse voltage applied to it changes by dv/dt is given by the displacement current $i = C_j(dv/dt)$ which is in excess of the I_0 of Eq. (6.20).

To calculate this depletion-layer capacitance in terms of device design parameters, refer back to Eqs. (6.9) and (6.18). The total charge on either the p- or n-side of the space-charge region in the depletion-layer approximation is given by

$$q_{sc} = qAN_d l_n = qAN_a l_p = A \left| \frac{2\epsilon_0 \,\epsilon_r q N_d N_a (\Delta V_0 - v)}{N_a + N_d} \right|^{1/2}. \tag{7.7}$$

Here v represents the applied voltage, negative for reverse bias, which may be time-dependent, and A is the junction cross-sectional area. Performing the differentiation as indicated in Eq. (7.6a) gives the junction space-charge capacitance:

$$C_j = A \left| \frac{\epsilon_0 \,\epsilon_r q}{2(1/N_a + 1/N_d)(\Delta V_0 - v)} \right|^{1/2}. \tag{7.6b}$$

Hence the reciprocal of the capacitance squared for an abrupt junction is seen to vary linearly as the applied voltage and the resistivity of the most lightly doped side of the diode. A graph of $1/C_j^2$ versus v should yield a straight line whose intercept at $1/C_j^2 = 0$ gives ΔV_0. In the case of a p-n junction which is not abrupt but has a linear grading of impurities, the exponent in Eq. (7.6b) has the value $\tfrac{1}{3}$. This represents a less rapid change of capacitance with voltage. A very sensitive variation of capacitance with voltage is obtained with a diode containing a hyperabrupt junction. This is designed with a very steep donor-impurity gradient so that the exponent in Eq. (7.6b) is 2.

Having derived an expression for the space-charge capacitance of a p-n junction, we are now in a position to point out a few of the applications of this device. A reverse-biased p-n junction diode is a voltage-variable capacitor (varactor) which may be used for frequency modulation or as a tuning capacitor whose capacitance can be varied electronically rather than mechanically. The device also has applications in harmonic-generator and parametric-amplifier circuits owing to its nonlinear variation of capacitance with voltage. The high-frequency limitation f_{max} of the varactor diode can be written as

$$f_{max} = \frac{1}{2\pi r_s C_j}, \tag{7.8}$$

where r_s is the resistance in series with the reverse-biased junction capacitance, which in a p^+-n diode is mainly the bulk resistance of the n-region. This cutoff frequency is defined as the frequency at which the capacitive reactance of the junction just equals the diode resistance. The impedance of this device is mainly capacitive below this frequency, and the electric energy then is stored in the capacitor and not dissipated in the series resistance. An important factor in the design of a varactor diode is the doping of the n-region. Light doping yields lower capacitance but results in higher series resistance. Heavy doping gives higher capacitance but much lower series resistance. The design of a varactor device is the subject of a problem at the end of this chapter.

B. Diode Diffusion Capacitance

The p-n junction space-charge capacitance under forward bias can be obtained from Eq. (7.6b). However, this forward bias also introduces an additional injected charge redistribution in the n-region which can be described by a capacitance known as the *diffusion* capacitance. This capacitance is especially of interest when the junction is the emitter of a transistor. In fact, the diffusion capacitance in forward bias in general is numerically greater than the depletion-layer capacitance and hence in this case can limit the high-frequency perform-ance of the device. This capacitance will now be calculated in terms of device parameters.

Under steady-state forward bias conditions in a p^+-n junction diode, there exists an excess charge in the form of injected holes in the bulk n-region (see

FIGURE 7.16

Minority carrier charge distribution in the n-region of a p^+-n junction diode before and after the application of an incremental voltage dv. The crosshatched area represents the additional hole charge stored in the n-region owing to the additional applied voltage dv.

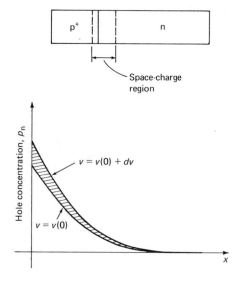

Fig. 6.7b). The total of these excess minority carrier charges may be obtained by integrating the expression for the excess hole distribution under forward bias, over this whole region. With Eq. (6.24) this gives, for the total excess minority hole charge diffused into the n-region,

$$q_{\text{diff}} = \int_0^\infty qA \, \Delta p_n(0) e^{-x/L_p} \, dx = qA \, \Delta p_n(0) L_p . \tag{7.9}$$

If now the forward bias is changed incrementally, this stored charge in the n-region will change by

$$dq_{\text{diff}} = \frac{dq_{\text{diff}}}{dv} \, dv = C_D \, dv, \tag{7.10}$$

where C_D represents the diffusion capacitance. The charging current required to accomplish this charge change is given by

$$i_{\text{diff}} = \frac{dq}{dv} \frac{dv}{dt} = C_D \frac{dv}{dt} . \tag{7.11}$$

The minority charge distribution in the n-region before and after the application of an incremental voltage dv, in a quasi-static manner, is indicated in Fig. 7.16. From Eqs. (7.9) and (7.10) the diffusion capacitance is given by

$$C_D = \frac{dq_{\text{diff}}}{dv} = qAL_p \frac{d\Delta p_n(0)}{dv} . \tag{7.12}$$

Now $\Delta p_n(0)$ can be obtained from Eq. (6.26) as

$$\Delta p_n(0) = \frac{i_p L_p}{qD_p A} . \tag{7.13}$$

Then on differentiating with respect to voltage, one obtains

$$\frac{d\Delta p_n(0)}{dv} = \frac{L_p}{qD_p A} \frac{di_p}{dv} . \tag{7.14}$$

Introducing the diode law (Eq. 6.20) into Eq. (7.14) and substituting into Eq. (7.12) gives, for the diffusion capacitance,

$$C_D = \frac{qI\tau_p}{kT} . \tag{7.15}$$

EXAMPLE 7.2

The capacitance of an abrupt p^+-n junction, 1.0×10^{-8} m^2 in area, measured at -1.0 V reverse bias is 5.0 pF. The built-in voltage ΔV_0 of this device is 0.90 V. When the diode is forward biased with 0.50 V, a current of 10 mA flows. The n-region minority hole lifetime is known to be 1.0 μsec at 300°K.

Calculate (a) the depletion-layer capacitance of this junction at 0.50 V forward bias and (b) the diffusion capacitance of the diode operating as in part a at 300°K.

SOLUTION   a) The depletion-layer capacitance of a p-n junction is given by Eq. (7.6b). Hence in the forward direction at 0.50 V the depletion-layer capacitance is

$$C_j = 5.0\ \text{pF} \left| \frac{-1.0 - 0.90}{0.50 - 0.90} \right|^{1/2} = 11\ \text{pF}.$$

b) From Eq. (7.15),

$$C_D = \frac{I\tau_p}{kT/q} = \frac{10 \times 10^{-3}\ \text{A}\ (1.0 \times 10^{-6}\ \text{sec})}{0.026\ \text{V}}$$

$$= 39 \times 10^{-8}\ \text{F} = 390,000\ \text{pF}.$$

This example illustrates that the diffusion capacitance of a forward-biased junction diode is typically four or more orders of magnitude greater than the depletion-layer capacitance of a p-n junction. Then the p-n junction minority carrier charging time depends on the product of the junction dynamic resistance and the capacitance. This product is given essentially by

$$r_{jct} C_D = \left(\frac{dv}{di} \right)_{jct} C_D = \frac{kT}{qI} \frac{qI\tau_p}{kT} = \tau_p, \tag{7.16}$$

where use is made of Eqs. (6.42) and (7.15). Hence the time necessary to incrementally change the injected charge distribution in a forward-biased p-n junction is about equal to the minority carrier lifetime in the n-region, when the width of this region is much greater than L_p.

When both the time for charge redistribution in the junction depletion layer and the redistribution of charge in the semiconductor bulk are considered, a small-signal equivalent circuit can be constructed to approximate the high-frequency diode performance. This is shown in Fig. 7.17. Note that the diffusion capacitance is negligible for a reverse-biased diode and dominates in forward bias. The current i through a p-n junction with a dc bias voltage V and a small impressed ac voltage v is given by

$$i = I_0(e^{qV/kT} - 1) + (C_j + C_D) \frac{dv}{dt}. \tag{7.17}$$

FIGURE 7.17

Approximate high-frequency equivalent circuit of a p-n junction diode including the diffusion capacitance C_D, the depletion-layer capacitance C_J, and the dynamic junction resistance, r_{jct}.

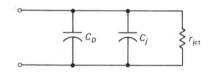

7.3   LARGE-SIGNAL SWITCHING OF A P-N JUNCTION DIODE

In computer-circuit applications the junction diode is often used as a switch which is ON when the diode is conducting in the forward direction and OFF or nonconducting when a reverse potential is applied to the device. Consider that voltage V_F is applied to the circuit so that the diode is in its normal conducting ON state, carrying a current $I_F \simeq V_F/R_L$, where the current is limited by a circuit resistance R_L, much greater than the junction diode resistance.[2] Now the potential is suddenly reversed in an attempt to place the diode instantaneously into the nonconducting OFF state. A circuit for demonstrating this type of large-signal switching is shown in Fig. 7.18.

Figure 7.19 gives a plot of the current through the load resistor R_L and the voltage across the diode as a function of time. Note that the current persists for a time longer than t_s after turn-off is attempted at $t = 0$. This effect is referred to as diode *charge storage* and is characteristic of all p-n junction diodes; however, it doesn't occur in Schottky barrier diodes, which turn off more rapidly. This demonstrates that minority carriers, injected while the device is forward biased, persist for a time after the forward bias is removed, owing to their finite lifetime. Reference to the previous section indicates that a finite time is also required to establish the steady-state minority carrier distribution characteristic of the forward current I_F. Schottky devices do not exhibit these phenomena normally since the current carriers in such diodes are majority carriers. A formulation will now be presented for the large-signal switching problem. This will point up the importance and the simplicity of the charge-control description of diode switching.

A.   Charge Control and p-n Junction Diode Switching

In the steady-state situation, the distribution of holes in the n-region of a forward-biased p^+-n diode (see Fig. 6.7b) is given by Eq. (6.24). The integral of these charges over the whole n-region yields a value for the number of excess injected charge carriers stored in this region.[3] If we take the ratio of this charge to the current flowing through the junction (Eqs. 6.26a and 7.9), we obtain

$$\frac{Q_p}{I_p} = \frac{qAL_p\,\Delta p_n(0)}{qAD_p\,\Delta p_n(0)/L_p} = \frac{D_p}{L_p^2} = \tau_p. \qquad (7.18)$$

An interpretation of this simple relation is obtained by noting that the hole

[2] Here we assume the junction contact potential is small compared with V_F.
[3] See Eq. (7.9). Note that only the holes stored in the n-region are considered since very few electrons are injected by this p^+-n junction.

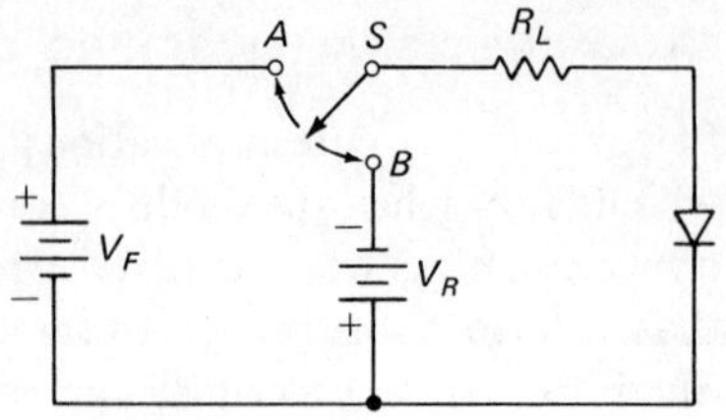

FIGURE 7.18

Circuit for demonstrating large-signal switching of a p-n junction diode. The diode is initially forward biased by having the switch S at terminal A. The diode is biased in the reverse direction by suddenly switching S to terminal B.

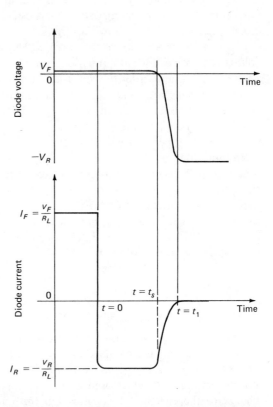

FIGURE 7.19

Graph of the voltage across and current through the diode in Fig. 7.18 versus time. The moment of switching from the forward to the reverse direction is at $t = 0$. A time greater than t_s is required for the diode to recover to its normal very low reverse current.

current through the junction simply serves the purpose of maintaining the exponential distribution of holes in the n-region, where they are constantly recombining. Since the average hole recombination time is τ_p, a charge Q_p must be supplied in a time τ_p by the current I_p. The normal definition of current as charge transmitted per unit time gives us Eq. (7.18). Note that the relationship between junction current and charge is linear, whereas the relation between current and junction voltage is exponential (see Eq. 6.20). This illustrates the simplicity of the so-called charge-control model in dealing with current flow in the p-n junction diode. It is also of considerable use in the description of transistor switching when applied to the emitter p-n junction. This will be discussed later.

Consider a p^+-n diode being used as an electronic switch, with a current I_F flowing through it at time $t < 0$ under steady-state ON conditions (Fig. 7.18 with the switch moved to terminal A). Now the diode is to be turned OFF by reversing the potential on the device, by moving the switch to terminal B at $t = 0$. After some transient recovery time during which appreciable reverse current flows, the diode finally attains the high-impedance state with only a very small leakage current flowing through it, typical of a reverse-biased junction diode. The finite time required for the diode to recover to its high-impedance state is the period necessary for the excess minority holes introduced into the n-region of the diode under forward bias conditions to disappear by recombination with majority electrons or by diffusing out of this region. The minority carrier distribution in the n-region of the diode just before and during switching is shown in Fig. 7.20.

FIGURE 7.20

a) The minority injected carrier distribution in the n-region of the p-n diode biased in the forward direction, just prior to switching. The excess hole density at the edge of the space-charge region at $x = 0$ is $\Delta p_n(0)$.
b) The minority carrier distribution just after switching at $t = 0^+$. Note that the slope of the carrier density at the edge of the space-charge region is proportional to the reverse current I_R, since it is a diffusion current.
c) Carrier distribution at a later time, less than t_s. Note that the carrier slope at $x = 0$ is the same as in part b, indicating the reverse current is still I_R.
d) Carrier distribution at t_s, when the excess hole density at the edge of the space-charge region becomes zero.

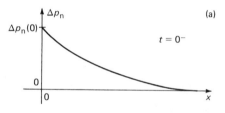

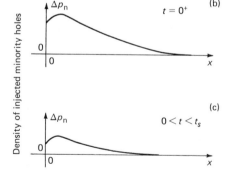

Under transient conditions the minority hole current in the diode n-region is given by[4]

$$i_p(t) = \frac{q_p(t)}{\tau_p} + \frac{dq_p(t)}{dt},$$ (7.19)

where $q_p(t)$ represents the hole charge integrated over x at a time t. The first term on the right-hand side of the equation refers to the supply of current necessary to compensate for the recombination of holes in the n-region under steady-state conditions. However, under transient conditions the net number of charges in the n-region must change with time. If the diode is being switched OFF, the net number of charges there must decrease with time, and hence dq_p/dt must be negative. In the circuit shown in Fig. 7.18, before turn-off (switch in position A), a steady-state current $I_F \simeq V_F/R_L$ flows in the diode if V_F is much greater than the forward diode drop. At $t = 0$ the switch is suddenly moved to B. Now the diode is reverse biased with a potential $-V_R$, and a current $I_R \simeq -V_R/R_L$ flows from time $t = 0$ to $t = t_s$. During this period the junction is so swamped with minority injected carriers that its impedance is negligibly small compared to the circuit resistance R_L, which then limits the current. After a time t_s a sufficient number of excess holes have recombined in, or flowed out of, the n-region into the p-region so that the junction begins to return to its high-impedance state. The recovery is essentially exponential in character and is effectively completed at time $t = t_1$ (see Fig. 7.19), when approximately the dc reverse steady-state leakage current I_0 is all that flows.

Equation (7.19) can now be written for the period $0 < t < t_s$ as

$$-I_R = -\frac{V_R}{R_L} = \frac{q_p}{\tau_p} + \frac{dq_p}{dt}.$$ (7.20)

The general solution of this equation is

$$q_p(t) = -I_R \tau_p + Be^{-t/\tau_p},$$ (7.21)

where B is an integration constant which will now be evaluated. Just prior to $t = 0$, the steady-state hole current is given by $I_F = Q_p/\tau_p$, where Q_p is the total charge stored in the n-region which cannot change instantaneously at $t = 0$. Hence introducing this charge into Eq. (7.21), taken at $t = 0^+$, yields $B = \tau_p(I_F + I_R)$. Now Eq. (7.21) becomes

$$q_p(t) = \tau_p[-I_R + (I_F + I_R)e^{-t/\tau_p}].$$ (7.22)

At $t = t_s$ the excess hole density at the junction becomes negligibly small, the diode voltage starts to reverse, and the junction impedance begins to regain control of the current. Depending on the details of diode doping design, the

---

[4] It is assumed here that the junction capacitance charging current, $i_c = C_j(dv/dt)$, is negligible compared to bulk charge storage terms, since the voltage change across the diode during switching is usually small during most of the turn-off time.

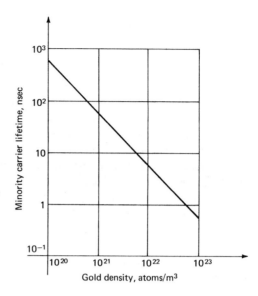

FIGURE 7.21

Logarithmic plot of the variation of minority hole lifetime with gold impurity density in a 0.01 Ω-m n-type silicon crystal doped with gold atoms, at 300°K.

excess junction current will then decay more or less rapidly and become negligible compared to I_R at a time $t = t_1$ when the stored charge $q_p(t_1)$ becomes negligibly small; then from Eq. (7.22) we get as an approximation, assuming $t_1 \sim t_s$ and hence falls within the interval in which I_R is essentially constant,

$$t_1 \simeq \tau_p \ln \left(1 + \frac{I_F}{I_R} \right). \qquad (7.23)$$

In some diode designs $t_1 \sim t_s$ and so this latter equation represents approximately the charge storage time t_s, which is related to the lifetime τ_p. Hence the storage time can be reduced by increasing I_R but only with additional energy expenditure. Note that the approximation of Eq. (7.23) is not useful for $t_1 \gg t_s$. The exact equation for obtaining t_s for a uniformly doped p-n junction is[5]

$$\mathrm{erf}(\sqrt{t_s/\tau_p}) = \frac{I_F}{I_F + I_R}, \qquad (7.24)$$

where erf is a tabulated function called the *error function*. Since $\mathrm{erf}(0) = 0$, $t_s \to 0$ as $I_R \to \infty$; since $\mathrm{erf}(0.45) = 0.5$, $t_s = 0.2\tau_p$ when $I_R = I_F$. See Fig. 11.4 for other values of the error function.

If the p-n diode behaved as an "ideal" diode this recovery time would be zero. In a real diode, though, the recovery time is of the order of the minority carrier lifetime in the lightly doped n-region if its width is greater than L_p. This time delay during which the diode returns to its blocking state and the time required to turn the diode on lengthen the total diode switching time and

[5] B. Lax and S. F. Neustader, "Transient Response of p-n Junctions," *J. Appl. Phys.,* Vol. 25, p. 1148 (1954).

hence reduce the maximum switching rate of the device in a computer application. In silicon junction diodes designed for rapid switching, gold impurity atoms are introduced into semiconductor bulk in order to reduce the lifetime there and hence improve the switching speed. However, there is a limit to the density of gold atoms which can be introduced metallurgically into the silicon, and hence the recovery time cannot be reduced much below 1 nsec. (Also gold traps out or immobilizes electrons, causing the silicon resistivity to increase.) The variation of hole lifetime with gold impurity density in a silicon crystal is plotted in Fig. 7.21. The storage time is also reduced by decreasing the n-region thickness, and hence reducing the total minority charge stored by reducing the volume in which this charge is contained.

For subnanosecond switching the Schottky-barrier diode is employed since the forward current in this device is not carried by injected minority carriers and the problem of minority carrier storage does not arise. It will be seen in the next chapter that the minority carrier storage effect also limits the switching speed of bipolar junction transistors. In integrated circuits, Schottky-barrier devices are sometimes used to speed up the switching of bipolar transistors. (See Section 11.1D, Integrated Schottky diodes.)

EXAMPLE 7.3

The steady-state current I_F carried by a silicon p^+-n junction diode biased in the forward direction is 10 mA. The hole lifetime in the 0.001 Ω-m n-region is adjusted by introducing 10^{22} atoms of gold per cubic meter of silicon. The diode is switched on and off in the circuit of Fig. 7.18.

a) Determine the integrated excess hole charge in the n-region which must be eliminated to turn this diode off.

b) How long will it take to essentially eliminate the charge calculated in part a if the reverse turn-off I_R is 10 mA?

c) Calculate the storage time under the conditions of part b.

SOLUTION a) From Fig. 7.21, 10^{22} atoms/m^3 of gold correspond to a lifetime of 6×10^{-9} sec. Hence from Eq. (7.18) the total excess hole charge in the n-region is

$$Q_p = I_p \tau_p = 10 \times 10^{-3} \text{ A } (6 \times 10^{-9} \text{ sec})$$
$$= 6 \times 10^{-11} \text{ C} \quad \text{or} \quad 60 \text{ pC.}$$

b) The time to remove this charge, by Eq. (7.23), is

$$t_1 \simeq \tau_p \ln \left(1 + \frac{I_F}{I_R}\right)$$

$$= (6 \times 10^{-9} \text{ sec}) \ln \left(1 + \frac{10 \times 10^{-3} \text{ A}}{10 \times 10^{-3} \text{ A}}\right)$$

$$= 4 \times 10^{-9} \text{ sec.}$$

c) The storage time, from Eq. (7.24), is given by

$$\text{erf}\left(\sqrt{\frac{t_s}{\tau_p}}\right) = \frac{I_F}{I_F + I_R}$$

$$= \frac{10 \times 10^{-3} \text{ A}}{(10 \times 10^{-3} + 10 \times 10^{-3}) \text{ A}}$$

$$= \tfrac{1}{2}.$$

Hence

$$\sqrt{\frac{t_s}{\tau_p}} = 0.45$$

and

$$t_s = \tau_p (0.45)^2 = 6 \times 10^{-9} \text{ sec } (0.45)^2$$
$$= 1.2 \times 10^{-9} \text{ sec}.$$

Note that this is less than the minority carrier recombination time (lifetime) since the reverse current extracts holes from the n-region.

PROBLEMS

7.1 A silicon diode is operating in a half-wave rectifying circuit such as is shown in Fig. 7.4. The voltage source is a 60 Hz sinusoid with a peak voltage of 165 V and negligible generator resistance. The load resistance R_L is 7.5 Ω. Assuming a piecewise linear model of the diode as shown in Fig. 7.3a, with $V_0 = 0.70$ V, a series resistance $R_s = 0.10$ Ω, and a reverse leakage of 1 nA, calculate:

a) the rms value of the generator voltage,
b) the required minimum avalanche breakdown voltage of the diode,
c) the average current through the load resistor,
d) the power dissipated in the load resistor,
e) approximately, the power dissipated in the diode,
f) approximately, the lowest value of capacitance for a filter capacitor which is connected across the load resistor.

7.2 If two diodes similar to that of Problem 7.1 are used in the full-wave rectifier circuit of Fig. 7.8a, and the same power as in Problem 7.1 is to be delivered to the 7.5 Ω load, determine:

a) the peak secondary transformer voltage required,
b) the minimum avalanche voltage required for the diodes,
c) the approximate power dissipated in each diode.

7.3 Repeat Problem 7.2 for the bridge rectifier circuit of Fig. 7.9.

7.4 Given a Zener-diode-regulated circuit as shown in the figure below. The diode rectifier D_1 is of the type described in Problem 7.1, while the Zener diode Z has a breakdown voltage of 20 V, a reverse leakage of 1 nA, and a Zener resistance of 10 Ω. The circuit is driven by an 18 V rms sinusoidal voltage source with negligible generator resistance. Calculate the current through the load resistance R_L of 200 Ω and through the Zener diode at various times when the generator voltage is, with respect to ground,

a) +10 V,   b) +20 V,   c) +25 V,   d) +35 V.
e) Give a plot of the current through the load versus time.

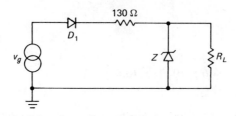

7.5 a) Repeat Problem 7.4 if the resistance in series with the Zener diode is reduced from 130 to 50 Ω.

b) Discuss the suitability of using one or the other resistor in series with the Zener diode.

7.6 The piecewise linear model of a p-n diode has the *I-V* characteristic shown in the figure below. Two such diodes, D_1 and D_2, are used in the circuit shown.

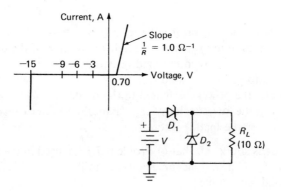

a) What voltage V is necessary so that the current through the load resistor R_L is 1.0 A?
b) Calculate the load current through R_L if the applied voltage V is 15 V.
c) What is the current through diode D_2?

7.7 Refer to the diode digital AND circuit shown in Fig. 7.14b, containing diodes of the type described in Problem 7.1, with a positive supply voltage of 3 V and a load resistance of 100 Ω.

a) Determine the current through diodes F and G when there is no input pulse. (Points F and G are at ground potential.)
b) Determine the potential at point H under the conditions of part a.
c) Determine the current through the diode at F if a 3 V positive-going pulse is applied at point F; point G is still at ground potential.
d) Determine the potential at point H under the conditions of part c.
e) Determine the potential at point H when $+3$ V pulses are applied to points F and G.
f) What is the current through the diodes under the conditions of part e?

7.8 a) Calculate the junction depletion-layer capacitance at a reverse voltage of -1.0 V of an abrupt p^+-n silicon junction diode, 1.0×10^{-8} m^2 in area, whose n-region resistivity is 0.0010 Ω-m and p-region resistivity is 0.000010 Ω-m.
b) How much charge is stored on the n-side of the space-charge region in part a, and what is the nature of this charge?
c) Repeat parts a and b at a reverse applied voltage of -10 V.
d) If the n-region is only 5.0×10^{-4} m thick, calculate the maximum operating frequency of the diode of part a used in a varactor diode application.
e) Determine the effect on the frequency cutoff of the diode of part d if the doping in the n-region is reduced so that the resistivity there is 0.01 Ω-m.

7.9 Derive an expression which will give the voltage delivered by the solar cell of Example 7.1 when it is delivering maximum power to a matched load. (*Hint:* Set $d(IV)/dV = 0$.)

7.10 Show that Eq. (7.21) is a solution of Eq. (7.20).

7.11 Explain the approximations used to obtain Eq. (7.23).

7.12 A p^+-n diode like that shown in Fig. 7.16 is used in a switching application in a circuit similar to that shown in Fig. 7.18. The minority hole lifetime in the n-region of this diode is determined by the gold doping there of 10^{21} gold atoms/m^3. The minority electron lifetime in the p^+-region is very low compared to that in the n-region because of the high impurity content there.

a) Why is the switching speed of this diode limited by the minority hole lifetime in the n-region?
b) If a forward current of 10 mA is initially flowing through the diode, determine the excess hole charges stored in the n-region of the diode.
c) If the diode is suddenly reverse biased so that a reverse current of 10 mA is drawn out of the diode, compute the time necessary to reduce the excess stored hole charge to a negligible amount.

d) Repeat part c for 100 mA of reverse current extracted.

e) Compute approximately the maximum switching rate possible with the diode operated as in part c.

7.13 A p^+-n junction diode biased in the forward direction is conducting a steady-state current of 100 mA. A reverse bias of 50 V is suddenly applied to the diode through a 5.0 kΩ resistor. The avalanche breakdown voltage of the diode is 100 V, it has a saturation current of 0.10 μA, and the minority carrier lifetime in the n-region of the diode is 1.0 μsec.

a) What is the current through the diode just after switching?

b) What is the current through the diode 1 msec after switching?

c) How long does it take for the diode current to reach an essentially steady-state value in this reverse direction?

8 Introduction to the Bipolar Junction Transistor

8.1 THE BIPOLAR JUNCTION TRANSISTOR

The analysis and modeling of the p-n junction diode provides the background for a discussion of the physics and modeling of the bipolar semiconductor junction transistor. As previously described in Chapter 1, this device consists of two back-to-back p-n junctions in series, connected by a thin region of semiconductor material. Although the individual diodes are passive devices, when they are coupled by a pure single crystal and a nearly structurally perfect thin semiconductor region, the device can become active and exhibit good power gain. This amplification results from the nearly undiminished transfer of minority current carriers from the emitter junction to the collector junction with the former forward biased (low impedance) and the latter reversed biased (high impedance). This is then the source of power gain (see Section 1.2B), with the device operating in the active region.

In the following sections the physics of the operation of the transistor as an electronic device will be developed; first dc or low-frequency operation will be described, and later some high-frequency limitations will be considered. The device will be modeled for ease in analyzing electronic circuits utilizing bipolar transistors. A few applications of transistors as amplifying elements and switches will also be presented.

A. Current Flow in a p-n-p Transistor

The n-p-n transistor is the most common type of transistor used in integrated circuits today. However, since the flow of positively charged current carriers in the p-n-p device is simpler to follow, that device will be discussed here. The n-p-n transistor description is identical with that of the p-n-p device if the roles played by holes and electrons are reversed. A schematic diagram for this device is given in Fig. 8.1 with currents and voltages specified in a conventional manner. Figure 8.2 shows the circuit symbol and typical bias arrangements for the transistor as an amplifier. Note that the directions of current and junction potential biases normally assumed are independent of whether the device is n-p-n or p-n-p. The transistor has three segments with three leads connected

FIGURE 8.1

a) Schematic diagram of a p-n-p transistor with exter-nally applied voltages and currents defined. Note that the first letter of the double-letter subscript represents the electrode which is the more positive in potential. For example, V_{CE} is the collector-to-emitter potential and is positive if the collector is positive in potential relative to the emitter. If the emitter electrode is more positive than than the collector, V_{CE} is a negative quantity $(=-V_{EC})$. b) Schematic diagram of an n-p-n transistor with extern-ally applied voltages and currents defined.

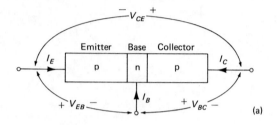

(a)

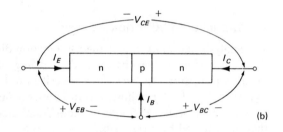

(b)

FIGURE 8.2

a) Circuit symbols and typical voltage bias arrange-ment for p-n-p transistor biased in the active region. b) Circuit symbols and typical voltage bias arrange-ment for an n-p-n transistor biased in the active region.

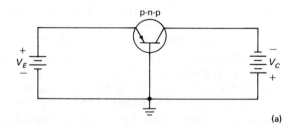

(a)

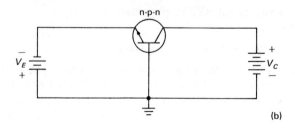

(b)

respectively to the emitter, base, and collector regions. The currents in these three leads are respectively called the emitter, base, and collector currents and are conventionally assumed to flow into the device. Since one of Kirchhoff's laws requires that

$$I_E + I_B + I_C = 0, \qquad (8.1)$$

only two of the three currents are independent of each other. The specification of any two currents fixes the third. The Kirchhoff law relating to voltages requires that

$$V_{EB} + V_{BC} + V_{CE} = 0, \qquad (8.2)$$

and again only two of the terminal voltages are independent. In modeling this transistor, we will see that the device operation is completely specified by a pair of input voltage and current values and a pair of output voltage and current values.

There are four basic modes of transistor operation:

1) active mode — emitter junction forward biased, collector reverse biased,
2) saturated mode — both emitter and collector junctions forward biased,
3) cutoff mode — both emitter and collector junctions reverse biased,
4) inverse mode — collector junction forward biased and emitter reverse biased.

The active mode is of primary interest when the transistor is utilized as an amplifier. The saturated and cutoff modes are of interest when the transistor is used as an electronic switch. In certain special applications the inverse mode is of interest, although it should be pointed out that the transistor is a completely symmetric device in principle. Hence the labeling of one junction as emitter and the other as collector is arbitrary if the device is designed symmetrically. In practice, however, the emitter region is most often more heavily doped with impurities than the collector region, and the collector junction cross-sectional area is normally greater than the emitter junction area (see Fig. 1.2). From this viewpoint the transistor to be discussed here should properly be labeled a p^+-n-p transistor, with the emitter most heavily doped. Such a device may be constructed by the impurity diffusion techniques which are used in integrated circuit technology.[1] Here the base region automatically is more heavily doped than the collector region. Certain discrete devices are of an "alloy" or "epitaxial" variety where the collector region can be as heavily doped as the emitter. In this case the junction with the larger cross-sectional area is called the collector.

Let us describe the bipolar transistor from the quantum mechanical energy

---

[1] The technology and processes for constructing these devices will be described in Chapter 11.

FIGURE 8.3

a) Energy band diagram for three separate pieces of a semiconductor crystal material doped p$^+$, n, and p.
b) Energy band diagram for a p$^+$-n-p transistor in thermal equilibrium. The Fermi energy E_F is constant throughout the crystal.
c) Energy band diagram for a p$^+$-n-p transistor biased in the active mode. The emitter is forward biased with a voltage V_E and the collector is reverse biased with a voltage V_C. The emitter space-charge region is thinner and the collector space-charge region is wider than the corresponding equilibrium case.

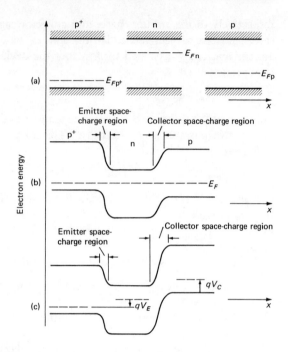

band viewpoint. First consider three separated sections of a semiconductor crystal material, the first heavily doped p-type, the second part n-type, and the last p-type, as shown in Fig. 8.3a. Now let us do a "thought" experiment in which we bring these three parts together. Majority holes from the p-sections spill over into the n-region where they are few in number; correspondingly, the majority electrons in the n-section flow into the p-regions where electrons are in the minority. This charge transfer results in the creation of two separate p-n junction space-charge regions containing ionized acceptors and donors and hence in electric fields therein which tend to limit the charge flow. This identical process is described for an individual p-n junction in Section 6.1A. Figure 8.3b shows the energy band diagram for this composite p$^+$-n-p structure after equilibrium has been established and the corresponding Fermi levels of the three sections are lined up. In equilibrium, minority electrons from the p$^+$-region can flow downhill into the n-region but, of course, an equal number of majority electrons from the n-region mount this p$^+$-n potential barrier so that the net electron flow is indeed zero. A corresponding description applies to the hole charge flowing across this junction and to electrons and holes flowing across the collector junction. Finally, Fig. 8.3c shows the energy band diagram for the p$^+$-n-p transistor biased in the active or amplifying mode as indicated in Fig. 8.2a. Note that the p$^+$-n emitter barrier is reduced in height by the forward bias, inducing a net flow of holes into the n-region. Note too that the collector p-n barrier is substantially increased in height owing to the reverse biasing of the collector junction. This barrier prevents electrons from entering the p-type collector region but is very effective in collecting holes (it is downhill for holes) from the n-region which have mainly been injected into this region from the emitter. The efficient (without recombination) transfer of holes from the emitting junction to the collecting junction is the basis of the transistor power gain as discussed in Section 1.2B.

We shall now describe analytically a p^+-n-p transistor operating in the active mode under small-signal low-current conditions. In this type of device and under these conditions, minority carrier flow is of primary importance. The analysis procedure will be to define the minority carrier concentrations at the edges of the emitter and collector p-n junction space-charge regions in terms of the applied voltages (see Section 6.2B). Then the emitter-injected minority carrier flow through the base region to the collector junction will be considered. Under low-level conditions the voltages applied to the emitter and collector junctions appear essentially across these junctions, and very little electric field penetrates into the transistor base region. Hence the minority carrier current flow through the base layer will be considered to have only a diffusion component; the field-dependent drift-current component is taken as negligible (see Section 6.2A and Appendix A).

B. Minority Carrier Flow in the Base Region

The junction transistor may be looked upon as a control device in which the potential V_{EB} applied across the input emitter-base junction controls the current I_C through the output collector junction. When the emitter junction is forward biased, excess carriers are produced at the edges of the emitter space-charge region (see Eq. 6.30). Again as for the p^+-n diode, the number of excess holes injected into the n-side of the junction is taken as much greater than the number of excess minority electrons injected into the p^+-side of the junction, this number being dependent on the minority carrier concentration in each region. So again, the minority carrier current through the p^+-n junction is comprised primarily of holes. However, in contrast to the case of the diode structure previously analyzed in which the n-region was taken as wide, the n-region base of the transistor is assumed very thin, for good gain. Specifically, the base-region width W is taken as much less than the minority hole diffusion length L_p. Consider this p^+-n-p transistor operating in the active mode with the emitter junction forward biased and the collector junction reverse biased. Under low-level, static, or slowly varying current conditions, the excess hole density at the base-region edge of the emitter space-charge region, $x = 0$ (see Figure 8.4), is given in analogy with Eq. (6.30) by

$$\Delta p_E(0) = p_{n0}(e^{qV_{EB}/kT} - 1) \approx p_{n0}\,e^{qV_{EB}/kT}, \tag{8.3}$$

FIGURE 8.4

Linear distribution of minority holes in the base region
and electrons injected into
the emittter region of a
typical p$^+$-n-p transistor
operating in the active mode,
where the base current is
negligible (solid line). Note
that hole injection through
the emitter-base junction
predominates over electrons
owing to the much heavier
doping of the emitter region,
since $p_{no} \gg n_{po}$ (see Eq.
6.30). The dashed line indicates the distribution of holes
in the base region when the
base current is not negligible.

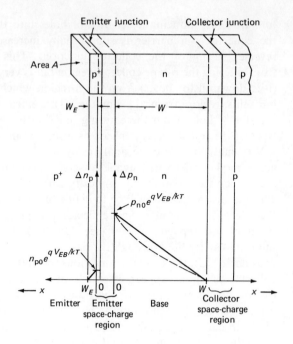

where it is assumed[2] that $V_{EB} \gg kT/q$, so that $e^{qV_{EB}/kT} \gg 1$. Here p_{no} represents
the minority hole density in the n-type base region at equilibrium.

At the base-region edge of the collector space-charge region, $x = W$, the
excess hole density is

$$\Delta p_C(W) = p_{no}(e^{qV_{CB}/kT} - 1) \approx -p_{no}, \tag{8.4}$$

since here too it is assumed that $|V_{CB}| \gg kT/q$, but V_{CB} is negative. The minus
sign indicates an excess hole concentration below the equilibrium value or a
total hole concentration there of approximately zero. Since p_{no} is normally
quite small, especially in silicon transistors, the excess hole density at the collector is nearly zero.

Hence to a good approximation the excess density of holes in the base
region varies from $p_{no} e^{qV_{EB}/kT}$ at the emitter to zero at the collector for a
device biased in the active mode. To a first approximation this variation is
linear as shown in Fig. 8.4 and is more nearly so as the base current becomes
more negligible compared with the emitter or collector current. The validity of
this approximation will now be shown. Note that the current through the base
region, from emitter to collector, is determined essentially by diffusion and
hence by the gradient of excess holes there. This results in a hole diffusion
current which is constant throughout the base region when the gradient is

---

[2] At $T = 300°K$ (room temperature), $kT/q = 26$ mV.

constant. Then the magnitude of the emitter current essentially must equal the collector current, which is given by Eq. (6.22) as

$$I_E \approx |I_C| = -qAD_p \frac{d(\Delta p_n)}{dx} \approx qAD_p \frac{\Delta p_E(0)}{W}. \tag{8.5}$$

Equation (8.1) indicates that this is only strictly true if the base current is negligible compared with I_E or I_C. Good current gain transistors have relatively small base currents, and hence this is the case in most practical devices. Combining Eqs. (8.3) and (8.5) yields an expression for the collector current under the condition of negligible base current:

$$I_C = \frac{-qAD_p p_{n0}}{W} (e^{qV_{EB}/kT} - 1) \approx \frac{-qAD_p p_{n0} e^{qV_{EB}/kT}}{W}, \tag{8.6}$$

where A is both the emitter and collector cross-sectional area. In the case where the base current is not negligible, the magnitude of the hole gradient at the emitter junction is greater than the gradient at the collector junction, and hence the collector current is less than the emitter current by the value of the base current. This is indicated by the dashed line in Fig. 8.4.

Base recombination current   Although the base current normally is small, it nevertheless plays an essential role in transistor operation. For example, in the "common-emitter" circuit connection, an input current supplied to the transistor base lead is amplified by the transistor and appears as the output collector current. Physically though, one function of the base current is to supply a majority electron for each minority hole which is lost by recombination in transit from emitter to collector. Hence the base lead must supply a current of majority electrons equivalent to the minority carrier recombination current. In the steady state this represents the electron flow which must be supplied to maintain the linear minority carrier distribution as indicated in Fig. 8.4. This is normally assumed for reasons of space-charge neutrality. To maintain this charge distribution, electrons must be supplied at a rate given in analogy with Eq. (7.18) by

$$I_B = -\frac{q_F}{\tau_p}, \tag{8.7}$$

where q_F represents the steady-state total charge of excess holes in the base region, and τ_p is the rate at which this charge tends to disappear. (Previously, τ_p was defined as the minority hole lifetime.) An approximate expression for q_F can be obtained by integrating the excess charge density distribution indicated in Fig. 8.4. By estimating the area under this curve, one obtains, in the base region,

$$q_F = \tfrac{1}{2}qAW \, \Delta p_E(0) \tag{8.8}$$

and hence the base current which must be supplied is, from Eq. (8.7),

$$I_B = -\frac{qAW\,\Delta p_E(0)}{2\tau_p}. \tag{8.9}$$

If in fact this base current is small compared to the current of Eq. (8.5), this would require that

$$\left|\frac{I_B}{I_C}\right| = \frac{qAW\,\Delta p_E(0)/2\tau_p}{qAD_p\,\Delta p_E(0)/W} \ll 1, \tag{8.10a}$$

or

$$\frac{W^2}{2D_p\tau_p} = \frac{W^2}{2L_p^2} \ll 1. \tag{8.10b}$$

Hence good current gain and a linear minority hole distribution in the base region require a very thin base region W compared with the minority carrier diffusion length L_p. In a typical silicon p-n-p transistor $W^2/2L_p^2 \ll 1$, owing to the low hole recombination rate in the base region. The role played here by the base current in supplying charge to operate the device is part of a more complete charge-control model of the transistor to be discussed later.

Emitter injection efficiency Actually the base current must also compensate for another source of inefficiency in the transfer of current from the emitter to the collector. This results from the fact that the emitter current is not made up entirely of hole current, but in practical devices about 1–5 percent of electron current is present.[3] This electron current is supplied through the base lead and injected back through the emitter junction, representing the lack of so-called emitter *injection efficiency*. To calculate this electron component of the emitter current, a procedure identical to that used in calculating the hole current can be used, assuming a very thin emitter width W_E. A linear distribution of electrons in the p$^+$-type emitter region is assumed as indicated in Fig. 8.4. This "defect" current, which is supplied through the base lead, then becomes

$$(I_E)_{\text{electron}} = I'_B \approx -\frac{qAD_n\,\Delta n_E(0)}{W_E} = \frac{qAD_n\,n_{p0}(e^{qV_{EB}/kT} - 1)}{W_E}, \tag{8.11}$$

in analogy with Eq. (8.5). Here $\Delta n_E(0)$ represents the density of injected electrons at the p$^+$-edge of the emitter space-charge region. One can define an emitter injection efficiency γ representing the ratio of hole current injected by the emitter p$^+$-n junction to the total emitter current flowing as

$$\gamma = \frac{(I_E)_{\text{holes}}}{(I_E)_{\text{holes}} + (I_E)_{\text{electrons}}}. \tag{8.12}$$

---

[3] No electron component of the emitter current may contribute to the collector current since the reverse-biased collector potential barrier rejects electrons and collects holes.

Using Eqs. (8.5) and (8.11) gives

$$\gamma = \frac{1}{1 + D_n n_{p0} W / D_p p_{n0} W_E}.$$ (8.13)

Multiplying the second term in the denominator of Eq. (8.13) by $n_{n0} p_{p0} / n_{n0} p_{p0}$, noting that $n_{p0} p_{p0} = p_{n0} n_{n0} = n_i^2$, and using $D_n / D_p = \mu_n / \mu_p$ from the Einstein relation give

$$\gamma = \frac{1}{1 + \sigma_n W / \sigma_p W_E};$$ (8.14)

here σ_p and σ_n are respectively the conductivities of the emitter and base regions. If the base width W and the emitter width W_E are comparable in magnitude and $\sigma_p / \sigma_n \gg 1$, the injection efficiency may be approximated from Eq. (8.14) as

$$\gamma \approx 1 - \frac{\sigma_n W}{\sigma_p W_E};$$ (8.15)

and so hole injection efficiency by a p^+-n junction emitter can be better than 0.99 or 99 percent.

There are two other sources of emitter "defect" current which must be supplied through the transistor base lead. These are particularly of consequence at low levels of emitter current and, in fact, tend to limit the current gain of a silicon transistor as the current is reduced to very low values. The first of these results from carrier recombination in the emitter space-charge region. This effect was discussed in some detail for the p-n junction diode in Section 6.2C. This recombination current was indicated to be significant compared to the diffusion current at low current levels for junctions fabricated from semiconductor materials with an energy gap equal to that of silicon or even greater. Hence the base lead must supply electrons for recombination with holes in the space-charge region of the emitter junction and, in fact, at the surface near the junction, where the recombination rate is also high.

The other source of low-level emitter defect current results from surface effects of the type discussed in Sections 5.6 and 6.3C (Voltage breakdown due to surface effects). It was pointed out in the latter section that for a planar p^+-n junction (such as the emitter junction of a p^+-n-p transistor), positive charges in the oxide and at the oxide-silicon interface will tend to attract electrons to the lightly doped n-region at the surface, essentially causing a p^+-n^+ junction there. Equation (8.15) indicates a low hole injection efficiency for that portion of the emitter junction, and hence an electron defect current which must be supplied by electrons through the base lead.

Finally, since the collector junction is reverse biased with the transistor operating in the active region, a component (albeit small) of the collector current must be the normal diode leakage currents of Eqs. (6.32) and (6.33) and the surface leakage. This leakage current is also supplied through the base

FIGURE 8.5

Schematic diagram identifying the various components
of the base current of a
p$^+$-n-p transistor, biased in
the active region, as well as
the emitter and collector
currents. $(I_E)_n$ represents the
emitter electron "defect"
current, $(I_E)_p - (I_C)_p$ represents the base hole recombination current, and I_{CBO}
represents the collector
"leakage" current which
includes both the diffusion
component and the space-
charge-generated current.

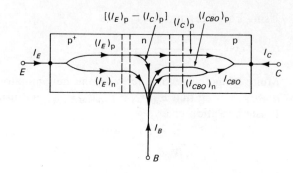

lead, but it is normally much smaller than the other defect currents, except at
very high temperatures in silicon transistors. The various sources of the
base-current components are summarized in Fig. 8.5.

EXAMPLE 8.1

A silicon p$^+$-n-p transistor has a base width of 2.0×10^{-6} m and base resistivity of 0.0010 Ω-m. The hole lifetime in the base region is 0.10 μsec. The
emitter region has thickness 1.0×10^{-6} m and resistivity 0.000010 Ω-m. The
emitter and collector areas are both 2.5×10^{-9} m^2.

a) Show that the component of the base current necessary for supplying electrons
for recombining holes in the base region is small compared to the collector current.
b) Calculate the injection efficiency of holes for the emitter of this transistor.

SOLUTION   a) According to Eq. (8.10b), to show that the electron current for
recombining holes is negligible, it is necessary to prove that $W^2/2L_p^2 \ll 1$,
where $L_p^2 = D_p \tau_p$. Now

$$\frac{W^2}{2L_p^2} = \frac{(2.0 \times 10^{-6})^2 \text{ m}^2}{2(1.25 \times 10^{-3} \text{ m}^2/\text{sec})(1.0 \times 10^{-7} \text{ sec})}$$

$$= 1.6 \times 10^{-2} \ll 1.$$

b) From Eq. (8.14),

$$\gamma = \frac{1}{1 + \sigma_n W/\sigma_p W_E}$$

$$= \frac{1}{1 + (1/0.0010)(\Omega\text{-m})^{-1}(2.0 \times 10^{-6} \text{ m})/(1/0.000010)(\Omega\text{-m})^{-1}(1.0 \times 10^{-6} \text{ m})}$$

$$= \frac{1}{1 + 2.0 \times 10^{-2}} = 0.98 \quad \text{or} \quad 98 \text{ percent.}$$

FIGURE 8.6

The common-base transistor circuit connection for a p-n-p transistor represented as a "black box," with two input ports and two output ports. Note that V_{CB} is positive when the collector potential is positive relative to the base. In addition, $V_{CB} = -V_{BC}$.

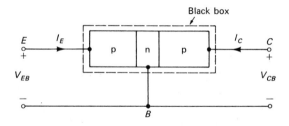

8.2   THE EBERS-MOLL TRANSISTOR EQUATIONS

To analyze electronic circuits containing the transistor it is convenient to model the device as a two-port circuit element. Two currents and two voltages are sufficient to specify the operation of this component modeled as a "black box." The *common-base* transistor circuit connection of a p-n-p transistor is given in Fig. 8.6, showing the input current and voltage I_E and V_{EB} and the output current and voltage I_C and V_{CB}. By analogy with the junction-diode current-voltage equations (6.32) and (6.33), one would expect similar types of current-voltage relationships for a transistor, which consists basically of two coupled diodes. These can in fact be written as

$$I_E = A_{11}(e^{qV_{EB}/kT} - 1) + A_{12}(e^{qV_{CB}/kT} - 1) \qquad (8.16a)$$

and

$$I_C = A_{21}(e^{qV_{EB}/kT} - 1) + A_{22}(e^{qV_{CB}/kT} - 1), \qquad (8.16b)$$

where A_{12} and A_{21} are coupling coefficients. Here the input current I_E and the output current I_C are specified in terms of the input and output voltages V_{EB} and V_{CB} as well as the A_{ij} coefficients which describe the internal design of the particular transistor. These equations plus the two Kirchhoff laws of Eqs. (8.1) and (8.2) comprise four equations with six unknown currents and voltages, once the transistor design parameters A_{ij} are specified. Hence if two of these currents or voltages are given, all others are determined in principle. In contrast to the description in the last section, which was confined to the transistor operating in the active mode, this formulation applies more generally to all four modes of transistor operation. Equations (8.16) are in a form which can be very useful for transistor modeling, particularly in computer-aided circuit analysis; they are not restricted to low-level conditions. They are of a type commonly referred to as the *Ebers-Moll equations.*[4]

By comparing Eqs. (8.3) and (8.4) and (8.16), Eqs. (8.16a) and (8.16b) may be rewritten as

$$I_E = a_{11}\,\Delta p_E(0) + a_{12}\,\Delta p_C(W) \qquad (8.17a)$$

[4] J. J. Ebers and J. L. Moll, "Large-Signal Behavior of Junction Transistors," *Proc. IRE,* Vol. 42, p. 1761 (1954).

and

$$I_C = a_{21}\,\Delta p_E(0) + a_{22}\,\Delta p_C(W). \tag{8.17b}$$

In this form the transistor currents are seen to be linearly dependent on the excess minority carrier concentrations at the emitter and collector edges of the base region respectively. This is in contrast to the nonlinear relations in terms of the emitter and collector junction voltages as expressed in Eqs. (8.16). Since physical processes related to current flow in the transistor, when expressed in terms of minority carriers, are linear, the principle of superposition may be applied. This will be of considerable value later in the formulation of the charge-control model of the transistor.

A. Ebers-Moll Parameter Formulation in Terms of Physical Device Parameters

Expressions for the transistor emitter and collector currents in terms of physical device parameters may be derived by considering the excess minority carrier flow in a manner similar to what has already been done for the p-n junction diode. Under low-level steady-state conditions the excess hole density distribution in the n-type base region of the transistor shown in Fig. 8.7 may be determined from the continuity equation

$$D_p \frac{d^2(\Delta p_n)}{dx^2} = \frac{\Delta p_n}{\tau_p}, \tag{8.18}$$

exactly as in the treatment of the diode in Eq. (6.23b). Again the solution of this second-order linear differential equation may be written as

$$\Delta p_n(x) = Be^{-x/L_P} + Ce^{x/L_P}, \qquad (0 \le x \le W) \tag{8.19}$$

as in Eq. (6.24). In contrast to the diode case, the transistor base width W, for reasons of good current gain, will be taken as very narrow; that is, $W/L_p \ll 1$. The boundary conditions for excess hole density at the emitter and collector edges of the base region are already expressed in Eqs. (8.3) and (8.4).

FIGURE 8.7

General type of low-level excess hole density distribution in the n-type base region of a p-n-p transistor.

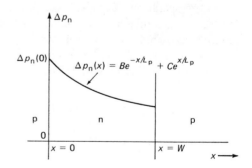

Introducing these expressions into Eq. (8.19) permits determination of the integration constants B and C as

$$B = \frac{\Delta p_E(0)e^{W/L_p} - \Delta p_C(W)}{e^{W/L_p} - e^{-W/L_p}}$$

and   (8.20)

$$C = \frac{\Delta p_C(W) - \Delta p_E(0)e^{-W/L_p}}{e^{W/L_p} - e^{-W/L_p}}.$$

The complicated general solution of Eq. (8.19) may be simplified by noting that $W/L_p < 1$ in practical cases. Then the exponentials may be series expanded and the solution for the excess hole density in the base region approximated as

$$\Delta p_n(x) \approx \frac{C - B}{L_p} x + B + C. \qquad (0 \le x \le W) \qquad (8.21)$$

This is an expression of the nearly linear distribution of excess holes in the base region for all values of emitter and collector junction potential. This has already been discussed for the special case of the transistor biased in the active region.

The minority hole diffusion current flow in the transistor base region can now be calculated, assuming the emitter current is exclusively hole current (injection efficiency $\gamma = 1$), as

$$I_E = -qAD_p \left. \frac{\partial(\Delta p_n)}{\partial x} \right|_{x=0} \qquad (8.22a)$$

and

$$I_C = qAD_p \left. \frac{\partial(\Delta p_n)}{\partial x} \right|_{x=W}. \qquad (8.22b)$$

Here the emitter and collector areas are both taken as A. After differentiating the exact expression for the excess hole density of Eq. (8.19), Eqs. (8.22) become

$$I_E = \frac{qAD_p}{L_p}(B - C) \qquad (8.23a)$$

and

$$I_C = \frac{qAD_p}{L_p}(Ce^{W/L_p} - Be^{-W/L_p}). \qquad (8.23b)$$

By introducing the previously derived values for B and C given in (8.20), Eqs. (8.23) may be written in the Ebers-Moll form of (8.17a) and (8.17b) as

$$I_E = a_{11} \Delta p_E(0) + a_{12} \Delta p_C(W), \qquad I_C = a_{21} \Delta p_E(0) + a_{22} \Delta p_C(W),$$

where[5]

$$a_{11} = a_{22} = \frac{qAD_p}{L_p} \coth \frac{W}{L_p} \quad \text{and} \quad a_{12} = a_{21} = \frac{-qAD_p}{L_p} \operatorname{csch} \frac{W}{L_p}.$$

(8.24)

Again if $W/L_p < 1$, the expression for the a_{ij} can be simplified by the approximations

$$\coth \frac{W}{L_p} \approx \frac{L_p}{W} + \frac{W}{3L_p} \quad \text{and} \quad \operatorname{csch} \frac{W}{L_p} \approx \frac{L_p}{W} - \frac{W}{6L_p}$$

(8.25)

to read

$$a_{11} = a_{22} = \frac{qAD_p}{W} + \frac{qAW}{3\tau_p} \quad \text{and} \quad a_{12} = a_{21} = \frac{-qAD_p}{W} + \frac{qAW}{6\tau_p}.$$

(8.26)

Then the base current can be obtained by combining Eqs. (8.17) and the Kirchhoff law of Eq. (8.1) as

$$I_B = -[(a_{11} + a_{21}) \Delta p_E(0) + (a_{12} + a_{22}) \Delta p_C(W)].$$

(8.27)

If values for the a_{ij} as given in Eq. (8.26) are introduced into Eq. (8.27) and the active region of transistor operation considered $[\Delta p_C(W) \approx 0]$, agreement with Eq. (8.9) is obtained; note that the latter equation was derived from a different viewpoint.

EXAMPLE 8.2

Assuming the transistor design of Example 8.1, calculate the magnitude and dimensions of a_{11}, a_{12}, a_{21}, and a_{22} using (a) the approximate expressions of Eq. (8.26) and (b) the exact expressions of Eq. (8.24).

SOLUTION   a) From Eq. (8.26),

$$a_{11} = a_{22} = \frac{qAD_p}{W} + \frac{qAW}{3\tau_p}$$

$$= \frac{1.6 \times 10^{-19} \text{ C } (2.5 \times 10^{-9} \text{ m}^2)(1.25 \times 10^{-3} \text{ m}^2/\text{sec})}{2.0 \times 10^{-6} \text{ m}}$$

$$+ \frac{1.6 \times 10^{-19} \text{ C } (2.5 \times 10^{-9} \text{ m}^2)(2.0 \times 10^{-6} \text{ m})}{3(1.0 \times 10^{-7} \text{ sec})}$$

---

[5] In this formulation, the emitter and collector junctions are not considered dissimilar. This accounts for the symmetry of the equations for I_C and I_E expressed by Eqs. (8.24). In real transistors this symmetry does not exist in general. The Ebers-Moll type equations will nevertheless still apply, although generally $a_{11} \neq a_{22}$ when the emitter and collector junctions are not identical. However, in general $a_{12} = a_{21}$, even in the case of lack of symmetry. This expresses the *reciprocity* of the transistor configuration.

$$= 2.5 \times 10^{-25} \text{ A-m}^3 + 2.7 \times 10^{-27} \text{ A-m}^3$$

$$= 2.5 \times 10^{-25} \text{ A-m}^3.$$

$$a_{12} = a_{21} = -\frac{qAD_\text{p}}{W} + \frac{qAW}{6\tau_\text{p}} = -2.5 \times 10^{-25} \text{ A-m}^3.$$

b) From Eq. (8.24),

$$a_{11} = a_{22} = \frac{qAD_\text{p}}{L_\text{p}} \coth \frac{W}{L_\text{p}}$$

$$= \frac{1.6 \times 10^{-19} \text{ C } (2.5 \times 10^{-9} \text{ m}^2)(1.25 \times 10^{-3} \text{ m}^2/\text{sec})}{[(1.25 \times 10^{-3} \text{ m}^2/\text{sec})(1.0 \times 10^{-7} \text{ sec})]^{1/2}}$$

$$\times \coth \frac{2.0 \times 10^{-6} \text{ m}}{[(1.25 \times 10^{-3} \text{ m}^2/\text{sec})(1.0 \times 10^{-7} \text{ sec})]^{1/2}}$$

$$= 4.5 \times 10^{-26} \text{ A-m}^3 \coth 0.18$$

$$= 2.5 \times 10^{-25} \text{ A-m}^3, \quad \text{within the approximation of part a.}$$

$$a_{12} = a_{21} = -\frac{qAD_\text{p}}{L_\text{p}} \operatorname{csch} \frac{W}{L_\text{p}}$$

$$= -(4.5 \times 10^{-26} \text{ A-m}^3)$$

$$\times \operatorname{csch} \frac{2.0 \times 10^{-6} \text{ m}}{[(1.25 \times 10^{-3} \text{ m}^2/\text{sec})(1.0 \times 10^{-7} \text{ sec})]^{1/2}}$$

$$= -(4.5 \times 10^{-26} \text{ A-m}^3) \operatorname{csch} 0.18$$

$$= -2.5 \times 10^{-25} \text{ A-m}^3, \quad \text{within the approximation of part a.}$$

B. Ebers-Moll Parameter Determination by Measurements on Devices

Some simple measurements can be made on a real transistor in the laboratory to determine the coefficients A_{ij} in Eqs. (8.16) when the device is to be represented by these equations:

1) Connect the base and collector leads together ($V_{CB} = 0$) and then measure the current-voltage (I-V) characteristic of the emitter-base diode. Equation (8.16a) then becomes

$$I_E|_{V_{CB}=0} = A_{11}(e^{qV_{EB}/kT} - 1), \tag{8.28}$$

which is the ordinary diode law of Eq. (6.32), where A_{11} is the emitter diode leakage current usually referred to[6] as I_{ES}. Experimentally, this is usually determined by fitting the measured forward-biased emitter diode characteristic to Eq. (8.28). It is also instructive to measure the ratio of collector current to emitter current under the shorted-collector condition.[7] In this situation Eqs. (8.16a) and (8.16b) can be combined to define a *short-circuit common-base current gain* as

$$\alpha_F \equiv -\frac{I_C}{I_E}\bigg|_{V_{CB}=0} = -\frac{A_{21}}{A_{11}}$$

$$= \frac{\text{csch}\,(W/L_p)}{\text{coth}\,(W/L_p)} \simeq 1 - \left(\frac{W}{L_p}\right)^2. \tag{8.29}$$

Here Eqs. (8.24) are used to give an expression for α_F in terms of device parameters, assuming unity injection efficiency.

2) Now connect the base and emitter leads together ($V_{EB} = 0$) and then measure the collector-base diode *I-V* characteristic. Equation (8.16b) then becomes

$$I_C\big|_{V_{EB}=0} = A_{22}(e^{qV_{CB}/kT} - 1), \tag{8.30}$$

which is the ordinary diode law for the collector junction, where A_{22} is the collector diode leakage current with the emitter shorted, usually referred to as I_{CS}. Except in the special case where the emitter and collector diodes are identical, in general $I_{CS} \neq I_{ES}$. Now the ratio of emitter current to collector current under the shorted-emitter condition may be written as

$$\alpha_R \equiv -\frac{I_E}{I_C} = -\frac{A_{12}}{A_{22}}, \tag{8.31}$$

where α_R is the short-circuit common-base current gain with the transistor operating in the reverse condition, i.e., collector forward biased and emitting and the emitter junction collecting.

With the definitions given above, Eqs. (8.16a) and (8.16b) coupled with (8.1) may now be expressed for a p-n-p transistor as

$$I_E = I_{ES}(e^{qV_{EB}/kT} - 1) - \alpha_R I_{CS}(e^{qV_{CB}/kT} - 1), \tag{8.32a}$$

$$I_C = -\alpha_F I_{ES}(e^{qV_{EB}/kT} - 1) + I_{CS}(e^{qV_{CB}/kT} - 1), \tag{8.32b}$$

$$I_B = -(1 - \alpha_F)I_{ES}(e^{qV_{EB}/kT} - 1) - (1 - \alpha_R)I_{CS}(e^{qV_{CB}/kT} - 1). \tag{8.32c}$$

---

[6] The subscript S here refers to the shorting of the collector junction. In small-signal transistors this quantity is of the order of picoamperes.

[7] Experimentally this condition can be approximated by placing a very low-resistance ammeter between collector and base to measure the collector current. However, more commonly the common-base current gain α is measured at a specified collector-base potential and is defined as $|I_C/I_E|_{V_{CB}=\text{constant}}$.

These are the original forms of the equations as published by Ebers and Moll.[8] In general, $I_{ES} \neq I_{CS}$ and $\alpha_F \neq \alpha_R$ but always

$$\alpha_R I_{CS} = \alpha_F I_{ES}. \tag{8.33}$$

This is a consequence of the fact that in general $a_{11} \neq a_{22}$ but always $a_{12} = a_{21}$ (see footnote 5). Eliminating I_{ES} from Eqs. (8.32a) and (8.32b) yields the useful equation

$$I_C = -\alpha_F I_E + I_{CBO}(e^{qV_{CB}/kT} - 1), \tag{8.34a}$$

where by definition

$$I_{CBO} \equiv (1 - \alpha_F \alpha_R)I_{CS}. \tag{8.35a}$$

Similarly,

$$I_E = -\alpha_R I_C + I_{EBO}(e^{qV_{EB}/kT} - 1), \tag{8.34b}$$

where

$$I_{EBO} \equiv (1 - \alpha_F \alpha_R)I_{ES}. \tag{8.35b}$$

Note that I_{CBO} may be determined[9] by measuring the reverse collector leakage current ($V_{CB} \ll 0$) with the emitter unconnected ($I_E = 0$), according to Eq. (8.34a). Similarly, I_{EBO} may be determined by measuring the reverse emitter leakage current with the collector unconnected ($I_C = 0$), according to Eq. (8.34b). In the active region ($V_{CB} \ll 0$) this states that the collector current is just the emitter current multiplied by the short-circuit common-base current gain plus the collector saturation current as measured with the emitter open-circuited. Since α_F in practical devices is normally between 0.95 and 1.0, the common-base connection yields nearly unity current gain. Since the collector junction is normally reverse biased with 10 V or more and the emitter junction requires less than 1 V to provide a practical emitter current, the device in this connection can have appreciable voltage and hence power gain (see Section 1.2B). Figure 8.8 shows a plot of the output characteristics of a typical common-base-connected p-n-p transistor, relating the collector current to collector-base voltage for different input values of emitter current.

In a corresponding manner the input characteristics of the transistor in the common-base connection may be derived. From Eq. (8.32a) the emitter-base I-V characteristic of the transistor operating in the active region becomes

$$I_E = I_{ES}(e^{qV_{EB}/kT} - 1) + \alpha_R I_{CS}. \tag{8.36}$$

This is essentially the expression for the ordinary junction diode except for an

---

[8] For an n-p-n transistor the Ebers-Moll equations are

$$I_E = -I_{ES}(e^{qV_{BE}/kT} - 1) + \alpha_R I_{CS}(e^{qV_{BC}/kT} - 1), \tag{8.32d}$$
$$I_C = \alpha_F I_{ES}(e^{qV_{BE}/kT} - 1) - I_{CS}(e^{qV_{BC}/kT} - 1), \tag{8.32e}$$
$$I_B = (1 - \alpha_F)I_{ES}(e^{qV_{BE}/kT} - 1) + (1 - \alpha_R)I_{CS}(e^{qV_{BC}/kT} - 1). \tag{8.32f}$$

[9] The subscript O refers to the open-circuiting of the emitter. In small-signal transistors this quantity is of the order of picoamperes.

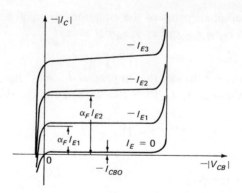

FIGURE 8.8

Common-base *output* current-voltage characteristic of a p-n-p transistor. Collector current versus collector-base voltage at a fixed emitter current is plotted for different values of emitter current ($I_{E3} > I_{E2} > I_{E1} > 0$). The forward gain factor α_F is used to define the collector output current in terms of the emitter input current. The sudden increase in current at high V_{CB} is due to avalanching of the collector junction. Note that in this text generally symbols with letter subscripts only (V_{CB}) refer to variable quantities, whereas those with a combination of letters and numbers refer to specific numerical values of a quantity.

FIGURE 8.9

Common-base *input* current-voltage characteristic of a p-n-p transistor for different values of collector output voltage. As the collector potential is increased ($|V_{CB3}| > |V_{CB2}| > |V_{CB1}|$), space-charge widening reduces the effective transistor base width causing additional current to flow through the emitter diode (I_{ES} and α_R increase).

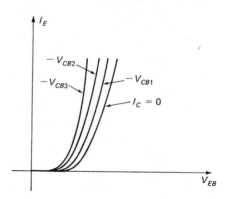

additional coupling term $\alpha_R I_{CS}$ which is normally small in practical devices. The typical input characteristics of a common-base-connected p-n-p transistor, relating the emitter current to the emitter-base voltage for different values of collector-base reverse voltage, are shown in Fig. 8.9. Here the effect of higher collector reverse voltage is to widen the collector space-charge region, bringing the collector and emitter space-charge regions closer together. This will affect the parameters I_{ES} and α_R (see Eqs. 8.16, 8.26, and 8.32), and hence I_E for each V_{CB}.

EXAMPLE 8.3

Assuming the transistor design of Example 8.1, calculate the short-circuit common-base current gain α_F using (a) the appropriate expressions of Eqs. (8.26) and (8.29), (b) the exact expressions of Eq. (8.24).

SOLUTION a) From (8.29),

$$\alpha_F = 1 - \frac{1}{2}\left(\frac{W}{L_p}\right)^2 = 1 - \frac{1}{2}\left[\frac{2.0 \times 10^{-6}\,\text{m}}{\sqrt{(1.25 \times 10^{-3}\,\text{m}^2/\text{sec})(1.0 \times 10^{-7}\,\text{sec})}}\right]^2$$

$$= 0.98.$$

b) Combining the expressions of Eq. (8.24) with the definition of Eq. (8.29) gives

$$\alpha_F = \frac{-(qAD_p/L_p)\,\text{csch}\,(W/L_p)}{(qAD_p/L_p)\,\text{coth}\,(W/L_p)} = \text{sech}\,\frac{W}{L_p}$$

$$= \text{sech}\,\frac{2.0 \times 10^{-6}\,\text{m}}{\sqrt{(1.25 \times 10^{-3}\,\text{m}^2/\text{sec})(1.0 \times 10^{-7}\,\text{sec})}} = \text{sech}\,0.18$$

$$= 0.98.$$

Hence both give the same value within the accuracy of the given data.

C. The Ebers-Moll Equivalent Circuit

The Ebers-Moll equations may be expressed in terms of circuit models to aid in the solution of circuit network problems which involve transistors. One of these models is shown in Fig. 8.10. This is the model for the Ebers-Moll equations expressed in the form

$$I_C = -\alpha_F I_E + I_{CBO}(e^{qV_{CB}/kT} - 1), \tag{8.34a}$$

$$I_E = -\alpha_R I_C + I_{EBO}(e^{qV_{EB}/kT} - 1), \tag{8.34b}$$

where I_{EBO} and I_{CBO} are respectively the emitter and collector leakage currents, when the opposite terminal is open-circuited. From reciprocity, $\alpha_F I_{EBO} = \alpha_R I_{CBO}$, and hence only three measurements are needed to determine the four parameters α_F, α_R, I_{EBO}, and I_{CBO} and hence to specify completely the I-V characteristics of a transistor.

FIGURE 8.10

Ebers-Moll common-base transistor model. The model consists of two current generators and two diodes coupled together. The diode labeled I_{EBO} has a current-voltage characteristic given by $I = I_{EBO}(e^{qV_{EB}/kT} - 1)$; the symbol I_{CBO} represents the I-V characteristic given by $I = I_{CBO}(e^{qV_{CB}/kT} - 1)$.

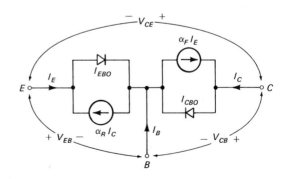

The most common connection of the transistor in amplifier and switching applications is the common-emitter configuration shown schematically in Fig. 8.11. Here the applied voltages V_{CE} and V_{BE} are specified relative to the common-emitter terminal. Now the input current I_B is supplied to the transistor base, and the output current is the collector current I_C. A typical set of I-V characteristics for a transistor in this connection is shown in Fig. 8.12. When the collector current is expressed in terms of a base-current input, one can write with the aid of Eqs. (8.34a), (8.34b), and (8.1)

$$I_C = \left| \frac{\alpha_F}{1 - \alpha_F} \right| I_B + \left| \frac{I_{CBO}}{1 - \alpha_F} \right| (e^{qV_{CB}/kT} - 1). \qquad (8.37)$$

This equation expresses the common-emitter output current I_C in terms of the input current I_B and the collector-base voltage V_{CB}. Since in this connection the applied voltage should be expressed in terms of V_{CE} (see Fig. 8.11), Eq. (8.2) may be used to give

$$V_{CB} = V_{CE} + V_{EB} = V_{CE} - V_{BE}$$

FIGURE 8.11

Common-emitter circuit configuration for a p-n-p transistor. This is sometimes referred to as the grounded-emitter connection.

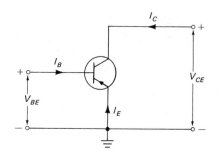

FIGURE 8.12

Common-emitter output current-voltage characteristic of a p-n-p transistor. Collector current versus collector-emitter voltage at a fixed base current is plotted for different values of base current ($|I_{B4}| > |I_{B3}| > |I_{B2}| > |I_{B1}| > 0$). The common-emitter forward gain factor β is used to define the collector output current in terms of the base input current. A set of specification sheets for a typical commercial n-p-n silicon planar transistor is included in Appendix D.1. Appendix D.2 gives specifications for a p-n-p device.

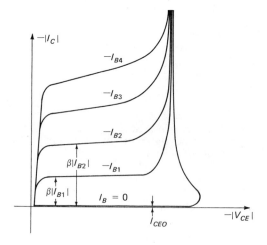

and Eq. (8.37) may be rewritten as

$$I_C = \beta_F I_B + (\beta_F + 1)I_{CBO}(e^{-qV_{BE}/kT}e^{qV_{CE}/kT} - 1), \qquad (8.38)$$

where, by definition, $\beta_F \equiv \alpha_F/(1 - \alpha_F)$. Here β_F is termed the *common-emitter short-circuit current gain* $(I_C/I_B$ with $V_{CB} = 0)$ and has practical values ranging from 10 to several hundred. From Eq. (8.29),

$$\beta_F \equiv \frac{\alpha_F}{1 - \alpha_F} = \frac{1 - \frac{1}{2}(W/L_p)^2}{1 - [1 - \frac{1}{2}(W/L_p)^2]} \simeq 2\left(\frac{L_p}{W}\right)^2 \qquad (8.39)$$

for $W/L_p \ll 1$. The expression for β_F in terms of device parameters assumes unity injection efficiency. In general practice, the common-emitter current gain is measured at a specified collector-emitter voltage and is defined as $\beta \equiv |I_C/I_B|_{V_{CE}=\text{const}}$. Note that when the input current I_B is zero the output current in this circuit connection has the value $-(\beta_F + 1)I_{CO} \equiv -I_{CEO}$ in the active region where $|V_{CE}| \gg |V_{BE}|$. $|I_{CEO}|$ can be a few orders of magnitude greater than $|I_{CBO}|$, the output leakage current in the common-base connection, and is indicated in Fig. 8.12. The common-emitter input I-V characteristics of a transistor can be derived from Eq. (8.32c) as

$$I_B = -[(1 - \alpha_F)I_{ES}(e^{-qV_{BE}/kT} - 1) + (1 - \alpha_R)I_{CS}], \qquad (8.40)$$

in the active region, where $V_{CB} \ll 0$. Here again this is the equation for the ordinary junction diode except for the additional term $(1 - \alpha_R)I_{CS}$, which is normally small. Some typical input characteristics for a common-emitter-connected p-n-p transistor, relating the base current to the base-emitter voltage, are shown in Fig. 8.13. Again the effect of higher collector voltage is to widen the collector space-charge region, increasing α_F but decreasing $(1 - \alpha_F)I_{ES}$. Hence the effect here is opposite that in the common-base case (Fig. 8.9).

The Ebers-Moll circuit model for the transistor in the common-emitter connection is shown in Fig. 8.14. Note that the collector current source can be specified in terms of the input emitter current through α_F or the base current through β_F.

8.3 THE CHARGE-CONTROL TRANSISTOR MODEL

Mention has already been made of the function of the base current in supplying charge to maintain a constant steady-state minority carrier distribution in the base region. It will now be pointed out how static or low-frequency transistor operation may be conveniently described in terms of the charge distributed in the transistor base region by relating the terminal currents to this charge. The concept will prove to be useful in extending the description of low-frequency transistor behavior to medium-frequency transistor performance. This method also is particularly powerful in explaining the operation of transistors in the switching mode. In this case use is made of the linear relationship between the terminal currents and the charge in the transistor base region, which permits the principle of superposition to be applied.

FIGURE 8.13

Common-emitter input
current-voltage characteristic
of a p-n-p transistor for
different values of collector
output voltage. As the
collector potential is
increased ($|V_{CE3}| > |V_{CE2}|$
$> |V_{CE1}|$), space-charge
widening reduces the effective
transistor base width, which
causes less current to flow
through the emitter diode.
This is in contrast to the
common-base input charac-
teristic of Fig. 8.9.

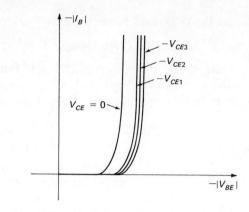

FIGURE 8.14

Ebers-Moll common-emitter
transistor model. Here the
collector current generator
may be specified in terms of
the emitter current I_E or in
terms of the input current I_B.
When I_E is specified the
diode symbol I_{CBO} has an I-V
characteristic given by
$I = I_{CBO}(e^{qV_{CB}/kT} - 1)$;
when I_B is specified the
diode symbol represents
$(\beta_F + 1)I_{CBO}(e^{qV_{CB}/kT} - 1)$.

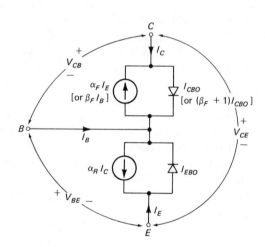

A. The Charge-Control Equations

Equation (8.7) is an expression relating the terminal base current to the total
charge q_F in the transistor base region in the steady-state active mode. By
combining Eqs. (8.5) and (8.8), an equation relating the collector current to this
total charge in the base region may be written for the transistor operating in the
active mode as

$$I_C = -\frac{q_F}{W^2/2D_p} \equiv -\frac{q_F}{\tau_t}. \qquad (8.41)$$

Here the quantity $W^2/2D_p$ has the dimensions of time and hence is defined as
τ_t. This time is in fact essentially the time for minority holes to traverse from
the emitter junction to the collector junction where they are collected. This may

be proved by solving the diffusion equation for minority carrier flow in the base region. Alternatively, this fact may be made plausible as follows: The average time for hole recombination in the base region is τ_p. To maintain a steady-state charge distribution in the base, the holes thus lost must be reinjected through the emitter with an accompanying supply of electrons through the base lead for reasons of space-charge neutrality. However, during the time for one hole recombination in the base, many more holes will have successfully reached the collector space-charge region, constituting the collector current. Hence the ratio of base current to collector current should reflect the ratio of the emitter-collector transit time to the recombination time. This is verified mathematically by combining Eqs. (8.7) and (8.41) to give

$$\frac{I_C}{I_B} = \frac{\tau_p}{\tau_t}, \tag{8.42}$$

where τ_t can be identified clearly as the transit time. From Eqs. (8.1), (8.7), and (8.41) the emitter current can now be expressed in the active mode as

$$I_E = q_F \left(\frac{1}{\tau_t} + \frac{1}{\tau_p} \right). \tag{8.43}$$

Finally the dc current gains of the transistor in the active region can be expressed in terms of the charge-control parameters τ_p and τ_t as

$$\beta_F = \frac{I_C}{I_B} = \frac{\tau_p}{\tau_t}$$

and   (8.44)

$$\alpha_F = -\frac{I_C}{I_E} = \frac{1}{1 + \tau_t / \tau_p}.$$

The formulation thus far has referred to the active region with the emitter forward biased and collector reverse biased. As explained in Section 8.1A, in the saturated or inverse mode of operation the collector junction is injecting minority carriers into the base region. Hence for a more general formulation of the charge-control model this "reverse" operation should be considered. It may be described by a set of equations analogous to that for the "forward" operation; i.e.,

$$I_B = -\frac{q_R}{\tau_p}, \tag{8.45a}$$

$$I_E = -\frac{q_R}{\tau_t}, \tag{8.45b}$$

$$I_C = q_R \left(\frac{1}{\tau_p} + \frac{1}{\tau_t} \right). \tag{8.45c}$$

Here q_R represents the total charge injected into the base by the collector junction. As discussed earlier, because of the linear relationship between the terminal currents and the charge in the base region, the principle of superposition may be applied. Therefore, the more general charge-control equations for the transistor in the dc or static situation may be obtained from Eqs. (8.7), (8.41), (8.43), and (8.45) as

$$I_B = - \left| \frac{q_F + q_R}{\tau_p} \right|, \tag{8.46a}$$

$$I_C = - q_F \left(\frac{1}{\tau_t}\right) + q_R \left(\frac{1}{\tau_p} + \frac{1}{\tau_t}\right), \tag{8.46b}$$

$$I_E = q_F \left(\frac{1}{\tau_p} + \frac{1}{\tau_t}\right) - q_R \left(\frac{1}{\tau_t}\right). \tag{8.46c}$$

Here the three terminal currents are expressed in terms of the charge-control parameters q_F, q_R, τ_p, and τ_t for the transistor in any of its four operating modes. If q_F and q_R are expressed in terms of the terminal voltages via equations of the type of (8.3) and (8.8), the analogy between these equations and the Ebers-Moll equations may be verified.

A plot of the minority charge distribution in a p-n-p transistor base region when both the emitter and collector junctions are in forward bias (as in the saturation mode) is shown in Fig. 8.15a. This linear distribution may be

FIGURE 8.15

a) Minority hole charge density distribution for a transistor in *saturation* (both emitter and collector junctions are forward biased). The solid line indicates the total hole distribution $q_F + q_R$.
b) Forward charge density distribution whose integral is q_F.
c) Reverse charge density distribution whose integral is q_R.

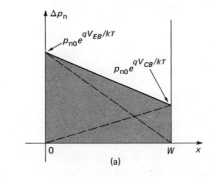

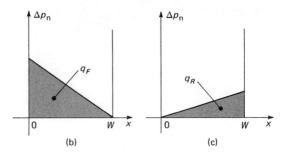

Transistor packages. (Courtesy of Unitrode Corp., 580 Pleasant Street, Watertown, Mass. 02172).

considered to be composed of the forward and reverse charge components, q_F and q_R, shown in Fig. 8.15b and 8.15c, and may be constructed by addition or superposition of these distributions.

Thus far only the minority charges in the base region have been considered, and the minority charges in the emitter and collector regions ignored. This is an excellent approximation as far as the emitter region is concerned, owing to the high doping and small injected minority carrier concentration there. However, bipolar transistors in integrated circuits usually have lightly doped collector regions, and hence a more general treatment must take into account the minority electron charges in the p-n-p transistor collector region.

The power of the charge-control model will be seen in the case of transient or ac transistor operation to be described later. Of interest then will be the change of the charge in the base Δq_F, with a signal applied to the emitter junction. Under ac conditions an additional alteration of charge will occur in the space-charge region of the emitter and collector junctions owing to charging of their space-charge capacitances. In the charge-control model this may be included simply by adding more charge parameters.

8.4 THE BEHAVIOR OF REAL TRANSISTORS

The discussion thus far has been confined to what may be termed the *intrinsic* transistor. In modeling a semiconductor diode, factors other than the intrinsic p-n junction characteristic had to be considered for effective representation of the behavior of a real device (see Section 6.3). In the case of the transistor even more factors are important in describing the behavior of a real transistor. Which of these additional factors are of particular importance depends greatly on the specific transistor design and operating conditions. Some of these additional effects are listed below:

1) The emitter, base, and collector regions have bulk series resistance, and the ohmic voltage drop due to these resistances in the presence of current flow must be accounted for in the external I-V characteristics of the transistor.

2) The current gain, α or β, for a given transistor design is not a constant but is dependent on the collector voltage and the collector current.

3) The collector current will increase abruptly above a certain collector voltage owing to avalanche multiplication in the collector junction space-charge region, in a manner similar to the diode (see Section 6.3C).

4) Minority carrier transport across the base region of a transistor is not exclusively by diffusion but is electric-field-aided in some transistor designs. Some of these effects can easily be incorporated into one of the transistor models.

A. Transistor Ohmic Voltage Drops

To accurately formulate the current-voltage characteristic of the transistor as, for example, in Fig. 8.12, possible voltage drops across resistances in the transistor bulk need to be considered. The voltage drop in the emitter bulk can normally be neglected in planar transistors, such as in Fig. 1.2, because of the heavy doping and hence high conductivity of the region. Also this region is usually very thin.

Since the base doping is only moderate and since the base current which flows transversely (perpendicular to the emitter current) has a rather long path, the voltage drop in the base resistance R_B is of consequence. Since V_{EB} in the Ebers-Moll equations (8.32) represents the actual voltage drop across the emitter junction space-charge region, the relation between this voltage and the actual voltage V'_{EB} applied externally between the emitter and base lead is given by

$$V_{EB} = V'_{EB} - I_B R_B. \qquad (8.47)$$

Since the transistor I-V characteristics are expressed in terms of terminal voltages, Eq. (8.47) should be introduced into Eqs. (8.32). Because this expression contains a terminal current and appears in the exponent of a term in the expressions relating current and voltage, explicit relations between the terminal currents and voltages become very complicated. Fortunately, in a properly designed transistor, the base resistance is small and the base current is small if the device has good current gain and operates at low level. Hence for many purposes this voltage drop can be neglected in comparison with the voltage drop across the emitter junction. At high current levels, however, this voltage drop increases in proportion to the current whereas the junction voltage hardly changes, owing to the exponential nature of the junction I-V characteristic. Under these conditions the base drop cannot be neglected.

Finally, the importance of the series voltage drop in the transistor collector bulk should be considered. In planar transistors the collector bulk is more lightly doped than the base region and since the collector current is often greater than the base current, this collector series resistance R_{CS} should be considered. In the active region, the collector junction is reverse biased and hence represents a rather high impedance. Then the collector voltage drop $I_C R_{CS}$ is normally small compared to the collector junction resistance and can be neglected. However, in the saturation region, where the collector junction is forward biased, in practice the collector bulk voltage drop even exceeds the potential drop across the junction. The effect of this extra collector series resistance can be seen in Fig. 8.16, where the common-emitter characteristic is plotted for a transistor with different values of R_{CS}. The effect is most pronounced at low collector-emitter voltages, where the transistor is in saturation and hence is collecting minority carriers inefficiently (the collector is back-injecting to the emitter). The collector-emitter potential drop represents the voltage across the transistor when operated as a switch. With the switch closed (the transistor ON), this voltage should ideally be zero and hence any drop across the series collector resistance represents power loss or switching inefficiency.

B. Current-Gain Variation with Collector Voltage and Current

The expression for transistor current gain α_F as given in terms of transistor parameters by Eqs. (8.29) and (8.24) involves the transistor base width W. This represents the distance the minority injected holes must travel before collection, or the distance between the edges of the emitter and collector space-charge regions. In Section 6.1C it was pointed out that the space-charge

FIGURE 8.16

Common-emitter transistor current-voltage characteristic showing the effect of series collector resistance R_{CS}. The initial straight-line slopes represent essentially the ohmic collector resistances. The solid line represents a typical characteristic, and the dashed lines represent the characteristic of the same device with either lower or higher R_{CS}.

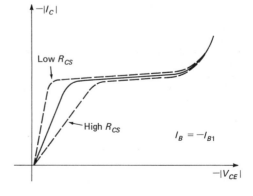

region of a p-n junction widens as increased reverse bias is applied to the junction. Hence since the collector junction of a transistor in the active region is reverse biased, the edge of its space-charge layer should extend toward the emitter junction as the collector voltage is increased. Then the transistor base width becomes voltage-dependent and thus so does the current gain. Hence for a given emitter current, the collector current will increase somewhat as the collector-base voltage V_{CB} is increased. This may be seen by examining Fig. 8.8 and, in a similar sense, Fig. 8.12, where I_C increases with V_{CE}. An extreme situation in this respect is called *punchthrough*, in which the collector space-charge region reaches through to the emitter region, causing the collector current to increase drastically, limited only by the external circuit. The extent to which this occurs may be seen by noting that the rate of increase of I_C with V_{CB} can be expressed by taking the differential of Eq. (8.34a) (neglecting the small I_{CBO} term) as

$$\left.\frac{\partial I_C}{\partial V_{CB}}\right|_{I_E=\text{const}} = -I_E \left.\frac{\partial \alpha_F}{\partial V_{CB}}\right|_{I_E=\text{const}}. \tag{8.48}$$

Therefore the slope of I_C versus V_C at constant emitter current is proportional to this current as well as the rate of change of α_F with V_{CB}. The latter effect results from the reduction in effective base width with collector space-charge widening due to increased collector potential. This will be large when the base doping is small (see Section 6.1C). The expression corresponding to Eq. (8.48) for the common-emitter transistor connection, obtained by differentiating Eq. (8.38), is

$$\left.\frac{\partial I_C}{\partial V_{CE}}\right|_{I_B=\text{const}} = I_B \left.\frac{\partial \beta_F}{\partial V_{CE}}\right|_{I_B=\text{const}}. \tag{8.49}$$

In effect, Eqs. (8.48) and (8.49) represent the transistor output conductance, which should be maintained at a low value for good power gain.

FIGURE 8.17

Variation of collector current with emitter current for a given collector potential in a typical silicon transistor. The linear portion of the curve represents constant current gain, whereas the curvature at both high and low currents indicates falloff of gain at these extremes.

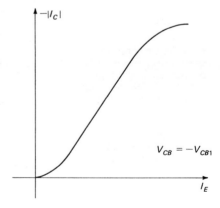

The variation of current gain with collector current at a given collector voltage is a somewhat more complex situation. Equation (8.34a) as stated indicates a linear variation of collector current with emitter current. A plot of I_C versus I_E should be a straight line. A typical variation of I_C with I_E for silicon transistors is indicated in Fig. 8.17. Deviation from linearity occurs at both very small values of emitter current and high values of I_E. In silicon transistors the loss of current gain at low emitter-current values results from the generation of carriers in the emitter space-charge region (see Section 6.2C). Since space-charge neutrality dictates the generation of pairs of holes and electrons, the high field in the emitter space-charge layer causes some electrons to be swept out, constituting a majority carrier current. This fraction of the emitter current is not collected and hence effectively constitutes a reduction in emitter efficiency. As the emitter current is increased, though, this space-charge-generated current becomes negligible compared to the normal minority carrier-injected diffusion current (see Section 6.2C), and the current gain rises. Surface effects also reduce low current gain (see Section 8.1B, Emitter injection efficiency).

At higher current levels the effective injection efficiency again falls off, but for another reason. This can be understood by referring to Section 8.1B. There it is pointed out that the doping on the p-side of the emitter p-n junction should be much greater than that on the n-type base side of the junction for good injection efficiency. The expression derived as Eq. (8.14) assumes that the carriers present on each side of the junction derive only from the impurity atoms there, as under equilibrium conditions. This is a good approximation at low current levels, but at high levels of injection a significant number of holes are injected into the n-type base region of a p-n-p transistor, comparable to or greater than the electrons present initially. Hence the conductivity of the base region will be significantly altered or *modulated*. However, the p-type emitter side of the junction is not similarly modulated, relatively speaking, since there are an enormous number of carriers there to begin with, owing to the high doping of this p^+-n junction. Hence the ratio of the conductivity of the p-region to that of the n-region is reduced at high injection levels and so the injection efficiency and the current gain are reduced.

At very high current levels the transistor gain falls off significantly owing to the modification of the collector depletion region by the large density of injected holes entering that region. These positive charges add to the positively charged donor atoms located in the normally depleted collector space-charge region, causing the latter to move toward the collector contact, effectively widening the base region. This base-width *stretching* reduces the current gain.

C. Avalanche Breakdown of the Collector Junction

Current multiplication in a p-n junction such as the collector junction of a transistor has been described in Section 6.3C. This results from avalanching caused by charge carriers diffusing into the space-charge region and

FIGURE 8.18

Common-emitter and common-base characteristics of a p-n-p transistor, emphasizing the breakdown region. BV_{CBO} represents essentially the collector diode avalanche characteristic with the emitter lead open. (The BV_{CBO} measuring circuit is indicated at bottom left.) BV_{CEO} represents the peak voltage which can be supported collector-to-emitter in the common-emitter connection with the base lead open. (The BV_{CEO} measuring circuit is indicated at bottom right.) BV_{CER} is the peak supported collector-emitter voltage with a resistor between emitter and base. The voltage supported at higher currents is less than at low currents, causing a negative-resistance portion in the curve. Also indicated is a third curve for the common-emitter connection when some base current is assumed to flow.

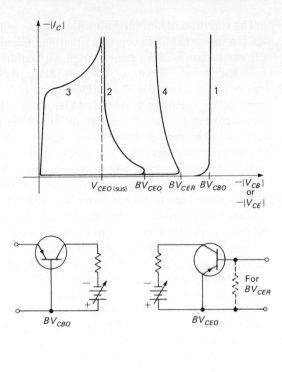

being accelerated to an energy high enough for impact ionization of some valence electrons of the semiconductor lattice atoms. However, in the case of the collector junction of a p-n-p transistor operating in the active region, the charge carriers entering the space-charge region may originate from injection by the emitter, which differentiates the transistor phenomenon from the diode effect. The *I-V* characteristics of the collector junction of a p-n-p transistor in avalanche breakdown can be seen by examining the transistor collector curves already drawn in Figs. 8.8 and 8.12. There the collector current increases rapidly above a certain collector voltage, for a given input current.

Note that the shape of the breakdown curve for a given transistor depends on the input current and the type of circuit connection, i.e., common-base or common-emitter. A few of these curves are repeated for clarity in Fig. 8.18. Curve 1 represents the normal collector-base p-n junction breakdown curve since it is measured with the emitter open-circuited. The breakdown voltage here is defined as BV_{CBO}. If, now, reverse potential is applied between collector and emitter leads, as in the common-emitter connection, and the base lead is left open-circuited, curve 2 results. Note that a breakdown voltage $BV_{CEO} <$

BV_{CBO} occurs here and the voltage across the device becomes lower, with avalanching still maintained, as multiplication causes the collector current to increase. The lowest voltage to maintain the breakdown is called the *sustaining* voltage $V_{CE(sus)}$ in analogy with the gaseous discharge which occurs, for example, in a fluorescent lamp. Curve 3 again represents the common-emitter connection, but in this case some external base current is applied and current gain is indicated. Carrier multiplication increases the collector current in the neighborhood of breakdown and the sustaining voltage is again approached asymptotically at high current values. This voltage, $V_{CE(sus)}$, typically is of the order of one-half BV_{CBO} and can be even less if the transistor intrinsically has high gain. BV_{CER} is the peak supported collector-emitter voltage with a resistor connected between emitter and base. The breakdown curve in this case is curve 4 in Fig. 8.18; it lies between the curves for BV_{CBO} and BV_{CEO}, since the transistor gain is reduced from the open-base case.

The relation between the sustaining voltage and the transistor current gain will now be formulated. The multiplication factor M of Section 6.3C is defined as the ratio of the current leaving the junction space-charge region to that entering this region. From Eq. (8.34a) the transistor collector current in the active region, with the collector avalanching, is given by

$$I_C = M(-\alpha_F I_E - I_{CBO}). \tag{8.50}$$

For the common-emitter configuration it is convenient to express this output current in terms of the input base current. With Eq. (8.1), Eq. (8.50) can be written as

$$I_C = \frac{M\alpha_F}{1 - M\alpha_F} I_B - \frac{M}{1 - M\alpha_F} I_{CBO}. \tag{8.51}$$

Hence it is clear that the collector current will increase, limited only by the circuit supplying the current, with or without base current, when

$$\alpha_F M = 1. \tag{8.52a}$$

In terms of the common-emitter gain β_F, this collector avalanche condition, using Eq. (8.39), becomes

$$\beta_F(M - 1) = 1. \tag{8.52b}$$

Introducing the expression for M from Eq. (6.44) gives an equation from which the transistor avalanche *I-V* characteristic can be computed in the common-emitter connection, if the current dependence of β_F $[=\beta_F(I_C)]$ is known. This is

$$\beta_F(I_C) \left[\frac{1}{1 - (V_{CB}/BV_{CBO})^n} - 1 \right] = 1. \tag{8.53}$$

Since the transistor current gain at low currents increases with current (see Fig. 8.17), this expression predicts the reduction in collector blocking voltage

V_{CB} with collector current. Hence the voltage capability of a transistor in the common-emitter connection is lower than in the common-base connection, particularly for high-gain devices.

In all the above expressions the collector junction leakage current with the emitter open-circuited, I_{CBO}, should be understood to be the leakage current characteristic of a p-n junction diode, as already described in Sections 6.2B, C, and D. In addition, surface-limited voltage-breakdown effects similar to those described in Section 6.3C (Voltage breakdown due to surface effects) apply here as well. In silicon planar transistor structures used in integrated circuits (see Section 11.1D, Integrated transistors and diodes), the collector region is usually more lightly doped than the emitter or base region. Hence the collector junction in these devices behaves like an n^+-p junction in a p-n-p transistor and a p^+-n junction in an n-p-n transistor. Hence positive charges in the oxide covering the collector junction will induce an n-type surface channel in a p-n-p transistor which can be limited by a p-type "guard-ring" to avoid premature voltage breakdown at the surface; n-p-n transistors can suffer from low-voltage breakdown due to the presence of a p^+-n^+ surface junction. Lateral p-n-p structures used in integrated circuits have heavily doped collectors as well as emitters and thus suffer surface-breakdown effects typical of p^+-n junctions.

D. The Built-In Field Due to Nonuniform Base Doping

Thus far it has been assumed that the three transistor regions are uniformly doped. In fact, the planar transistors in integrated circuits have a substantially greater base doping near the emitter junction than near the collector junction. This is a natural consequence of the fact that these devices are fabricated by diffusing sequentially, from one surface, first p-type impurities and then n-type impurities into lightly doped n-type silicon.[10] The impurity profile of the transistor so obtained is shown in Fig. 8.19. The gradient of the impurity density in the transistor base region is particularly of interest. This donor variation introduces an electric field in the base region which is in a direction that accelerates minority electrons through the p-type base region, reducing the emitter-to-collector transit time, and hence increasing the current gain and transistor high-frequency capability. That is, these latter values are greater than the values obtained by assuming that carrier transport is solely by diffusion, as has been discussed thus far. However, the effect is only moderate since the voltage drop across the graded base region is typically only about $5kT/q$ or 130 mV, and hence the built-in field is small. This effect is only important at low current levels, for at high current density the injected carriers similarly produce an aiding field which "swamps out" the built-in field due to impurities. Then the diffusion-flow approximation is certainly no longer valid.

[10] N-P-N transistors are most common in integrated circuits but similar arguments apply for p-n-p devices with the role of electrons and holes reversed.

FIGURE 8.19

The impurity density distribution in a diffused n-p-n
transistor. The emitter
doping is n-type and consists
of about 10^{25} donors/m^3 at
the edge of the emitter. The
base doping is p-type and
consists of about
10^{21} acceptors/m^3 at the
base edge nearest the
emitter, falling off to about
10^{20} donors/m^3 at the
collector edge. This impurity
grading produces a built-in
electric field in the base
region.

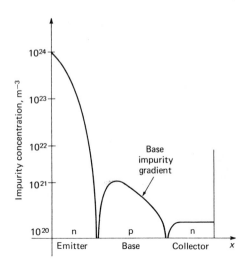

PROBLEMS

8.1 The transistor of Example 8.1 is operating in the active region under
steady-state conditions.

a) Calculate the total charge of holes stored in the base region when the transistor base current is 100 μA.

b) Determine the collector current under the conditions of part a.

c) Calculate the hole density at the edge of the emitter space-charge region, in
the n-region.

d) Calculate the emitter base voltage V_{EB} in this case.

8.2 Determine the relationship between the coefficients a_{ij} and A_{ij} of Eqs.
(8.16) and (8.17).

8.3 Show that the integration constants B and C of Eq. (8.19) may be expressed as given in Eq. (8.20).

8.4 a) Derive Eq. (8.21).

b) Under what conditions is the linear approximation of Eq. (8.21) valid?

8.5 Show that Eqs. (8.24) derive from Eqs. (8.23).

8.6 Show how Eqs. (8.26) derive from (8.24).

8.7 Using the a_{ij} expressions from Eqs. (8.26), show that Eq. (8.27) reduces to
(8.9).

8.8 Introduce the expressions of (8.26) into (8.17) and determine the physical
significance of the terms in these equations for I_E and I_C.

8.9 A transistor with undetermined parameters is to be used in the circuit shown below. Data from which the transistor parameters can be computed are taken as follows: (1) V_{CB} is made zero and V_{EB} is made -10 V (reverse bias). Then I_E is measured to be -50 nA and I_C is 48 nA. (2) V_{EB} is made zero and V_{CB} is made -10 V (reverse bias). Then I_C is measured to be -64 nA, and I_E is 48 nA.

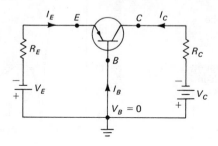

a) Determine α_F.
b) Determine α_R.
c) Determine I_{CBO}.
d) Determine I_{EBO}.
e) When the transistor is used in the active mode, determine the forward emitter voltage needed to produce a collector current of -5 mA at $300°$K.

8.10 By combining Eqs. (8.32a) and (8.32b) under emitter open-circuit conditions, prove Eq. (8.35a) is true.

8.11 a) Calculate the common-base current gain α_F for the transistor of Problem 8.1.
b) Calculate the common-emitter current gain β_F for the transistor of Problem 8.1.

8.12 Draw the approximate Ebers-Moll equivalent circuit for a transistor operating in the active region, where $\alpha_F \gg \alpha_R$ and $I_{CBO} \ll \alpha_F |I_E|$.

8.13 Derive Eq. (8.37) from Eqs. (8.34) and (8.1).

8.14 For a transistor with strongly reverse-biased collector junction and with emitter base open-circuited, show that the floating emitter-base voltage is $kT/q \ln (1 - \alpha_F)$.

8.15 Derive an expression for the "offset voltage" V_{offset}, for a p-n-p transistor in the grounded-emitter connection when $I_C = 0$, with the emitter junction forward biased. Show that this can be expressed as $V_{CE} = (kT/q) \ln \alpha_R$ under certain conditions. (*Hint*: Use Eqs. (8.32) and (8.37) and assume that I_{EBO}, $I_{CBO} \ll I_B$.)

8.16 Calculate the transit time for holes across the base region of the transistor of Problem 8.1.

8.17 The collector leakage current of a transistor in the grounded-emitter connection, with a resistor R connected between base and emitter, is called I_{CER}. Show that

$$|I_{CER}| = \frac{I_{CBO}(1 + qI_{EBO}\,R/kT)}{1 - \alpha_F\alpha_R + (qI_{EBO}\,R/kT)(1 - \alpha_F)}.$$

8.18 A silicon p-n-p transistor with a base width of 5.0 μm and base doping of 5.0×10^{20} donors/m^3 has heavily doped emitter and collector regions. With the device operating in the saturated mode, the collector potential is -0.10 V, and the base potential -0.70 V relative to the grounded emitter. Its emitter area is 6000 μm^2. The hole lifetime in the base is 1.0 μsec.

a) Calculate the total hole charge stored in the base region of this transistor.
b) If the base width is reduced to 1.0 μm, calculate the new value of the stored charge.
c) What does part b suggest about reducing the turn-off time of the transistor used as a switch?

8.19 Show the relationship between the charge-control equation (8.46b) and the Ebers-Moll equation (8.32b).

8.20 For the transistor of Problem 8.18 operating in the active region, determine:

a) the collector-to-emitter punchthrough voltage,
b) the output conductance $(\partial I_C/\partial V_{CE})_{I_B = \text{const}}$ for the transistor operating at a base current $I_B = 0.10$ mA. (*Hint*: Use Eq. (8.39) for β_F and Eq. (6.18) for the variation of base width with collector voltage.)

8.21 The common-base current gain α_F of a transistor increases from 0.90 at low currents to 0.98 at higher currents. The collector breakdown voltage BV_{CBO} is 100 V. Determine the collector sustaining voltage at higher currents for the transistor in the common-emitter connection. Assume that $n = 3$ in Eq. (8.53).

9 Bipolar Transistor Applications and Frequency Performance

9.1 THE TRANSISTOR AS A CIRCUIT ELEMENT

In Section 1.2B the ability of a junction transistor to provide power gain or amplification in an appropriate circuit was discussed. That is, when biased in the active region, the transistor is capable of delivering more signal power to the load than is supplied to its input circuit. This input may be in the form of a current or voltage signal; it may be sinusoidal in time or have some other time dependence such as the voltage pulse input described in Section 1.2C. In any case the usefulness of the device derives most often from its ability to take a small signal from a transducer or a communication channel and, after a number of stages of amplification, provide enough power to drive an output device such as a loudspeaker or other electromechanical driver. The manner in which this is accomplished will be described next with the aid of the load-line concept already introduced in Section 7.1B, to facilitate the description of junction diode circuit behavior. This will be done for a p-n-p junction transistor connected in the common-emitter circuit configuration. Note that the emitter here is normally connected to a metallic ground plane, as shown in Fig. 9.1a, and is normally taken as zero reference potential.

A. Common-Emitter Amplifier Load-Line Analysis

Because of the nonlinear character of the transistor I-V characteristics (see Fig. 8.10), signal amplification is most easily studied using graphical methods. Consider the elementary common-emitter transistor amplifier circuit in Fig. 9.1a. The I-V characteristics of the transistor shown are given in Fig. 9.1b along with a load line representing the output load resistance R_L. Again, as in the case of the diode, the circuit dc operating point is established by the

259

intersection of the load line and one of the I-V curves representing the transistor behavior, for a given base bias current. Assume that the dc base current is established by the battery V_B supplying base current $-I_{B2}$ through resistance R_B. This input current gives rise to a collector output current (curve labeled $-I_{B2}$), which is plotted against collector-emitter voltage v_{CE} as shown in Fig. 9.1b. Now v_{CE} in this circuit is established by the battery V_{CC} supplying current through load resistance R_L and is given by

$$v_{CE} = -(V_{CC} + i_C R_L). \tag{9.1}$$

The load line, which is shown as a straight line with negative slope equal to $-1/R_L$ in Fig. 9.1b, is a plot of Eq. (9.1). Note that the intersection of this line with the v_{CE} axis is at $v_{CE} = -V_{CC}$ since, should the collector current i_C be nearly zero, essentially all the battery voltage will appear across the transistor. Also the intersection of the load line with the i_C axis represents the maximum collector current possible, which occurs when all the battery voltage V_{CC} appears across the load resistor; this maximum current is then given essentially

FIGURE 9.1

a) Common-emitter transistor amplifier driven by a signal current. Note the grounding of the emitter in this common-emitter configuration.
b) Current-voltage collector characteristics for a p-n-p transistor for various values of base current. Also shown is the load line for a collector voltage $-V_{CC}$ and a load resistor R_L. The operating point for a base current $-I_{B2}$ is marked 0. Also shown is the maximum power curve $(p_C)_{max}$.

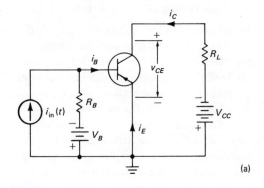

(a)

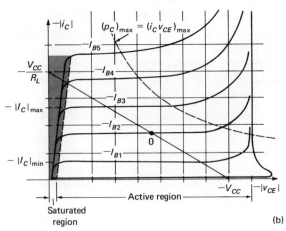

(b)

FIGURE 9.2

Graph of the sinusoidal signal current superimposed on a base current $-I_{B2}$.

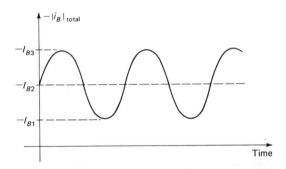

by $-V_{CC}/R_L$. The intersection of this load line and the transistor characteristic for base current equal to $-I_{B2}$ is the dc *operating point*, labelled 0 in Fig. 9.1b.

We have not yet included the effect of the signal current $i_{in}(t)$, which is, of course, the primary question. However, the effect of this input current can easily be ascertained with the aid of Fig. 9.1b. Consider that $i_{in}(t)$ is a small sinusoidal signal current, alternating in time. This current superimposed on the dc bias base current $-I_{B2}$ represents the input current to the transistor. Assume that this total input base current oscillates between the extremes $-I_{B1}$ and $-I_{B3}$ as shown in Fig. 9.2. The corresponding extremes of collector current through the load resistor R_L are $-|I_C|_{max}$ and $-|I_C|_{min}$ as shown in Fig. 9.1b. Since the ratio of the incremental increase in collector current resulting from the incremental increase in base current due to the signal is much greater than 1 for a good transistor, the device exhibits current gain.[1] In fact, this ratio is often referred to as h_{fe}, the common-emitter ac current gain,[2] and is defined by

$$h_{fe} \equiv \left[\frac{\partial i_C}{\partial i_B} \right]_{v_{CE}=\text{const}} \tag{9.2}$$

If this quantity were constant and independent of base current, a sinusoidal input signal current would yield an exact replica as output current, but increased in magnitude by the current gain. However, this is never quite the case except in a small current range as is indicated by the discussion in Section 8.4B. Hence in general a sinusoidal input signal will not result in an exact sinusoidal amplified output signal. That is, the transistor as an amplifier introduces *distortion* in processing the signal. In small-signal processing, where the current gain is reasonably constant over the range of current variation with signal, the device is said to be operating in the linear region. This is truer

---

[1] The argument presented in Section 1.2B indicates why the device also gives power gain. Of course, it can be shown that, owing to conservation-of-energy considerations, the power supplied by the energy sources (batteries) in the circuit exactly equals the power dissipated in the transistor and the remainder of the circuit. This is the subject of a problem at the end of this chapter. (See J. F. Gibbons, *Semiconductor Electronics*, McGraw-Hill Book Company, New York, 1966, p. 379.)

[2] By analogy the dc common-emitter current gain $[I_C/I_B]_{v_{CE}=\text{const}}$ is normally called h_{FE} or β.

FIGURE 9.3

Sketch of a typical mounting of a transistor package onto a metal plate for thermal radiation into the surrounding atmosphere, resulting in cooling of the transistor chip.

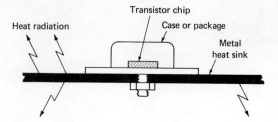

over an intermediate range of current values and not so true at the high and low extremes of collector current (see Fig. 8.17).

This is no longer so in the case of large input signals; for example, if the peak value of the input current reaches $-I_{B4}$ and is increased still further to $-I_{B5}$ (see Fig. 9.1b), the value of the output current hardly increases. (The collector currents obtained for these base currents are determined by the intersection of the load line and the I-V characteristic, for $-I_{B4}$ and then $-I_{B5}$.) Now the transistor is no longer operating totally in the active region, and the device is said to be driven into *saturation* or *bottoming*. The saturation region is indicated in Fig. 9.1b. The concept of current gain obviously has no meaning in this situation and can be applied only in the small-signal or linear region of operation. The dc operating point is generally chosen with regard for the nature of the input signal and the extent of distortion allowable in amplification.

B. Temperature and Frequency Limitations of Transistors

The maximum power level at which the transistor can operate, usually specified by the device manufacturer, also restricts the dc operating point. This power, handled by the transistor, causes internal heating, raising the collector junction temperature above the package or case temperature. The maximum power dissipation, $(p_C)_{max}$ $[=(i_C v_{CE})_{max}]$, graphs as a hyperbola and is indicated in Fig. 9.1b. Operating points below this curve are thermally stable while those above are unstable. Here internal thermal dissipation limits the current and voltage that the transistor can handle.[3] The efficiency with which the heat generated in a given transistor is removed (by mounting the transistor on a metal heat sink) can seriously affect the power-handling capability of the device. A typical thermal mounting for a packaged transistor is shown in Fig. 9.3. The transistor package itself limits the efficiency with which heat can be removed from the transistor chip and transferred to the surrounding atmosphere. This limitation to heat removal is usually expressed as

$$\theta_{jc} \equiv \Delta T_C / p_C , \qquad (9.3)$$

[3] In practice it is found that transistors can handle more power reliably at low voltage and high current, at a given power, than at high voltage and low current so that the hyperbolic power-limitation curve is only approximate. This results from a transistor thermal failure mechanism known as "second breakdown."

Power transistor package
with heat sink. (Courtesy of
RCA Solid State Division,
Somerville, N.J.)

where θ_{jc} is the transistor collector junction-to-case thermal resistance (usually specified by transistor manufacturers in degrees centigrade per watt), and ΔT_C is the rise in the transistor collector temperature above the package temperature when the transistor power p_C is dissipated in the device, mainly in the collector space-charge region. An additional case-to-ambient thermal resistance θ_{CA} must be added to θ_{jc} to account for the efficiency limitation of heat transfer from the case and heat sink to the ambient.

It should be understood that the ability of a transistor to function as an active device and produce power gain depends on its maintaining the nonlinear behavior characteristic of its p-n junctions. It is to be noted that when the device is raised to excessively high temperatures (greater than 300°C for silicon transistors), the large number of electron-hole pairs thermally generated in the semiconductor material tend to dominate current flow, and the p-n junctions tend toward conducting in an ohmic fashion, causing the power gain to fall off. Also when the input signal to a transistor is of a sufficiently high frequency, the capacitive character of the p-n junctions of the transistor tends to dominate the behavior of the device. The high-frequency current then passes through the emitter-junction capacitance as a displacement current, which is a majority carrier current, and little minority carrier injection results. The junctions in this case become linear passive elements and hence the gain of the transistor falls off significantly with frequency. The hybrid-π model of the transistor will be discussed later to explain this high-frequency behavior of bipolar transistors.

FIGURE 9.4

A typical medium-frequency amplifier circuit driven by a voltage signal v_{in}, and supplying an amplified signal v_{out}. The capacitors are for the purpose of isolating the dc bias circuit from the input and output circuits. Bypass capacitor C_E ensures that the dc bias is unaffected by the signal current. All potentials are specified with respect to the ground reference potential taken as zero.

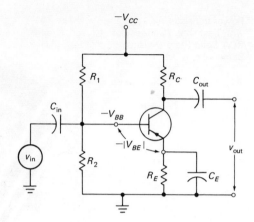

C. Transistor Circuit Applications

Some typical circuit applications of transistors will now be presented to illustrate more generally the utility of the device. First, typical dc biasing circuits will be discussed for transistors in amplifier applications. Differential-amplifier and emitter-follower circuits will then be described briefly. Finally, a digital application of the transistor used as a switch will be presented. The discussion will be confined mainly to discrete transistor circuits. These are circuits which are typically constructed by interconnecting individual (discrete) packaged devices on a printed circuit board. This is in contrast to a monolithic integrated circuit which contains all the necessary devices on a single block of silicon. However, some reference will be made to differences in the design of these circuits in the integrated form as compared to the discrete version. A more complete description of modern forms of integrated circuits will be presented in Chapter 11. No detailed analysis of these circuits will be given here, and results will sometimes be stated without proof since the intent is only to illustrate the usefulness of the transistor device.[4]

Transistor bias circuits   The first step in designing a transistor amplifier circuit is to establish a proper dc operating point by an appropriate emitter junction voltage bias arrangement. A standard circuit for accomplishing this, utilizing one voltage supply, is shown in Fig. 9.4. Here the resistors R_1, R_2, and R_E are effective in establishing $-V_{BB}$, the base voltage bias point. For small-signal silicon transistors the emitter-base junction is usually biased by more than half a volt in order that this diode can conduct a current of the order of a milli-ampere. (In germanium transistors only a quarter of a volt is needed, because of the smaller emitter junction barrier height due to the narrower energy gap of germanium.)

---

[4] This being a text on materials and devices, circuits are merely discussed to provide insight into some final uses of the transistor. For a more complete analysis of the circuits described here, see J. Millman and C. C. Halkias, *Integrated Electronics*, McGraw-Hill Book Company, New York (1972).

Photomicrograph of a linear
integrated circuit operational
amplifier, the CA 3130.
(Courtesy of RCA Solid
State Division, Somerville,
N.J.)

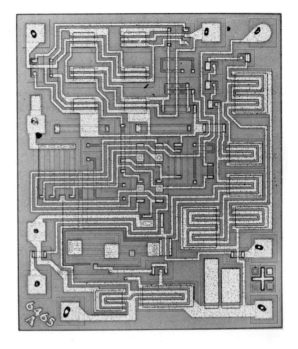

The circuit of Fig. 9.4 is designed to provide stability of the dc bias condition with respect to temperature changes. (The sensitivity of the emitter diode I-V characteristic with respect to temperature was discussed in Section 6.3A.) In addition, the bias current of Fig. 9.4 is designed in such a way that the current and voltage operating point is relatively insensitive to transistor design-parameter variations. For example, it is convenient that circuit operation is not seriously affected by the value of the current gain of the transistor used, because it is not unusual for a transistor manufacturer to specify the β of a particular transistor type only within a factor of two.

The bias circuit of Fig. 9.4 will next be shown to establish a value of dc collector current which is relatively insensitive to temperature changes as well as variability in the current gain β of the transistor used. The bias technique utilized here is very useful in designing discrete transistor amplifier circuits. Another technique is used for the purpose of biasing transistors in integrated circuit chips. Integrated circuit technology lends itself to the use of differential-amplifier type circuits which are automatically stable with respect to temperature and device-parameter variations. This will be discussed later.

Let us now consider in moderate detail the operation of the amplifier circuit of Fig. 9.4. The capacitors C_{in} and C_{out} act to isolate the dc biasing circuit from the input and output circuits respectively. It is assumed, of course, that they provide low-impedance paths for the high-frequency signal current.

Similarly, the capacitor C_E serves as a low-impedance path to ground for the emitter signal current, bypassing the resistor R_E which serves to establish the dc bias point. An expression for the dc collector current of the transistor in the quiescent condition is

$$|I_C| = \frac{(|V_{BB}| - |V_{BE}|)\beta}{R_E(\beta + 1) + R'} ,$$ (9.4)

where $|V_{BB}| = V_{CC} R_2/(R_1 + R_2)$, and $R' = R_1 R_2/(R_1 + R_2)$. This expression is derived by the standard application of Kirchhoff's laws of circuit analysis. Consider a circuit design in which $R_E(\beta + 1) \gg R'$ and the transistor $\beta \gg 1$. Then Eq. (9.4) reduces to

$$|I_C| = \frac{|V_{BB}| - |V_{BE}|}{R_E} .$$ (9.5)

Note that the collector quiescent current is now independent of the transistor β, which can vary from device to device and change with temperature. To ensure that the bias point is insensitive to the strong variation with temperature of the base-emitter voltage V_{BE} at a fixed current requires that $|V_{BB}| \gg |V_{BE}|$.

FIGURE 9.5

a) Schematic diagram of a differential-amplifier circuit utilizing a constant current source I_2.
b) Typical current source used in integrated circuits for supplying a differential amplifier.

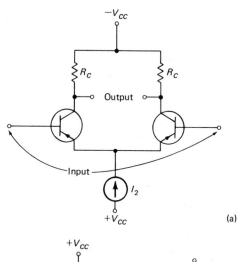

(a)

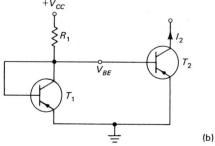

(b)

This explains the usefulness of the emitter resistor R_E, which unfortunately dissipates power and reduces the amplifier gain.

The type of amplifier bias circuit used in integrated circuits is somewhat different from that presented here for a discrete-device amplifier. The standard form of the amplifier in the integrated version is the *differential amplifier*. A schematic form of this circuit is shown in Fig. 9.5a. The evident mirror symmetry of the circuit makes it ideal for use in integrated circuits for reasons to be discussed in Chapter 11. The dc bias in this amplifier is established by the constant current source I_2 in Fig. 9.5a. One form of this current source is shown in Fig. 9.5b. Just why this type of bias circuit is insensitive to variations in the transistor β and in temperature is discussed next.

Because the same base-emitter voltage V_{BE} is applied to transistors T_1 and T_2, both must have the same base current and hence also the same collector current, if the devices are well matched. Since all devices are produced simultaneously in integrated circuit technology, the nearly identical nature of these transistors ensures that their emitter junctions and their current gains are nearly the same. Hence

$$I_2 = \frac{\beta}{\beta + 2} \frac{V_{CC} - |V_{BE}|}{R_1} \approx \frac{V_{CC} - |V_{BE}|}{R_1}, \tag{9.6}$$

when the transistor current gains $\beta \gg 1$. Here again as in Eq. (9.5) the dc bias current is nearly independent of the transistor current gain; the latter can vary from circuit to circuit and with temperature. The bias current I_2 can be temperature sensitive, however, owing to the variation of V_{BE} and R_1 with temperature. Although the temperature variation of discrete resistor values is usually negligible, integrated circuit resistors, composed of silicon, change in value by several thousand parts per million per degree centigrade. Fortunately, near room temperature the value of R_1 will increase with temperature, while V_{BE} decreases with temperature. Hence the increase in the numerator of Eq. (9.6) is somewhat compensated for by the increase in the denominator with increasing temperature, so that the bias current I_2 is maintained essentially constant.

Emitter follower circuit A low output resistance for an amplifier is often required to match a low-resistance load. The high collector impedance of the common-emitter connected transistor provides a high-resistance output. The *emitter follower* circuit of Fig. 9.6, however, presents a low output resistance to a load connected across R_E. Its name derives from the fact that the emitter (output) potential follows by one diode drop the base (input) potential. Hence a voltage signal applied to the base input appears essentially as an equal change in potential across the load driven by the emitter current, so that there is no voltage gain. However, the circuit does give current gain. This circuit not only provides a low-resistance output, but it also has a high input resistance and hence acts as a resistance transformer with power gain in addition.

FIGURE 9.6

Typical emitter-follower transistor amplifier.

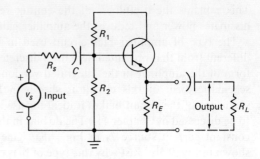

Small-signal circuit analysis gives for the output resistance

$$r_{\text{out}} \approx \frac{R'}{h_{fe} + 1} + r_e, \tag{9.7}$$

where R' represents the parallel combination of R_1, R_2, and R_S; and r_e is the dynamic emitter resistance at the dc bias operating point. The small-signal input resistance when a load resistor R_L is connected across the output terminals is given by

$$r_{\text{in}} \approx (h_{fe} + 1) \left(\frac{R_E R_L}{R_E + R_L} + r_e \right). \tag{9.8}$$

The ratio of input resistance to output resistance is then

$$\frac{r_{\text{in}}}{r_{\text{out}}} \approx \frac{(h_{fe} + 1)\,[R_E R_L/(R_E + R_L) + r_e]}{R'/(h_{fe} + 1) + r_e}. \tag{9.9}$$

If $R_E \sim R_L \gg r_e$ and $R'/(h_{fe} + 1) \sim r_e$, this ratio reduces to

$$\frac{r_{\text{in}}}{r_{\text{out}}} \approx \frac{(h_{fe} + 1)R_E}{4 r_e}, \tag{9.10}$$

which can be made appropriately large since in practical transistors $h_{fe} \gg 1$.

EXAMPLE 9.1

Consider the emitter follower circuit of Fig. 9.6. Take $R_1 = R_2 = 100$ kΩ, $R_S = 2.0$ kΩ, $R_E = 5.0$ kΩ, $h_{fe} = 80$, $r_e = 26$ Ω, and $R_L = 5.0$ kΩ. Find, approximately, (a) the output resistance, (b) the input resistance, and (c) the ratio of the input resistance to the output resistance.

SOLUTION   a) Using Eq. (9.7), we have

$$r_{\text{out}} \approx \frac{R'}{h_{fe} + 1} + r_e \approx \frac{2.0 \times 10^3\,\Omega}{81} + 26\,\Omega$$

$$= 51\ \Omega.$$

b) From Eq. (9.8),

$$r_{in} \approx (h_{fe} + 1) \left(\frac{R_E R_L}{R_E + R_L} + r_e \right)$$

$$\approx 81 \left[\frac{5.0 \times 10^3 \, \Omega \, (5.0 \times 10^3 \, \Omega)}{5.0 \times 10^3 \, \Omega + 5.0 \times 10^3 \, \Omega} + 26 \, \Omega \right]$$

$$= 210{,}000 \, \Omega.$$

c) From the results of parts a and b,

$$\frac{r_{in}}{r_{out}} \approx \frac{210{,}000}{51} = 4100.$$

The flip-flop Section 1.2C contains a brief description of the transistor as a switch. That is, the device can be placed in the conducting state by the application of a current pulse to its base. When the pulse shuts off, the transistor becomes essentially nonconducting. There is a need in computer circuits for a device which provides a *memory function* — one which when pulsed into the ON condition, maintains this state after the triggering pulse is removed. At some subsequent time such a device can be triggered into the OFF state by the application of another pulse. A circuit for performing this information-storage function consists basically of two nearly identical cross-coupled transistors in a configuration known as a *flip-flop*, which is shown schematically in Fig. 9.7a. In this arrangement, when the p-n-p transistor T_1 is ON (conducting), transistor T_0 must be OFF (nonconducting) and this represents one state of the device, say state **1**. The other state, with T_1 OFF and T_0 ON, is defined as state **0** in digital counting terminology. These are the only two stable states of the device.

To understand the operation of this flip-flop consider that it is in the **1** state; that is, p-n-p transistor T_1 is ON and transistor T_0 is OFF. Moreover, let us assume that when T_1 is ON and conducting it is in saturation or bottoming. A self-consistent argument will now be presented to show under what conditions this is indeed the case. Assume T_1 is ON and in saturation. Since the normal saturation voltage of a silicon transistor is only a few tenths of a volt, V_{C1}, the collector potential of T_1, is a few tenths of a volt below ground potential, causing V_{B0}, the base potential of T_0, to be somewhat less negative. This is insufficiently negative for a silicon transistor to maintain T_0 in the ON condition and hence T_0 is cutoff. With T_0 OFF, its collector potential V_{C0} must be essentially $-V_{CC}$, since there is hardly any current through, or potential drop across, resistor R_{C0}. This makes the base potential V_{B1} quite negative with respect to ground. It is now necessary to prescribe the conditions under which V_{B1} is sufficiently negative in potential to maintain transistor T_1 stably in the saturated ON condition, as originally assumed.

FIGURE 9.7

a) Elementary form of a
transistor flip-flop circuit.
b) Elementary version of a
transistor flip-flop showing
an output V_{C1} for sensing
the state of the device and
two inputs via transistors T_3
and T_4 for changing the state
of the flip-flop.

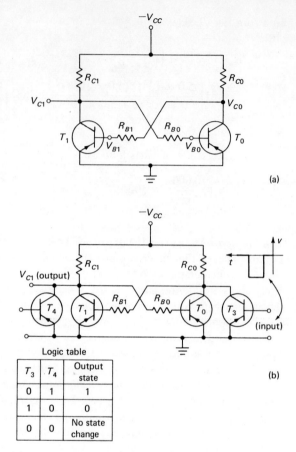

(a)

(b)

Logic table

T_3	T_4	Output state
0	1	1
1	0	0
0	0	No state change

If T_0 is OFF, the base current of transistor T_1 is given by

$$|I_{B1}| = \frac{V_{CC} - |V_{BE1}|}{R_{C0} + R_{B1}}. \tag{9.11}$$

The collector current of T_1 is correspondingly

$$|I_{C1}| = \frac{V_{CC} - |V_{C1}|}{R_{C1}}. \tag{9.12}$$

If β is the transistor grounded-emitter current gain, then the existence of T_1 in the saturated condition requires that $I_{C1}/I_{B1} < \beta$. From the assumed symmetry of the circuit, $R_{C0} = R_{C1}$, and from the fact that generally $|V_{B1}| \ll V_{CC}$, the saturation of transistor T_1 requires, in accordance with Eqs. (9.11) and (9.12), that

$$\left(1 - \frac{|V_{C1}|}{V_{CC}}\right)\left(1 + \frac{R_{B1}}{R_{C1}}\right) < \beta. \tag{9.13}$$

Since $1 - |V_{C1}|/V_{CC}$ is always less than one, this indicates that transistor saturation may be ensured by making $R_{B1}/R_{C1} + 1 < \beta$.

When this is done, the ON or **1** state is indicated when V_{C1} is near ground potential. When V_{C1} is nearly $-V_{CC}$, the OFF or **0** state is indicated. Hence the state of the device may be determined by sensing the output voltage V_{C1}. If the device is in stable equilibrium in the **1** state it may be changed to the **0** state by pulse triggering of transistor T_3 in Fig. 9.7b. If a negative-going pulse is applied to the base of T_3 it will draw collector current through R_{C0}, bringing the potential V_{C0} toward ground potential. This will cause potential V_{B1} to become less negative, tending to reduce the collector current of the previously conducting transistor T_1. The reduction of current through this transistor reduces the potential drop across R_{C1}, causing the potential V_{C1} to become more negative and to approach $-V_{CC}$. This negative potential will supply base current for transistor T_0, turning it ON. Hence the device has been caused to shift from state **1** to **0** by the application of a pulse to input transistor T_3. This pulse must be long enough for charge readjustment to take place enabling transistor T_1 to shut OFF and T_0 to turn ON. All the stored charge must be removed from the transistor T_1 so that it can turn OFF.[5] The time taken for this charge removal will limit the speed with which this device can change its state. A completely symmetrical argument can be presented to show that the original state **1** of the device may now be restored by pulsing input transistor T_4.

Hence a device has been described which can remain stably in either of two possible states. The state can be ascertained by noting whether output V_{C1} is nearer ground potential or $-V_{CC}$. The state of the device may be shifted by supplying a negative current pulse to input transistor T_3 or T_4. A shift back to the original state may be induced by supplying a negative pulse to the other input transistor. The switching rate is limited by the time necessary to remove charge stored in the saturated transistor. A logic table for this flip-flop is given in Fig. 9.7b.

9.2 SMALL-SIGNAL LINEAR TRANSISTOR MODELS

Useful ac circuit models for bipolar transistors will now be described. In Chapter 8 the Ebers-Moll and charge-control transistor models were discussed. The former is a rather general large-signal model intended primarily for dc and low-frequency analysis. The charge-control model is mainly useful for transistor switching applications. Consider now the transistor operating strictly in the active region with a small ac signal applied, in a typical linear amplification application. The device will be treated from a circuit viewpoint with the model parameters calculable from measured transistor characteristics. Initially a low-frequency model will be developed but afterward the model will be extended to high-frequency small-signal operation.

---

[5] Transistor turn-off and turn-on will be discussed in some detail at the end of this chapter.

FIGURE 9.8

Two-port transistor model utilizing a black-box transistor representation.

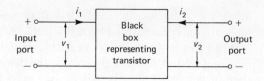

A. Two-Port Hybrid Transistor Model

The transistor may be considered as a black box with one input and one output port. The terminal characteristics are completely determined by the specification of two currents and two voltages. Two of the quantities may be specified as independent variables; the other two are then determined from these and certain device parameters. Such a transistor model is given in Fig. 9.8. It shows the small-signal input current and voltage, i_1 and v_1 respectively, and the output current and voltage, i_2 and v_2, as well as some arbitrarily chosen voltage polarities and current directions.[6] If input current i_1 and output voltage v_2 are chosen as the independent variables and linear, small-signal conditions are assumed, the remaining voltage and current can be written as

$$v_1 = h_{11}i_1 + h_{12}v_2, \qquad\qquad (9.14a)$$

$$i_2 = h_{21}i_1 + h_{22}v_2. \qquad\qquad (9.14b)$$

Here the quantities h_{11}, h_{12}, h_{21}, and h_{22} are parameters which describe the transistor characteristics at one particular dc operating point.

The letter h is chosen to describe these parameters since they are *hybrid* in the sense that they are not all dimensionally the same; two have the dimensions of ohms and the other two are dimensionally reciprocal ohms or mhos. They can be determined for a particular transistor using a small sinusoidal input signal, in a manner somewhat like that outlined in Section 8.2B for the determination of the Ebers-Moll device parameters. For example, Eqs. (9.14) give $h_{11} = v_1/i_1]_{v_2=0}$, which represents the transistor input resistance with the output short-circuited; $h_{12} = v_1/v_2]_{i_1=0}$ is the reverse voltage gain with the input open-circuited; $h_{21} = i_2/i_1]_{v_2=0}$ is the negative of the forward current gain with the output short-circuited; and $h_{22} = i_2/v_2]_{i_1=0}$ is the output conductance with open-circuited input. At low frequency the h parameters are real numbers and the currents and voltages are in phase. However, in general they will be complex numbers and frequency dependent so that the currents and voltages will bear certain phase relationships to each other with respect to time. These device parameters are considered constant over the small signal excursion about a dc operating point and are the ones normally specified on manufacturer specification sheets (see Appendixes D.1 and D.2).

A circuit model may now be constructed for the transistor for use in analyzing the performance of transistor amplifier circuits. This hybrid circuit model is

---

[6] Note that these small signals are superimposed on the dc current and voltage which bias the device into the active region.

FIGURE 9.9

Hybrid circuit model for a
transistor.

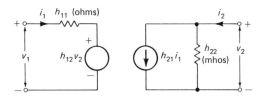

FIGURE 9.10

Common-emitter transistor
circuit utilizing one battery.
The input signal is v_{BE} and
the output is taken across
load resistor R_L.

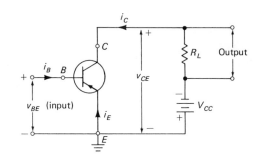

shown in Fig. 9.9. Note that it contains a voltage source $h_{12} v_2$ (open circle) as
well as a current source $h_{21} i_1$ (circle with arrow).[7] That this model can be
derived from Eqs. (9.14) is the subject of a problem at the end of this chapter.

B. The Transistor Hybrid Model for the Common-Emitter Connection

Consider the common-emitter transistor circuit of Fig. 9.10 containing a
p-n-p transistor. It is desired to derive the hybrid-model parameters for the
device in this circuit configuration. If i_B and v_{CE} are chosen as the independent
variables, it follows that v_{BE} and i_C are functions of these variables and can be
written, for the transistor operating in the active region, as

$$v_{BE} = f(i_B, v_{CE}), \tag{9.15a}$$

$$i_C = g(i_B, v_{CE}). \tag{9.15b}$$

Small-signal considerations permit a Taylor expansion of these equations about
an operating point I_C and V_{CE}. This gives, if we consider only first-order terms,

$$\Delta v_{BE} = \frac{\partial f}{\partial i_B}\bigg]_{V_{CE}} \Delta i_B + \frac{\partial f}{\partial v_{CE}}\bigg]_{I_B} \Delta v_{CE} = \frac{\partial v_{BE}}{\partial i_B}\bigg]_{V_{CE}} \Delta i_B + \frac{\partial v_{BE}}{\partial v_{CE}}\bigg]_{I_B} \Delta v_{CE}, \tag{9.16a}$$

$$\Delta i_C = \frac{\partial g}{\partial i_B}\bigg]_{V_{CE}} \Delta i_B + \frac{\partial g}{\partial v_{CE}}\bigg]_{I_B} \Delta v_{CE} = \frac{\partial i_C}{\partial i_B}\bigg]_{V_{CE}} \Delta i_B + \frac{\partial i_C}{\partial v_{CE}}\bigg]_{I_B} \Delta v_{CE}. \tag{9.16b}$$

---

[7] See the Ebers-Moll equivalent circuit (Section 8.2C) for an example of a current source.

Since the symbols v_{be}, v_{ce}, i_b, and i_c represent actual incremental or small-signal quantities, Eqs. (9.16) may be written as

$$v_{be} = h_{ie} i_b + h_{re} v_{ce} \qquad\qquad (9.17a)$$

and

$$i_c = h_{fe} i_b + h_{oe} v_{ce}, \qquad\qquad (9.17b)$$

where

$$h_{ie} \equiv \frac{\partial f}{\partial i_B} = \frac{\partial v_{BE}}{\partial i_B}\bigg]_{V_{CE}},$$

$$h_{re} \equiv \frac{\partial f}{\partial v_{CE}} = \frac{\partial v_{BE}}{\partial v_{CE}}\bigg]_{I_B},$$

$$h_{fe} \equiv \frac{\partial g}{\partial i_B} = \frac{\partial i_C}{\partial i_B}\bigg]_{V_{CE}},$$

FIGURE 9.11

Typical curves of transistor h parameters as a function of bias point and temperature.

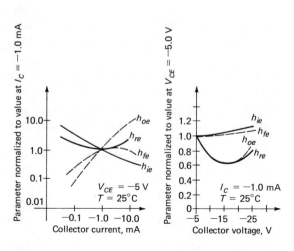

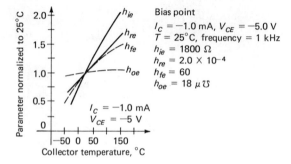

and

$$h_{oe} \equiv \frac{\partial g}{\partial v_{CE}} = \frac{\partial i_C}{\partial v_{CE}}\bigg]_{I_B}.$$

Equations (9.17) are of the same form as Eqs. (9.14) and hence they confirm that the equivalent circuit of Fig. 9.9 may be used to represent the common-emitter-connected transistor. Here h_{ie} is the commonly accepted symbol for the transistor input resistance, h_{re} represents the feedback voltage ratio, h_{fe} the small-signal current gain, and h_{oe} the output conductance for the common-emitter transistor connection. The definitions of the h parameters in Eqs. (9.17) suggest a method for finding their numerical values. A method for determining these numbers by measurements on a given transistor is the subject of an example which follows. Typical values for these h parameters and how they vary with bias point and temperature are indicated in Fig. 9.11 for a small-signal p-n-p transistor. (See also Appendixes D.1 and D.2 for typical manufacturer specification sheets.)

EXAMPLE 9.2

a) Estimate the h_{fe} of the transistor represented by the characteristic curves of Fig. 9.1b at the operating point 0 indicated, assuming $I_{B1} = 0.10$ mA, $I_{B2} = 0.20$ mA, $I_{B3} = 0.30$ mA, $V_{CC} = 16$ V, and $R_L = 800\ \Omega$. Find also $h_{FE} = \beta = (I_C/I_B)]_{V_{CE}}$.
b) Using Fig. 9.1b, estimate h_{oe} at the operating point 0.
c) Determine the power dissipated in the device operating in the quiescent condition at operating point 0.

SOLUTION

a) $\dfrac{V_{CC}}{R_L} = \dfrac{16\text{ V}}{800\ \Omega} = 20 \times 10^{-3}$ A,

$$h_{fe} = \frac{\partial i_c}{\partial i_B}\bigg]_{V_{CE}} = \frac{(11.0 - 6.0) \times 10^{-3}\text{ A}}{(0.25 - 0.15) \times 10^{-3}\text{ A}} = 50,$$

$$\beta = h_{FE} = \frac{I_C}{I_B}\bigg]_{V_{CE}} = \frac{8.7 \times 10^{-3}\text{ A}}{0.2 \times 10^{-3}\text{ A}} = 43.$$

b) $h_{oe} = \dfrac{\partial i_c}{\partial v_{CE}}\bigg]_{I_B} = \dfrac{(8.8 - 8.6) \times 10^{-3}\text{ A}}{(11.2 - 7.2)\text{ V}}$

$\qquad = 5.0 \times 10^{-5}\ \mho \quad$ or $\quad 50\ \mu\mho.$

c) $P_{\text{diss}}]_0 = I_C V_{CE}]_0 = 8.7 \times 10^{-3}\text{ A }(9.2\text{ V})$

$\qquad = 80 \times 10^{-3}\text{ W} \quad$ or $\quad 80\text{ mW.}$

9.3 A HIGH-FREQUENCY BIPOLAR TRANSISTOR MODEL

The two-part low-frequency small-signal transistor model of Fig. 9.9 contains current and voltage sources as well as resistive elements. In this type of circuit the current and voltages are always in phase; there are no time delays involved. In an actual transistor the time required for the charge carriers injected by the emitter (input) junction to reach the collector (output) junction is often quite short owing to the narrow base width; in fact the criterion for low-frequency operation is that this time constant is much smaller than the period of the input signal. However, when the period of the input signal becomes comparable to or less than the transport time for the injected carriers, provision must be made for the representation of this time delay in any valid high-frequency transistor model. In addition, a finite time is required to charge the junction capacitance of the emitter junction before carrier injection can begin, and this time delay will also enter into the high-frequency behavior of a transistor. It is the purpose of the next section to indicate how a high-frequency bipolar transistor model can be derived which will accurately represent the behavior of the transistor as a function of frequency. One such simple equivalent circuit which has proved successful in representing the circuit performance of a common-emitter-connected transistor over a broad range of frequencies is the *hybrid-π* model. Here reactive circuit elements are introduced to provide proper phase relationships between the currents and voltages. A separate set of circuit parameter values must be known for each dc operating point in this ac small-signal model of the transistor operating in the active region.

A. The Hybrid-π Common-Emitter Transistor Model

The hybrid-π small-signal common-emitter transistor model is shown in Fig. 9.12. A convenience of this model is that all passive circuit parameters are taken as independent of frequency. Also the resistive components in the circuit can be derived from the low-frequency h parameters already discussed. Additional parasitic resistances and capacitances can easily be added to this model to represent transistor circuit performance even more accurately over a broad range of frequencies.

In this model, $r_{bb'}$ represents the small-signal ohmic, transverse (parallel to the emitter junction) base resistance of the device in series with the base lead (see Fig. 9.12a). The resistor r_π represents the dynamic emitter resistance reflected[8] into the base input circuit configuration. That is,

$$r_\pi = (h_{fe} + 1)r_e, \qquad (9.18)$$

where r_e is the dynamic emitter resistance at the set operating point, and h_{fe} is the normal common-emitter ac current gain. The capacitance c_π takes into

---

[8] This reflection may be understood by remembering that the base and emitter current are related by $i_b = -i_e/(h_{fe} + 1)$.

FIGURE 9.12

a) Hybrid-π small-signal common-emitter transistor model useful in the megahertz frequency range. The various circuit parameters have a particular set of values for one value of bias point I_C, V_{CE}, and I_B.
b) Sketch of a typical planar transistor cross section showing the origin of the base resistance $r_{bb'}$, the intrinsic collector capacitance C_{ci}, and the extrinsic collector capacitance C_{ce}.

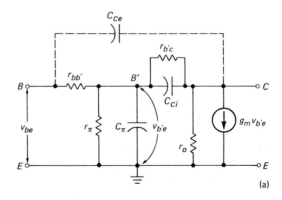

(a)

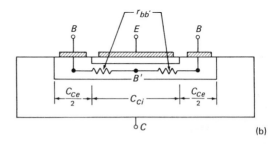

(b)

account the incremental change of minority carrier charge stored in the transistor base region owing to the ac input signal. This is essentially of the nature of the diffusion capacitance already introduced in Section 7.2B in connection with the high-frequency behavior of the junction diode.[9] In parallel with the diffusion capacitance is the emitter junction depletion-layer capacitance, but this is often negligibly small compared to the diffusion capacitance and can be neglected when the emitter is forward biased (see Section 7.2B).

For small voltage changes $v_{b'e}$ across the emitter junction, excess minority carriers are injected into the base region toward the collector junction where they give rise to collector signal current. This component of the collector current is denoted by the current generator labeled $g_m v_{b'e}$, where g_m is the small-signal transistor *transconductance*, defined by

$$g_m = \frac{\partial i_C}{\partial v_{b'e}}\Bigg]_{V_{CE}}. \tag{9.19}$$

The intrinsic part of the collector junction transition or depletion-layer capacitance under the emitter junction is denoted by C_{ci} (see Fig.9.12b), and this is shunted by the collector junction dynamic resistance, which derives from

---

[9] An expression for c_π in terms of transistor structure parameters is given in Section 9.4C.

the variation in transistor base width (and hence current gain) with base-to-collector voltage variation. This has already been discussed in Section 8.4B, and an expression for this resistance can be derived from Eq. (8.48). The change in base width reflects a change in the slope of the minority carrier distribution in the base, which determines the emitter diffusion current. This feedback effect is taken into account by a resistor $r_{b'c}$ connected between the internal base point B' and the collector terminal. The collector-to-emitter resistance is denoted r_o and this together with $r_{b'c}$ reflects the slope of the output i_C-v_{CE} characteristic. The collector junction capacitance not under the emitter junction is called *extrinsic* or *overlap* and is represented by C_{Ce} in Fig. 9.12.

It can be shown that the high-frequency hybrid-π circuit parameters can be expressed approximately in terms of the low-frequency hybrid parameters[10] (see Section 9.2A) at the transistor operating point I_C, V_{CE} as

$$g_m = \frac{q|I_C|}{kT}, \tag{9.19a}$$

$$r_\pi = \frac{h_{fe}}{g_m}, \tag{9.19b}$$

$$r_{bb'} = h_{ie} - r_\pi, \tag{9.19c}$$

$$r_{b'c} = \frac{r_\pi}{h_{re}}, \tag{9.19d}$$

and

$$\frac{1}{r_o} = h_{oe} - \frac{1 + h_{fe}}{r_{b'c}} \approx h_{oe} - g_m h_{re}. \tag{9.19e}$$

This model in practice is found to be useful in the hundreds of megahertz frequency range.

EXAMPLE 9.3

Given a p-n-p transistor operating at $I_C = -1.0$ mA and $V_{CE} = -5.0$ V at 25°C, with low-frequency design parameters as indicated in Fig. 9.11. Determine the hybrid-π circuit parameters: (a) g_m, (b) r_π, (c) $r_{bb'}$, (d) $r_{b'c}$, and (e) r_o.

SOLUTION   Using Eqs. (9.19a) to (9.19e), we have
a)

$$g_m = \frac{|I_C|}{kT/q} = \frac{1.0 \times 10^{-3}\ \text{A}}{0.026\ \text{V}} = 0.038\ \mho,$$

b)

$$r_\pi = \frac{h_{fe}}{g_m} = \frac{60}{0.038\ \mho} = 1600\ \Omega,$$

[10] See J. Millman and C. C. Halkias, *Integrated Circuits*, McGraw-Hill Book Company, New York (1972).

c)

$$r_{bb'} = h_{ie} - r_\pi = 1800 - 1600 = 200 \ \Omega,$$

d)

$$r_{b'c} = \frac{r_\pi}{h_{re}} = \frac{1600 \ \Omega}{2.0 \times 10^{-4}} = 8.0 \times 10^6 \ \Omega \quad \text{or} \quad 8.0 \ \text{M}\Omega,$$

e)

$$r_o = \frac{1}{h_{oe} - (1 + h_{fe})/r_{b'c}} = \frac{1}{18 \times 10^{-6} \mho - 61/8.0 \times 10^6 \ \Omega} = 96{,}000 \ \Omega.$$

9.4 COMMON-EMITTER SHORT-CIRCUIT TRANSISTOR CURRENT GAIN VERSUS FREQUENCY

It is of interest to determine the *intrinsic* upper frequency limitation of a transistor operating in the common-emitter connection. This can be accomplished with the aid of the hybrid-π high-frequency transistor model. The analysis will lead to the definition of the device parameters f_β and f_T which are often given on transistor specification sheets supplied by the manufacturer. These parameters represent respectively the bandwidth of the device and the frequency at which the short-circuit common-emitter gain drops to unity. Of course, a practical amplifier circuit containing this transistor will never achieve these frequency capabilities owing to limitations provided by circuit capacitances and inductances.

A. The Transistor Short-Circuit Bandwidth f_β

Consider the hybrid-π transistor equivalent circuit of Fig. 9.12 with the output short-circuited. This revised circuit is shown in appropriate form in Fig. 9.13. In this version $r_{b'c}$, which should appear between terminals B' and C, is neglected since generally $r_{b'c} \gg r_\pi$. Also r_o is eliminated since it is short-circuited. In addition, the output current supplied by C_{Ci}, C_{Ce}, and $r_{b'c}$ is assumed negligibly small compared with that supplied by the current generator $g_m v_{b'e}$. This output current, i_{out}, is then simply $-g_m v_{b'e}$, where

$$v_{b'e} = \frac{i_{in}}{1/r_\pi + j\omega(C_\pi + C_c)}. \tag{9.20}$$

Here $\omega \equiv 2\pi f$ is the angular frequency of the input signal, and $C_c = C_{Ci} + C_{Ce}$, the total collector junction capacitance. The short-circuit current gain A_i then can be written as

$$A_i \equiv \frac{i_{out}}{i_{in}} = -\frac{g_m}{1/r_\pi + j\omega(C_\pi + C_c)} \tag{9.21a}$$

FIGURE 9.13

Hybrid-π transistor equivalent circuit with the output short-circuited.

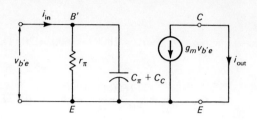

or

$$A_i = -\frac{h_{fe}}{1 + j\omega(C_\pi + C_C)h_{fe}/g_m},$$
(9.21b)

using the result of Eq. (9.19b).

This expression indicates correctly that the current gain numerically approaches h_{fe} as the signal frequency becomes vanishingly small. On the other hand, the current gain falls to $1/\sqrt{2}$ of its low-frequency value (3 dB point) at a frequency f_β given by

$$f_\beta = \frac{1}{h_{fe}}\frac{g_m}{2\pi(C_\pi + C_C)}.$$
(9.22a)

The range of frequencies 0 to f_β is sometimes defined as the transistor *bandwidth*, which expresses the intrinsic ability of the transistor to maintain its current gain over this frequency range. With Eq. (9.19b), Eq. (9.22a) can be rewritten as

$$t_{\text{ch}} = \frac{1}{2\pi f_\beta} = r_\pi(C_\pi + C_C).$$
(9.22b)

This can be simply interpreted as the time necessary to charge the diffusion and collector capacitances through the input resistor r_π, which accounts for the frequency limitation in this simplified model.

B. The Short-Circuit Common-Emitter Gain-Bandwidth Product

It is often convenient to specify the maximum useful high-frequency capability of a transistor as the frequency at which the absolute value of the short-circuit common-emitter current gain reduces to one. This frequency is usually labeled f_T on transistor specification sheets (see Appendixes D.1 and D.2) supplied by manufacturers and is that operating frequency beyond which the transistor has little value as an amplifier. Equation (9.21b) can be used in order to derive an expression for f_T which, assuming $|h_{fe}| \gg 1$, is

$$f_T = \frac{g_m}{2\pi(C_\pi + C_C)}.$$
(9.23)

Comparision of Eq. (9.22a) with (9.23) gives

$$f_T = h_{fe} f_\beta .$$ (9.24)

This latter equation permits the interpretation of f_T as the *short-circuit current gain-bandwidth product*. Figure 9.14 is basically a log-log plot of Eq. (9.21), where the ordinate is expressed in decibels (dB), as is normally the practice. For values of frequency somewhat in excess of f_β this graph exhibits a linear falloff in dB of the current gain, with a negative slope of 6 dB/octave. (An octave drop in frequency corresponds to a falloff in frequency by a factor of 2.)

The frequency cutoff f_T in some transistors extends into the hundreds of megahertz range. Special equipment is required to measure the transistor gain at these high frequencies, and the measurement is difficult to perform. Hence common practice is to measure the transistor current gain at some moderate frequency f_M, somewhat higher than f_β but much lower than f_T. Since f_M is along the straight-line falloff of 6 dB/octave, f_T can be calculated by making use of this linearity as

$$f_T = f_M (A_i)_M ,$$ (9.25)

where $(A_i)_M$ is the current gain measured at f_M. This expression may be derived by first combining Eqs. (9.21b) and (9.22a) as $A_i = -h_{fe}/[1 + j(f/f_\beta)]$, and is offered as an exercise at the end of this chapter. Now it remains only to measure the transistor short-circuit current gain at some moderate frequency f_M. The experimentally determined value of the diffusion capacitance C_π can be calculated from Eq. (9.23) once f_T, C_C, and g_m are known.

C. The Transistor Diffusion Capacitance in Terms of Device Design Parameters

In Section 9.3A it was stated that the diffusion capacitance reflects the incremental change of minority carrier charge stored in the transistor base region due to the ac input signal v_{be}. Now the total charge stored in the base region when

FIGURE 9.14

Log-log plot of transistor gain in dB versus operating frequency.

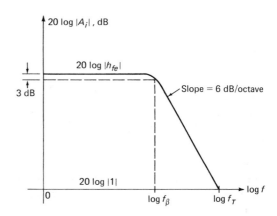

the transistor is normally biased in the forward direction, q_F, has been shown in Eq. (8.41) to be

$$q_F = \frac{|I_C| W^2}{2D_p}, \qquad (9.26)$$

where W is the transistor base width, and D_p is the minority hole diffusion constant in the p-n-p transistor considered. The rate of change of this charge with respect to a voltage signal v_{BE} applied to the emitter junction is defined as C_{DE}, the transistor diffusion capacitance, which by differentiating Eq. (9.26) can be written as

$$C_{DE} \equiv \frac{\partial q_F}{\partial v_{BE}} = \frac{W^2}{2D_p} \frac{d|I_C|}{dv_{BE}} \approx \frac{W^2}{2D_p} \frac{dI_E}{dv_{BE}}, \qquad (9.27)$$

assuming negligible recombination of these carriers in transport from the emitter to the collector junction. The differential in the last term of Eq. (9.27) can be evaluated starting with Eq. (8.36), the emitter junction i-v relationship, approximately as

$$\frac{dI_E}{dv_{BE}} = \frac{qI_E}{kT}. \qquad (9.28)$$

This differential can be recognized as the reciprocal of the dynamic emitter resistance r_e. Introducing Eq. (9.28) into (9.27) gives, for the diffusion capacitance,

$$C_{DE} = \frac{qW^2 I_E}{2D_p kT}. \qquad (9.29)$$

At normal emitter currents for the transistor operating in the active region, this diffusion capacitance is much greater in value than the emitter transition-layer capacitance (but in parallel with it) and hence $C_{DE} \approx C_\pi$ of Section 9.3A. (This is the subject of a problem at the end of this chapter.) Also $C_\pi \gg C_C$, the collector transition capacitance, so that Eqs. (9.23) and (9.19a) can be used with Eq. (9.29) to yield the following approximate expression for f_T in terms of transistor design parameters:

$$f_T = \frac{D_p}{\pi W^2}. \qquad (9.30)$$

This confirms that the high-frequency capability of a transistor requires that it be designed with narrow base width.

EXAMPLE 9.4

The p-n-p transistor of Example 9.3 has a bandwidth $f_\beta = 1.0\,\text{MHz}$. Determine approximately the following design parameters for this device: (a) C_π, (b) W, the transistor base width, (c) τ_p, the minority hole lifetime in the base region.

SOLUTION a) Using Eq. (9.22a) and the results of Example 9.3, we have

$$C_\pi \approx \frac{g_m}{2\pi h_{fe} f_\beta} = \frac{0.038 \, \mho}{2\pi(60)(10^6 \text{ Hz})} = 1.0 \times 10^{-10} \text{ F}.$$

b) From Eq. (9.24),

$$f_T = h_{fe} f_\beta = 60(10^6) = 60 \times 10^6 \text{ Hz} \quad \text{or} \quad 60 \text{ MHz},$$

and using Eq. (9.30) gives

$$W^2 = \frac{D_p}{\pi f_T} = \frac{1.25 \times 10^{-3} \text{ m}^2/\text{sec}}{\pi(60 \times 10^6 \text{ Hz})} = 6.9 \times 10^{-12} \text{ m}^2$$

and

$$W = 2.6 \times 10^{-6} \text{ m} \quad \text{or} \quad 2.6 \ \mu\text{m}.$$

c) Using Eq. (8.39) and noting that Eq. (8.39) with (9.2) gives $h_{fe} = \beta_F$, if β_F is independent of current, we get

$$h_{fe} = \beta_F \approx \frac{2D_p \tau_p}{W^2}$$

or

$$\tau_p \approx \frac{h_{fe} W^2}{2D_p} = \frac{60(6.9 \times 10^{-12} \text{ m}^2)}{2(1.25 \times 10^{-3} \text{ m}^2/\text{sec})} = 1.6 \times 10^{-7} \text{ sec} \quad \text{or} \quad 0.16 \ \mu\text{sec}.$$

9.5 THE BIPOLAR TRANSISTOR AS A SWITCH

The use of the bipolar transistor as an electronic switch was briefly discussed in Section 1.2B. The transistor switch is used extensively as an electronic gate to perform logic functions in digital computers, much more so than the diode gate described in Section 7.1D. In Fig. 1.6 a schematic diagram of a typical common-emitter switching circuit was shown. A redrawing of Fig. 1.6 is given in Fig. 9.15. It also shows the *I-V* characteristic of the transistor switch, including a load-line representation of the resistive load R_L, as well as the circuit current response for a voltage input pulse to the base terminal. In this case, however, a p-n-p transistor is considered. Before the input pulse comes on (goes negative) the device is cut off by a positive potential relative to ground; the transistor then conducts only a very small current, and most of the battery voltage V_{CC} appears across the transistor collector junction. This is indicated as $-V_{OFF}$ in Fig. 9.15b. After the input pulse turns on, the voltage and current of the device change along the load line until the steady-state collector current $-I_{C1}$ is reached. The voltage across the device then, marked $-V_{ON}$ in the figure, is often referred to as the *collector saturation voltage* and the device is then said to be *in saturation* (or *bottoming*). In this condition both the emitter and collector junctions are forward biased as illustrated in Fig. 8.15a. After a

FIGURE 9.15

a) Schematic diagram of a typical common-emitter switching circuit utilizing a p-n-p bipolar transistor.
b) Typical current-voltage characteristic of a transistor switch showing a load line. The voltage drop across the device when it is switched ON and OFF is indicated by $-V_{\text{ON}}$ and $-V_{\text{OFF}}$, respectively.
c), d) Base current and collector current versus time. The delay times for switching are indicated.

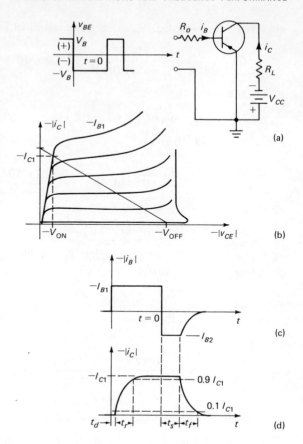

(a)

(b)

(c)

(d)

time the input voltage pulse goes positive, cutting off the transistor and causing the collector current to return eventually to zero. The base current and collector current versus time are praphed in Fig. 9.15c and d.

Hence when the switch is OFF practically all the voltage V_{CC} appears across the transistor and almost none appears across the load R_L. When the transistor is switched ON, a small saturation voltage V_{ON} appears across the transistor and nearly all the voltage V_{CC} appears across the load. Ideally this voltage V_{ON} should be as small as possible in order that the transistor may dissipate as little power and heat up as little as possible while it is conducting the maximum current I_{C1} and delivering power to the load. This saturation voltage is normally a few hundred millivolts.

Also indicated in Fig. 9.15d are the delay and distortion of the output pulse in the collector circuit relative to the signal input pulse. There is an initial time delay t_d after the pulse is applied to the emitter-base junction before 10 percent of the ultimate current I_{C1} begins to flow in the collector circuit. Then there is a rise time t_r before the collector current reaches 90 percent of its ultimate value

I_{C1}. After the base pulse shuts off there is a time delay t_s, normally referred to as the *storage time*, during which the collector current reduces again to 90 percent of I_{C1}. During the fall time t_f, the current reduces to only 10 percent of I_{C1}. The sum of the first two time constants is referred to as the *turn-on time*; the sum of the latter two time constants is called the *turn-off time*. These time delays are indicated in Fig. 9.15d and they limit the speed with which the transistor can operate successfully as an electronic switch.

It is the purpose of the next section to describe the physical basis of these time constants. This will permit the calculation of the maximum repetition rate at which a transistor switch of a given design can operate.

A. The Transistor Turn-on Time

Before the collector current in the transistor of Fig. 9.15a can begin ro rise, the input base current must charge the emitter-base junction capacitance as well as the collector-base capacitance. Hence there is a time delay t_d between the time the base pulse is turned on and the time the transistor enters the active region with the emitter junction injecting and the collector junction collecting minority carriers. Since the emitter junction, initially reverse biased by a small voltage V_B, has to be brought into forward bias to initiate injection, its transition capacitance C_e under these bias conditions is relatively high compared to the reverse-biased collector junction. This is true in spite of the fact that the collector area is generally larger than the emitter junction area, since in planar transistors the doping in the base region near the emitter junction is much greater than that near the collector junction. Therefore the time delay can be approximated by calculating the time necessary to alter the charge on the emitter junction, to bring it to the edge of the active region at 0 V. This delay time,

$$t_d = \frac{C_e \, \Delta V}{\Delta I} \approx \frac{C_e V_B}{I_{B1}},\tag{9.31}$$

represents the approximate time for the collector current to begin injecting or reach 10 percent of its ultimate value $-I_{C1}$ when driven by a base current $-I_{B1}$.

To calculate the rise time t_r for the collector current to reach approximately 90 percent of $-I_{C1}$, the charge-control model discussed in Section 7.3A in connection with diode switching will be utilized. The transistor charge-control model of Section 8.3A will also be useful. It has already been shown in Section 8.1B that to maintain a steady-state hole distribution in the transistor base region, a majority electron base current $-I_B = q_B/\tau_p$ must be supplied to ensure charge neutrality in the face of minority hole recombination. Here q_B is the total charge stored in the base region. Under transient conditions we must have, at any time t,

$$-i_B(t) = \frac{q_B(t)}{\tau_p} + \frac{dq_B(t)}{dt}.\tag{9.32}$$

The last term on the right represents the rate of deviation of base charge from the steady state. During the initial rise of collector current, the forward charge distribution of holes is established in the base region by emitter injection, with the collector junction, still in reverse bias, not injecting carriers into the base region. The total base charge in this case is $q_B(t) = q_F(t)$. During this time interval the base-current pulse value is constant at $-I_{B1}$ and Eq. (9.32) becomes

$$I_{B1} = \frac{q_F(t)}{\tau_p} + \frac{dq_F(t)}{dt}. \tag{9.33}$$

The solution of this differential equation, under the assumption that q_F is zero at $t = 0$, is

$$q_F(t) = I_{B1}\tau_p(1 - e^{-t/\tau_p}), \tag{9.34}$$

for $0 < t < t_r$. The charge-control model of Section 8.3A for the transistor in forward operation at any time t is given in Eq. (8.41) as

$$i_C(t) = \frac{-q_F(t)}{\tau_t}. \tag{9.35}$$

Just at time t_r, when the collector junction is at the edge of saturation (forward bias), the collector current is approximately

$$i_C(t_r) \approx I_{C1} \approx \frac{V_{CC}}{R_L}. \tag{9.36}$$

Introducing Eq. (9.36) into (9.35) and combining this result with Eqs. (9.34) and (8.44) gives finally

$$t_r = -\tau_p \ln \left(1 - \frac{I_{C1}}{\beta_F I_{B1}}\right). \tag{9.37}$$

The time variation of collector current obtained by combining Eqs. (9.35) and (9.34) is shown in Fig. 9.16. The collector current rises exponentially toward the value $\beta_F I_{B1}$ since the base input current is $-I_{B1}$. However, the switching circuit of Fig. 9.15 constrains the collector current to rise no higher than the value $-I_{C1}$ given by Eq. (9.36) at time t_r, at which time the transistor current saturates and so the device is no longer in the active region of operation. Since normally $I_{C1}/I_{B1}\beta_F \ll 1$ the logarithmic term of Eq. (9.37) can be approximated, and a simple expression for the transistor rise time is

$$t_r \approx \tau_p \frac{I_{C1}}{\beta_F I_{B1}}. \tag{9.38}$$

This indicates that the transistor may be caused to rise more quickly to collector current $-I_{C1}$ by driving it "harder" with a higher pulse base current $-I_{B1}$. In the next section it will be shown that this is not necessarily a way to "speed up" the transistor since a long storage time may then result.

FIGURE 9.16

Time variation of the collector current of a transistor switch. $-I_{C_1}$ represents the collector current of a p-n-p transistor driven into saturation by a base current $-I_{B_1}$. The dashed line represents the time variation of the collector current of the unsaturated transistor.

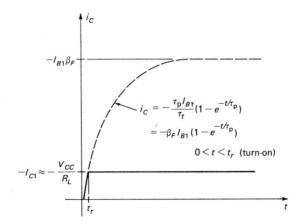

$$i_c = -\frac{\tau_p I_{B\uparrow}}{\tau_t}(1 - e^{-t/\tau_p})$$
$$= -\beta_F I_{B1}(1 - e^{-t/\tau_p})$$

$$0 < t < t_r \quad \text{(turn-on)}$$

B. The Transistor Turn-off Time

In Fig. 9.15 it is seen that the collector current persists at a constant value $-I_{C_1}$ for a time interval t_s, called the *storage time*, after the base drive pulse shuts off and reverses polarity. During this period the collector junction is operating in saturation; that is, it is forward biased and injecting minority holes into the transistor base region. This occurs when the collector potential rises above the base potential. Since both emitter and collector junctions are now forward biased, the device has typically only a few hundred millivolts across it, is in the low-impedance state, and no longer limits the current, so that I_{C_1} is circuit determined as essentially V_{CC}/R_L. This low-impedance state will be maintained after the base drive is removed until most of the excess minority carriers in the base undergo recombination or are extracted by a reverse base current supplied by the positive pulse voltage V_B in Fig. 9.15, and the device reenters the active region.

To calculate the storage time consider again Eq. (9.32), under the condition that the base current is suddenly reduced to zero, which can be written as[11]

$$0 = \frac{q_B}{\tau_p} + \frac{dq_B}{dt}. \tag{9.39}$$

The solution of this equation, under the condition that $t = 0$ and $q_B = I_{B1}\tau_p$ at the beginning of turn-off, is given by

$$q_B(t) = I_{B1}\tau_p e^{-t/\tau_p}. \tag{9.40}$$

---

[11] In this treatment minority carriers are taken to be stored only in the base region, which is the case when the collector region is heavily doped and hence the minority carrier lifetime there is very low. Charge storage in a lightly doped collector region will increase the storage time above that calculated here (see Section 8.3A).

FIGURE 9.17

Exponential decay of the total charge in the base region of a transistor in time after the base drive is turned off at time $t = 0$. t_s represents the transistor storage time.

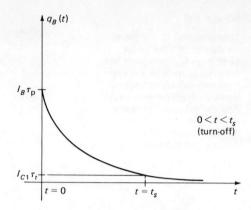

This exponential decay of base charge is graphed in Fig. 9.17. Between $t = 0$ and $t = t_s$ the collector current remains constant at a value $-I_{C1}$ until the base charge $q_B(t)$ reaches the value

$$q_B(t_s) = I_{C1}\tau_t, \qquad (9.41)$$

at which time, $t = t_s$, the transistor enters the active region (collector just reverse biased) and Eq. (9.35) applies again. Evaluating Eq. (9.40) at $t = t_s$ and using Eq. (9.41) gives, for the storage time,

$$t_s \approx \tau_p \ln \frac{\beta_F I_{B1}}{I_{C1}}. \qquad (9.42)$$

This equation indicates that should I_{B1} be increased to reduce the current rise time, the transistor will be driven "deeper" into saturation and the storage time will increase correspondingly. The storage time is directly proportional to the minority carrier lifetime in the base region since, with the base current reduced to zero in this calculation, the transistor will reenter the active region in a time depending on the carrier recombination rate in the base region. If reverse base current I_{B2} is drawn on turn-off, the storage time is reduced below that given by Eq. (9.42) since then carriers can be removed by this current as well.

Let us assume that the minority carrier charge stored in the transistor base region while the device is in saturation ($0 < t < t_s$) is entirely removed by a reverse base current I_{B2}. Then this base current can be written approximately as

$$I_{B2} \approx \frac{\Delta q}{\Delta t} = \frac{q_B(0) - q_B(t_s)}{t_s}. \qquad (9.43)$$

With Eqs. (9.40) and (9.41), this can be expressed as

$$I_{B2} \approx \frac{I_{B1}\tau_p - I_{C1}\tau_t}{t_s}. \qquad (9.44)$$

From Eq. (8.44), $\beta_F = \tau_p/\tau_t$; introducing this into Eq. (9.44) gives for the storage time

$$t_s \approx \tau_p \left(\frac{I_{B1}}{I_{B2}} - \frac{I_{C1}}{\beta_F I_{B2}} \right). \tag{9.45}$$

Hence if the reverse base current $I_{B2} \gg I_{B1}$, then the storage time can be reduced substantially below τ_p, but at the expense of additional energy. In this case fast rise time and short storage time can be achieved simultaneously.

The fall time t_f can be estimated by noting that the collector current decays after the storage time according to

$$i_C(t) = \beta_F I_{B1} e^{-t/\tau_p}, \tag{9.46}$$

which can be derived by combining Eq. (9.40) with (9.41). From this equation it can be shown that the fall time $t_f \approx \tau_p I_{C1}/\beta_F I_{B1}$, which is numerically the same as the value for the rise time given by Eq. (9.38).

EXAMPLE 9.5

A switching transistor of the design described in Example 9.4 delivers a collector current $-I_{C1}$ when driven by a base current $-I_{B1}$. The device is turned OFF by reducing the base current to zero. Assuming $\beta_F > I_{C1}/I_{B1}$, determine the change in rise time and storage time for this transistor if the base drive is increased by a factor of 4; originally $t_r = 53.3$ nsec and $t_s = 176$ nsec.

SOLUTION From Eq. (9.38),

$$\frac{t_{ro}}{t_{rf}} = \frac{\tau_p I_{C1o}/\beta_{Fo} I_{B1o}}{\tau_p I_{C1f}/\beta_{Ff} I_{B1f}} = \frac{I_{B1f}}{I_{B1o}},$$

where the subscripts o and f refer to original and final conditions respectively, assuming that $\beta_{Fo} = \beta_{Ff}$ and $I_{C1o} = I_{C1f}$ (the collector current is determined by the load resistance). Then

$$t_{rf} \approx \frac{I_{B1o}}{I_{B1f}} t_{ro} = \frac{1}{4} (53.3) \text{ nsec} = 13.3 \text{ nsec}$$

and

$$\Delta t_r = t_{rf} - t_{ro} \approx (13.3 - 53.3) \text{ nsec} = -40.0 \text{ nsec}.$$

Now

$$\Delta t_s = t_{sf} - t_{so} = \tau_p \ln \frac{\beta_{Ff} I_{B1f}}{I_{C1f}} - \tau_p \ln \frac{\beta_{Fo} I_{B1o}}{I_{C1o}}$$

$$= \tau_p \ln \frac{I_{B1f}}{I_{B1o}} = 0.16 \times 10^{-6} \text{ sec ln } 4$$

$$= 0.22 \times 10^{-6} \text{ sec} = 220 \text{ nsec}.$$

Hence increasing the base drive reduces the rise time by 40 nsec but increases the storage time by 220 nsec.

PROBLEMS

9.1 Show that the slope of the load line of Fig. 9.1b is $-1/R_L$.

9.2 a) Show that in the circuit of Fig. 9.1a the power supplied by the battery exactly equals the power dissipated in the transistor and the other circuit elements.

b) A transistor operating at a case temperature of 350°K has a junction-to-case thermal resistance θ_{jc} of 1.0°C/W. What is the maximum power which can be dissipated in this transistor if the collector junction temperature may never exceed 200°C?

9.3 Derive the transistor circuit model of Fig. 9.9 starting with Eqs. (9.14).

9.4 a) Draw a hybrid-π low-frequency transistor model starting with Fig. 9.12.

b) Estimate the frequency below which the capacitor C_π across r_π may be neglected in deriving the low-frequency version of the hybrid-π transistor model of part a.

9.5 Derive the hybrid-π circuit resistive components from the low-frequency h parameters.

9.6 Prove the validity of the approximation of Eq. (9.19e). Under what conditions is this approximation good?

9.7 Consider a planar p-n-p transistor as shown in Fig. 1.2 with emitter-region doping 5.0×10^{25} acceptors/m³, base-region doping 1.0×10^{22} donors/m³, and collector doping of 5.0×10^{25} acceptors/m³. The emitter junction area is 1.0×10^{-8} m², and the transistor base width is 2.0×10^{-5} m. For a transistor operating in the active region at 300°K with an emitter current of 10 mA and an emitter-base potential of 0.60 V, find the emitter depletion-layer capacitance and the emitter diffusion capacitance; compare these values.

9.8 Given a p-n-p transistor with an f_T of 50 MHz, low-frequency $h_{fe} = 50$, and collector capacitance of 5.0 pF, operating at a collector current of 2.0 mA. Determine its diffusion capacitance and f_β at 300°K.

9.9 Show, beginning with Eq. (9.21b), that $f_\beta = g_m/2\pi h_{fe}(C_\pi + C_c)$.

9.10 Show that $f_T = f_M(A_i)_M$, as in Eq. (9.25), after proving that $A_i = -h_{fe}/[1 + j(f/f_\beta)]$.

9.11 Show that Eq. (9.28) follows from Eq. (8.36).

9.12 Show that the falloff of gain is 6 dB/octave (or 20 dB/decade) in Fig. 9.14 by determining the slope of the curve in the falloff region.

9.13 Show that $C_\pi r_\pi \approx \tau_p$.

9.14 Consider a p-n-p silicon switching transistor with $C_c = 3.0$ pF, $C_e = 6.0$ pF, $\tau_p = 1.0$ μsec, $I_{C1} = 10$ mA, $I_{B1} = 1$ mA, and $\beta_F = 50$. Calculate approximately:

a) the time delay t_d before the rise of collector current after a negative current pulse $-I_{B1}$ is applied to the transistor base,
b) the rise time t_r,
c) the storage time t_s after the base current pulse is turned off,
d) the storage time t_s if a reverse base current pulse, $I_{B2} = 2.0$ mA, is applied to the transistor base.

9.15 Show that Eq. (9.34) is a solution of Eq. (9.33).

9.16 Show that Eq. (9.37) follows from Eqs. (9.34), (9.35), (9.36), and (8.44).

9.17 Show that Eq. (9.42) follows from Eqs. (9.40) and (9.41).

9.18 Show that Eq. (9.44) follows from Eqs. (9.43), (9.40), and (9.41).

10 The Field-Effect Transistor

The *unipolar* or *field-effect* transistor is another type of three-terminal device, capable of power gain like the bipolar transistor, but which operates on a basically different physical principle. It is the function of this chapter to discuss the principle of operation of this device, used extensively in digital integrated circuits, electronic memories, and amplifier circuits. Comparisons will be drawn with the bipolar transistor.

The term "unipolar" derives from the fact that only one type of current carrier is required to describe this transistor's mode of operation. In contrast, both electron and hole flow must be taken into account to analyze the operation of the bipolar transistor. Hence there are no minority carrier storage effects to slow down the operation of unipolar devices. The term "field-effect" describes the manner in which control of current flow is achieved in this device, namely by adjusting the electric field in the capacitorlike "gate" input electrode. This permits modulation of charge flow in the transistor without drawing any appreciable input current. Hence the field-effect transistor is voltage driven rather than current driven as is the bipolar transistor.

There are basically two versions of the unipolar field-effect transistor. Both *insulated-gate* and *junction* field-effect transistors are used in electronic circuits. However, the insulated-gate device is used much more extensively than the junction device because of its simplicity of fabrication. It is the insulated-gate field-effect transistor (IGFET) which will be described here in some detail. The sandwichlike metal-oxide-semiconductor (MOS) input structure of this transistor is typical of the most popular version of the device. This gives rise to a dc input resistance in excess of $10^{14}\ \Omega$. Schematic drawings of a typical

293

FIGURE 10.1

a) Cross-sectional view of a p-channel metal-oxide-semiconductor enhancement-type insulated-gate field-effect transistor showing the basic device structure.
b) Two versions of a symbol for a p-channel enhancement-type MOSFET with source S, drain D, and gate electrode G. The symbol to the left has an arrow indicating an n-type substrate. The symbol to the right has an arrow indicating the p-type source region. The arrows are often left off since the battery polarity indicates whether p- or n-channel devices are used.

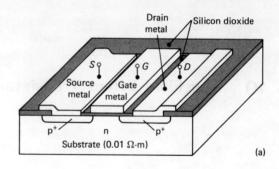

(a)

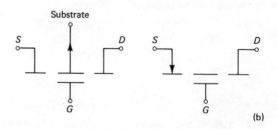

(b)

MOS-IGFET device and a junction field-effect transistor (JFET) are shown in Figs. 10.1 and 10.2 respectively.

The principle of operation of the JFET simply involves the control of the source-to-drain channel current by the narrowing of this current path by p^+-n junction space-charge widening into the n-type channel. The operation of the MOS device is a bit more complex to describe. The electrical characteristics of the JFET shown in Fig. 10.3 are very similar to those of the MOSFET shown in Fig. 10.7a. A cursory look at the device structures indicates strong similarities between the construction of field-effect devices and that of the planar bipolar transistor of Fig. 1.2. For example, both devices contain two p-n junctions and some associated metal contacts. However, when the fabrication of each of these devices is described in detail in the next chapter in connection with integrated circuit technology, it will be seen that the processing of MOS devices is significantly simpler and the steps less critical than for the comparable bipolar transistor or junction field-effect device. In addition, MOS-IGFETs can be made several times smaller in surface area than bipolar devices; hence their extensive use in large electronic memory circuits which employ thousands of devices in a small-area silicon chip.

A. The Junction Field-Effect Transistor (JFET)

Consider the junction field-effect structure of Fig. 10.2a. The two p^+-regions connected together form the JFET gate. Assume the gate is initially at the same electrical potential as the source metal electrode; this metal contact as well as the drain metal electrode make low-resistance ohmic connections to the n-type silicon region. If now the drain electrode is raised to a potential above that of the source, an electron current will flow from drain to source along the n-channel sandwiched between the space-charge regions of the two p^+-n junctions. The current does not flow into these space-charge regions owing to the electron potential barriers provided by the p^+-n junctions. Initially, a further increase in drain-source voltage V_{DS} will result in a linear increase in drain current I_D, since the channel conducts current like a linear resistor. In addition, the increased drain potential will tend to reverse bias the gate p^+-n junction more and more near the drain electrode. This will cause additional space-charge widening of the depletion regions there, tending to *pinch off* the n-type channel at the drain end, as illustrated in Fig. 10.2b. Now the rate of drain-current increase with voltage can no longer remain linear owing to this current-flow constriction, and the rate of rise of I_D becomes smaller with increasing V_{DS}.

Still further increase of the drain voltage results in a *saturation* of drain current, which tends to increase only very slowly with increased voltage. Figure 10.3 shows the output current-voltage characteristic of an n-channel JFET in

FIGURE 10.2

a) Cross-sectional view of an n-channel junction field-effect transistor with gate shorted to source and a small applied drain voltage. The darkened regions represent the junction space-charge regions.

b) Cross-sectional view of an n-channel junction field-effect transistor biased just at the point of channel pinchoff. The darkened regions represent the junction space-charge regions.

c) Circuit symbols for the junction field-effect transistor with source electrode S and drain D. The symbol to the left has an arrow indicating a p-type gate region G for this n-channel device. The symbol to the right has an arrow indicating an n-type gate for this p-channel device.

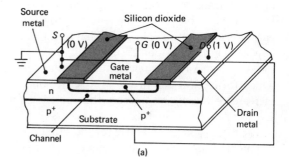

(a)

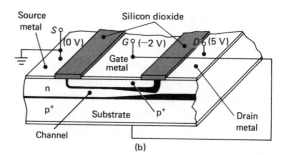

(b)

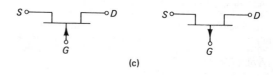

(c)

FIGURE 10.3

Typical set of output characteristics for an n-channel
junction field-effect transistor with a pinchoff voltage
of −4 V.

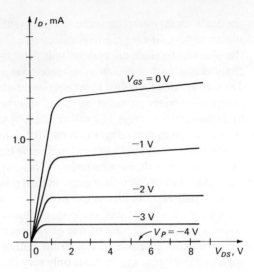

the common-source connection with zero gate-to-source potential V_{GS}. The figure indicates the linear I_D-V_{DS} region at low values of V_{DS} and the sharp bend in this curve due to the tendency toward drain pinchoff, followed by the current-saturation region. Note that as the drain voltage continues to increase beyond the pinchoff point,[1] the additional voltage cannot be absorbed by the the drop along the channel but instead appears across a space-charge region which forms near the drain, in a sense somewhat similar to the case of the increase of reverse bias across the depletion region of a p-n junction diode. Eventually, avalanche breakdown occurs just as in the case of the diode.

The discussion given above for the situation in which there is zero difference of potential between gate and source can be repeated for the case of the gate at a negative potential relative to the source. Owing to this reverse bias across the gate junctions, the channel width is now narrower to begin with so that pinchoff near the drain occurs at a lower drain-to-source voltage, causing the drain current to saturate at a lower value of drain current. This is illustrated in Fig. 10.3. Increasing this reverse gate potential still further until the two gate-junction space-charge regions essentially touch reduces the channel width to zero, preventing any appreciable current from flowing from drain to source other than a small depletion-region leakage current, regardless of the value of the drain potential. The gate-source voltage required for this pinchoff along

[1] Detailed analysis shows that complete pinchoff or the coming together of the gate space-charge regions doesn't always occur to yield current saturation. In certain cases there can remain a very narrow, long channel which supports a large voltage drop and hence includes a very high field. Carriers accelerated in this field may achieve a "scattering-limited" maximum velocity (saturation velocity) accounting for a limiting or saturation of current in spite of increased voltage. Eventually, avalanching occurs. For details see D. P. Kennedy and R. R. O'Brien, "Computer-Aided Two-Dimensional Analysis of the JFET," *IBM J. Res. Develop.*, Vol. 14, p. 95 (1970).

the total channel length is called the *pinchoff voltage* and is labeled V_P on the graph of Fig. 10.3.

Note the similarity between the general shape of the curve of Fig. 10.3 and the common-emitter output characteristics of the bipolar transistor given in Fig. 8.12. There are a number of differences, of course. The unipolar transistor is voltage driven rather than current driven, as is the case for the bipolar device. The FET often exhibits better current saturation and hence higher dynamic output resistance.

The above discussion indicates that the saturation value of the drain current $(I_D)_{sat}$ is a function of the gate-source voltage V_{GS}. A simple relationship has been found to predict the variation of $(I_D)_{sat}$ with V_{GS} rather accurately; it is

$$(I_D)_{sat} = I_{DSS}\left(1 - \frac{V_{GS}}{V_P}\right)^2, \tag{10.1}$$

where I_{DSS} is the saturated drain current with the gate shorted to the source $(V_{GS} = 0)$, and V_P is the gate-source voltage required to pinch off the entire drain-to-source channel. This square-law variation of drain saturation current with gate voltage is typical of field-effect devices and illustrates how the input gate voltage is used to modulate the output drain saturation current. An indication of the gain of this transistor may be obtained by specifying the rate of change of output current with input voltage by defining a *transconductance* $g_m \equiv [\partial(I_D)_{sat}/\partial V_{GS}]_{V_{DS}=const}$. By performing the appropriate differentiation on Eq. (10.1), g_m may be written as[2]

$$g_m = -\frac{2I_{DSS}}{V_P}\left(1 - \frac{V_{GS}}{V_P}\right) = -\frac{2}{V_P}[I_{DSS}(I_D)_{sat}]^{1/2}. \tag{10.2}$$

Since V_P is always at a negative potential relative to the source for an n-channel device, g_m is always a positive quantity.

The input to the JFET is in the form of a reverse biased p^+-n junction which has been described in Chapter 6 as exhibiting a very small leakage current of the order of picoamperes. This small input current ensures the high power gain of the device at low frequencies. The leakage current of the input electrode of the MOSFET (a capacitor with a pure SiO_2 dielectric layer), however, is typically several orders of magnitude lower than that of the JFET and hence is useful when extremely high input impedance is required. In contrast, the extremely low noise properties typical of JFETs dictate the use of this type of device when the circuit application requires that a very minimum of noise be contributed by the transistor.

A set of specification sheets for a typical commercial silicon n-channel junction field-effect transistor is included in Appendix E.1; Appendix E.2 gives specifications for a p-channel junction device.

---

[2] See L. J. Sevin, *Field Effect Transistors*, McGraw-Hill Book Company, New York (1965). In the literature g_m is often referred to as g_{fs}, the forward transconductance (see Appendix E.2).

B. The Silicon MOS Enhancement-Type Transistor

The metal-oxide-semiconductor device is normally fabricated with silicon as the semiconductor, its oxide SiO_2 as the oxide insulator, and a metal such as aluminum to provide ohmic contact to each of the three transistor regions. These three electrodes contact the three main sections of the device — the source, the gate, and the drain — as shown in Fig. 10.1. If a rough analogy is drawn with the bipolar transistor, the source would correspond to the emitter, the gate to the base, and the drain to the collector. The current which flows from source to drain is modulated by a potential of the order of a few volts applied to the gate electrode, insulated from the silicon chip by a thin silicon dioxide layer about 1000 Å thick. In many such devices the structure is completely symmetric so that either p^+-n junction may be labeled the source while the other becomes the drain.

To understand the current flow in this device, assume that the gate electrode of the MOS device pictured in Fig. 10.1 is biased a few volts negative with respect to the n-type silicon substrate beneath. Because the capacitorlike gate structure of this device consists of an insulating oxide layer sandwiched between the conducting metal gate electrode and the conducting (about 0.01 Ω-m) n-type silicon substrate, a negatively charged gate will induce a positive charge at the oxide-substrate interface.[3] This positive charge in the n-type silicon must of course consist of minority carrier holes which are collected at the oxide-silicon interface, just under the gate electrode, as well as some positively charged ionized donors there. When a large enough gate potential is applied, the significant number of these holes collected at the interface will outnumber the electrons there, and the originally n-type region will essentially convert to p-type. This p-region will then produce a conducting path between the p-type source and the p-type drain, permitting majority holes to flow between these latter regions, should a potential be applied between them. This gives rise to the designation of this transistor as a majority carrier device.

This situation is sketched in Fig. 10.4. There, for convenience, the source is electrically connected to the substrate and both are assumed to be at a reference potential of 0 V. Consider the gate electrode also connected momentarily to ground potential and the drain at -5 V relative to ground. Then the situation is simply one of 5 V connected across two back-to-back p-n junction diodes, with the drain junction reverse biased and conducting perhaps only a few nanoamperes of current. However, when the gate potential is made -2 V, the induced p-layer so produced will provide a p-type bridge or channel between source and drain, and typically milliamperes of current will flow. This "enhancement" of source-to-drain current, by about six orders of magnitude, by the charging of the gate electrode in this structure accounts for the designation of this device as an *enhancement-type* MOS transistor. In addition, it is

---

[3] This discussion will not make specific reference to charges which appear at the oxide-substrate interface due to surface states and charges in the oxide. This topic has already been considered in Section 5.6. Those charges will be superimposed on the ones discussed here.

FIGURE 10.4

P-channel enhancement-type
MOSFET with gate bias
showing the positive hole
and ionized donor charges
which form the p-type
channel.

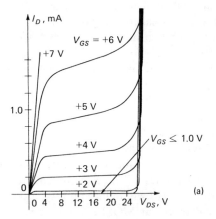

FIGURE 10.5

a) Typical set of output
characteristics for an
n-channel enhancement-
type MOSFET.
b) Typical set of output
characteristics for a
p-channel enhancement-
type MOSFET.

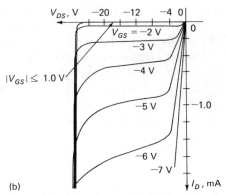

referred to as a *p-channel* device. It is also possible to construct a device
similar in every respect except that the source and drain will be n^+-regions
and the substrate low-conductivity p-type silicon. In this case a sufficiently
large positive potential applied to the gate electrode will collect electrons under
the gate, forming an n-type region at the oxide-substrate interface, providing a
conducting n-type bridge between source and drain. This device is referred to
as an *n-channel* enhancement-type device. A typical set of output characteristics
for both p- and n-channel MOSFETs is shown in Fig. 10.5. For the p-channel
device, while the gate potential is raised in a negative sense from 0 to -1 V,

the drain current remains in the nanoamperes range. A sufficient number of holes have not been attracted to the oxide-substrate interface to convert the n-type silicon beneath the gate to p-type, to form a p-type conducting channel. The few nanoamperes of current then represent essentially the leakage current of the reverse biased drain junction. When the gate potential is made somewhat more negative than -1 V, a drain current of the order of a milliampere begins to flow, and the device becomes active in the sense that increasing the negative gate potential results in higher output drain current. This critical voltage of -1 V is called the gate *threshold* voltage and is normally of the order of a volt or two in a well-designed MOS silicon device which has few surface states and charges in the oxide. Scrupulous care must be exercised in the preparation of the gate oxide and gate metalization for a p-channel enhancement-type device to avoid a high surface density of negative charges at the silicon surface induced by some positive charges in the oxide and the silicon-oxide interface (see Section 5.6). Proprietary cleaning procedures, annealing steps, and aluminum gate depositions using electron-beam heating yield p-channel MOSFETs with low threshold voltage. The replacement of the aluminum gate metalization with amorphous silicon can result in threshold voltages of 1 V or less. These steps tend to avoid sodium contamination. Positive oxide charges will induce an n-type skin on the surface of an intended n-channel enhancement MOSFET, giving rise to drain current even without applied gate voltage and limiting the usefulness of the device.

Note that the output characteristic of this unipolar transistor, in analogy with the bipolar transistor, includes an active region as well as a bottoming (or saturation) region and a drain junction avalanche region. Note also that the drain current in the active region increases at an increasing rate as the gate potential is made more negative, for a p-channel device. In fact, it will be shown in the next section that the drain current I_D increases essentially as the square of the gate voltage increase, giving rise to the term square-law device to describe this type of transistor. No such simple analytic type of description can be made for the bipolar transistor.

C. The Silicon MOS Depletion-Type Transistor

A *depletion-type* MOSFET is pictured in Fig. 10.6. There are two basic differences between this type of device and the enhancement-type MOSFET:

1) The p-channel (a lightly doped layer created by introducing acceptor impurities) is already present without the application of gate potential.

2) Generally the gate electrode covers only a small fraction of the channel length (see Fig. 10.6a).

Hence there is no gate threshold voltage necessary to begin source-to-drain current conduction. Source-to-drain current is conducted even at zero gate

FIGURE 10.6

a) Cross section of a p-channel depletion-type MOSFET. The crosshatched region represents the silicon dioxide insulator.
b) Sketch of a cross section of the p-type channel of a depletion-type MOSFET showing the manner in which a positive gate potential creates a depleted region which tends to pinch off the channel.
c) Circuit symbols for the MOS depletion-type field-effect transistor with source S, gate G, and drain D. The symbol to the left has an arrow indicating an n-type substrate region for this p-channel device. The symbol to the right has an arrow indicating a p-type substrate region for this n-channel device. The arrows are often left off since the battery polarity indicates whether p- or n-channel devices are used.

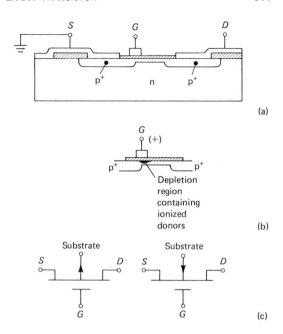

potential. In fact, the manner of device operation is to supply positive gate potential which, when sufficiently high, tends to cut off drain current. Typical output characteristics of this n-channel depletion type of device are shown in Fig. 10.7a. Characteristics for a p-channel device are shown in Fig. 10.7b. These curves are very similar to those for a JFET (Fig. 10.3).

In the case of the p-channel device a positive applied gate potential with respect to the substrate will tend to repel majority carrier holes from the channel, leaving behind negatively charged, ionized acceptor impurities, as shown in Fig. 10.6b. This "depletion" of holes from the channel is entirely analogous to the formation of the depletion layer of a reverse biased p-n junction. Since the mobile carriers tend to be depleted from this type of region, the semiconductor p-layer becomes intrinsic (nearly insulating) and the source-to-drain current is reduced nearly to zero. This resembles the pinchoff mechanism described for FETs. A gate potential of more than $+3$ V is required for the device whose characteristics are shown in Fig. 10.7b to achieve this condition, known as *current cutoff* or *channel pinchoff*. Since the pinchoff of any portion of the channel will interrupt source-to-drain current, the gate electrode need only cover a small fraction of the channel length.

The depletion type of device is used primarily in high-frequency amplifier

FIGURE 10.7

a) Typical set of output
characteristics for an
n-channel depletion-type
MOSFET.
b) Typical set of output
characteristics for a
p-channel depletion-type
MOSFET.

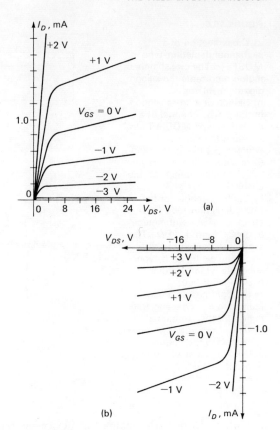

applications. This usage results directly from the small-area gate electrode required in this transistor, compared to the full-size gate needed for the enhancement-type device. The input gate-electrode capacitance can hence be significantly lower in value in the depletion-type MOSFET. It will be shown later that this input capacitance limits the high-frequency performance of the IGFET in a sense similar to the frequency limitation of the bipolar transistor described in Section 9.4C. It is common practice to fabricate the depletion device with the gate electrode only near the source region. Any overlap of the gate and drain regions tends to introduce direct capacitive coupling between input and output. This capacitance substantially reduces the high-frequency gain of the device.

As in the case of the enhancement-type MOSFET, the depletion-type device can be fabricated either with a p-channel or an n-channel.[4] Since the n-channel device conducts by electrons and the p-channel MOSFET operates by hole conduction, the former type tends to have a higher frequency capability. This is true since the mobility of electrons exceeds that of holes in silicon and so

[4] The n-channel device will, of course, have n+ source and drain regions.

FIGURE 10.8

Cross-sectional view of an n-channel enhancement-type MOSFET showing the negative electron charges forming the conducting channel between drain and source.

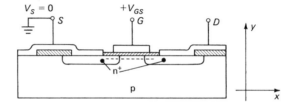

the source-to-drain transit time is shorter in the n-channel device. It will be shown later that this carrier transit time limits the high-frequency performance of the IGFET.

10.2 THE STATIC *I-V* CHARACTERISTIC OF THE ENHANCEMENT-TYPE MOSFET

Consider the n-channel enhancement-type MOSFET drawn in cross section in Fig. 10.8. For convenience, the source and substrate are taken at ground potential ($V_S = 0$). The calculation of the device *I-V* characteristic which will be presented here assumes the "shallow-channel" approximation. That is, the exact computation of the source-to-drain characteristic involves the solution of a two-dimensional current-flow problem even when only a cross section of the device is considered.[5] However, this complex analysis is approximated by reducing it to two simpler one-dimensional problems. First, it is assumed that nearly all of the gate control voltage appears across the oxide so that the control field is in the y-direction, normal to the oxide-substrate interface. Second, it is assumed that source-to-drain current conduction takes place in a thin (shallow-channel) surface sheet and that the electric field driving this current is in the x-direction, parallel to the oxide-substrate interface. The approach is that the control electric field induces the collection of mobile charges at the surface of the silicon, between the source and drain regions; these serve to conduct current between these regions. The more positive the gate control voltage, the greater the conductance of the channel and the larger the drain current conducted when a potential is applied between drain and source.

A. Derivation of the MOS *I-V* Output Characteristic

Consider the n-channel enhancement-type MOSFET structure shown in Fig. 10.9. If all of the gate control voltage V_{GS} appears across the oxide, the electric field in the oxide in the y-direction at any point x is given by

$$\mathcal{E}_{ox}(x) = \frac{V_{GS} - V(x)}{t_{ox}}, \qquad (10.3)$$

---

[5] This reduces the basically three-dimensional problem to a two-dimensional problem. See, for example, D. Frohman-Bentchkowsky and A. S. Grove, "Conductance of MOS Transistors in Saturation," *IEEE Trans. Electron. Devices*, Vol. ED-16, p. 108 (1969).

FIGURE 10.9

Sketch of an n-channel
enhancement-type MOSFET
showing a cross section
used in the derivation of the
I-V output characteristics of
the device. The *x*-direction
is taken along the channel
and the *y*-direction is normal
to the oxide layer.

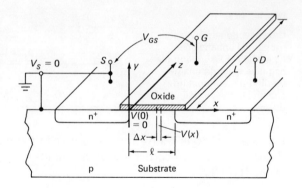

where $V(x)$ is the voltage in the channel (relative to the source-substrate taken as zero potential) at a distance x from the source, and t_{ox} is the oxide thickness. The surface charge density induced under the gate, ω_I (sheet charge in coulombs per square meter), is given by Gauss's law as

$$\omega_I(x) = \epsilon_{ox}\, \mathcal{E}_{ox}(x) = \frac{\epsilon_{ox}}{t_{ox}}\,[V_{GS} - V(x)], \qquad (10.4)$$

where ϵ_{ox} is the dielectric permittivity of the insulating oxide. However, a channel containing mobile charge carriers does not exist if $V_{GS} - V(x) \leqq V_{GST}$ for any value of x between source and drain. Here V_{GST}, the threshold voltage, is the minimum potential just required to cause sufficient mobile surface charge to be induced so that appreciable channel current can begin to flow. The mobile surface charge density is then given by[6]

$$\omega_m(x) = \frac{\epsilon_{ox}}{t_{ox}}\,[V_{GS} - V(x) - V_{GST}] \qquad (10.5a)$$

for $V_{GS} - V(x) > V_{GST}$, and

$$\omega_m(x) = 0 \qquad (10.5b)$$

for $V_{GS} - V(x) \leqq V_{GST}$.

The incremental conductance $\Delta G(x)$ of an infinitesimally thin channel of width L (*z*-direction) and length Δx is given by

$$\Delta G(x) = \frac{\omega_m(x)\mu L}{\Delta x}, \qquad (10.6)$$

where μ is the electron carrier mobility in the channel at the silicon surface just

---

[6] This derivation neglects the effect of induced ionized impurity charges in the substrate and is valid if the substrate is lightly doped. It also ignores oxide-semiconductor interface states and charges in the oxide (see Section 5.6), except as included in the gate threshold voltage V_{GST}.

under the oxide. Applying Ohm's law to the incremental portion of the channel Δx gives the source-to-drain current there which, with Eq. (10.6), is

$$I_D = \Delta G(x)\,\Delta V(x) = \omega_m(x)\mu L\,\Delta V/\Delta x. \tag{10.7}$$

Now this drain current must obviously be the same at every point x along the channel. Integrating Eq. (10.7) and taking into account this constancy of I_D gives

$$I_D \int_0^l dx = \frac{\epsilon_{ox}\mu L}{t_{ox}} \int_0^{V_{DS}} [V_{GS} - V(x) - V_{GST}]\,dV, \tag{10.8}$$

where the increment Δx has been assumed arbitrarily small (approaching zero), l is the channel length, and V_{DS} is the channel voltage at $x = l$, which is characteristic of the source-drain potential. Carrying out the integration gives

$$
\begin{aligned}
I_D &= \frac{\epsilon_{ox}\mu L}{lt_{ox}}\left[(V_{GS} - V_{GST})V_{DS} - \frac{1}{2}V_{DS}^2\right] \\
&\equiv \frac{C_{ox}\mu}{l^2}\left[(V_{GS} - V_{GST})V_{DS} - \frac{1}{2}V_{DS}^2\right],
\end{aligned} \tag{10.9}
$$

where C_{ox} is the gate oxide capacitance.

For $V_{DS} = 0$, $I_D = 0$. As V_{DS} increases in value, with the gate positively biased at some value V_{GS1}, I_D will correspondingly increase. However, when V_{DS} reaches the value given by $V_{DS} = V_{GS1} - V_{GST}$, the current must cease to increase. This is true since for this latter value of V_{DS}, termed $(V_{DS})_{sat}$, there is zero electric field in the gate oxide just at the drain, and hence there is no induced mobile charge there (see Eq. 10.5b). This might seem to indicate that the current will cease to flow due to the nonconducting end of the channel. However, this conclusion is absurd since it is the IR voltage drop in the channel which permits the voltage at the drain end of the channel to rise to a value $(V_{DS})_{sat}$. Hence the drain current $(I_D)_{sat}$ at this drain voltage must be maintained, and for values of $V_{DS} > (V_{DS})_{sat}$ the drain current remains essentially constant at a value $(I_D)_{sat}$, called the *saturation* current.[7] The output characteristic of an n-channel MOSFET, indicating the locus of the voltages $(V_{DS})_{sat}$ and $(I_D)_{sat}$ for various values of V_{GS}, is shown in Fig. 10.10. The value of this saturation current can be calculated by setting

$$\partial I_D / \partial V_{DS}]_{V_{GS}} = 0$$

for $V_{DS} = V_{GS} - V_{GST}$ to obtain

$$(I_D)_{sat} = \frac{C_{ox}\mu(V_{DS})_{sat}^2}{2l^2} = \frac{C_{ox}\mu}{2l^2}(V_{GS} - V_{GST})^2. \tag{10.10a}$$

---

[7] This saturation region should not be confused with the bipolar transistor bottoming or saturation region. As V_{DS} is increased beyond $(V_{DS})_{sat}$ excess voltage can no longer contribute to this ohmic drop along the channel but must appear near the space-charge region formed at the drain. See Section 10.2B.

FIGURE 10.10

Output characteristic of an
n-channel enhancement-
type MOSFET showing the
locus of drain-source voltage
saturation points for different
values of gate-source
voltage.

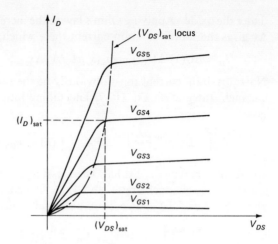

Multiplying numerator and denominator by V_{GST}^2 gives

$$(I_D)_{sat} = I_{DSS} \left(\frac{V_{GS}}{V_{GST}} - 1 \right)^2, \tag{10.10b}$$

where I_{DSS} represents $(I_D)_{sat}$ with gate shorted to source ($V_{GS} = 0$). [Note the similarity of this expression with Eq. (10.1) for the JFET.] Actually the drain current rises slightly with drain voltage so that the device output conductance $\partial I_D/\partial V_{DS}$ is not zero but has a finite though small value. The "square-law" nature of the field-effect transistor characteristics is indicated in Eqs. (10.10). No such simple relationship between output current and voltage can be written for the bipolar transistor.

A measure of the field-effect transistor gain is given by the transconductance g_m, which is obtained by differentiating Eq. (10.10a) with respect to gate voltage as

$$g_m \equiv \left[\frac{\partial (I_D)_{sat}}{\partial V_{GS}} \right]_{V_{DS}} = \frac{C_{ox}\mu}{l^2} (V_{GS} - V_{GST}) = \frac{C_{ox}\mu}{l^2} (V_{DS})_{sat}. \tag{10.11}$$

This represents the change in output drain current per unit change in input gate potential, at a fixed value of drain voltage, in the active region. This definition emphasizes the voltage-control nature of the IGFET. Note the direct dependence of the gain on C_{ox} and μ and the inverse relationship to the square of the channel length.

The device characteristic in the transistor bottoming region [$V_{DS} < (V_{DS})_{sat}$] can be studied by approximating the drain current I_D from Eq. (10.9), for $V_{DS} \to 0$, as

$$I_D = \frac{C_{ox}\mu}{l^2} (V_{GS} - V_{GST})V_{DS}. \tag{10.12}$$

The channel conductance in this bottoming region where I_D increases linearly with V_{DS} is given by

$$g_{ds}|_{V_{DS} \to 0} = \frac{\partial I_D}{\partial V_{DS}}\bigg|_{V_{GS}} = \frac{C_{ox}\mu}{l^2}(V_{GS} - V_{GST}). \qquad (10.13)$$

Hence the "bottoming conductance" increases with increasing gate voltage as well as with the inverse square of the channel length. The reciprocal of this quantity is usually called r_{ds} (see Appendix F.1).

EXAMPLE 10.1

Calculate the electric field in the gate oxide of a silicon n-channel enhancement-type MOSFET with gate voltage $V_{GS} = 5.0$ V and drain voltage $V_{DS} = 4.0$ V (a) at the source ($x = 0$), (b) at the drain ($x = l$). (c) Compare these fields with the avalanche field in silicon. The oxide thickness t_{ox} is 1.0×10^{-7} m, and the channel length l is 1.0×10^{-5} m.

SOLUTION   From Eq. (10.3),

$$\mathcal{E}(x) = \frac{V_{GS} - V(x)}{t_{ox}}.$$

a) At $x = 0$,

$$\mathcal{E}(0) = \frac{(5.0 - 0)\ \text{V}}{1.0 \times 10^{-7}\ \text{m}} = 5.0 \times 10^7\ \text{V/m}.$$

b) At $x = l$,

$$\mathcal{E}(l) = \frac{(5.0 - 4.0)\ \text{V}}{1.0 \times 10^{-7}\ \text{m}} = 1.0 \times 1.0^7\ \text{V/m}.$$

c) These fields are of the same order as the electric field at which avalanching occurs in silicon.

EXAMPLE 10.2

a) Derive an expression for the potential $V(x)$ at any point x along the conducting channel of the MOSFET of Example 10.1, at the edge of saturation.
b) Write an expression for the electric field $\mathcal{E}(x)$ along the channel.
c) Calculate the field parallel to the channel halfway between source and drain, if $(V_{DS})_{sat} = 5.0$ V.

SOLUTION   a) Integrating Eq. (10.8) gives

$$I_D \int_0^x dx = \frac{\epsilon_{ox}\mu L}{t_{ox}} \int_0^{V(x)} [(V_{GS} - V_{GST}) - V(x)]\, dV,$$

which gives

$$I_D x = B\{(V_{GS} - V_{GST})V(x) - \tfrac{1}{2}[V(x)]^2\},$$

where $B = \epsilon_{ox}\mu L/t_{ox}$ and which satisfies the boundary condition $V(0) = 0$. Now at the point of saturation, since $V_{GS} - V_{GST} = (V_{DS})_{sat}$,

$$[V(x)]^2 - 2(V_{DS})_{sat}V(x) + \frac{2(I_D)_{sat}\,x}{B} = 0.$$

Solving this quadratic equation for $V(x)$ yields

$$V(x) = (V_{DS})_{sat}\left[1 \pm \sqrt{1 - \frac{2(I_D)_{sat}\,x}{B(V_{DS})_{sat}^2}}\,\right].$$

Introducing Eq. (10.10a) for $(I_{DS})_{sat}$ then gives

$$V(x) = (V_{DS})_{sat}(1 - \sqrt{1 - x/l}),$$

where the negative sign of the radical is chosen to ensure that $V(0) = 0$.
b) The electric field is obtained by differentiation as

$$\mathcal{E}(x) = -\frac{\partial V(x)}{dx} = \frac{(V_{DS})_{sat}}{2l(1 - x/l)^{1/2}}.$$

c)

$$\mathcal{E}(\tfrac{1}{2}) = \frac{5.0\text{ V}}{2(1.0 \times 10^{-5}\text{m})\,(1 - \tfrac{1}{2})^{1/2}}$$

$$= 2.5\sqrt{2} \times 10^5 \text{ V/m} \quad \text{or} \quad 3.5 \times 10^5 \text{ V/m},$$

which is large but much smaller than the field in the oxide.

B. Calculation of the MOSFET Output Conductance in the Active Region

For drain voltages beyond the knee in the I-V curve (see Fig. 10.10), the output conductance is determined by the change of channel length with drain voltage. In this current saturation region, the channel may be considered to be made up of two parts as shown in Fig. 10.11. The normal channel ohmic conduction region discussed thus far and adjoining the source is labeled l_s. The portion of the channel near the drain, l_d, is a small section which represents the space-charge depletion region of the reverse-biased drain junction. This region

FIGURE 10.11

Sketch of the cross section of an n-channel enhancement-type MOSFET showing the two sections of the channel. Ohmic conduction takes place in the section marked l_s. The section marked l_d is the drain space-charge region. The dashed line indicates the boundary of the space-charge layer.

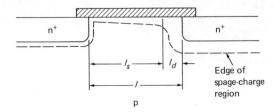

begins to form as the drain voltage is increased to $(V_{DS})_{sat}$ and then continues to expand as this voltage continues to increase. In fact, the drain voltage in excess of $(V_{DS})_{sat}$ appears across this depletion region in a manner very similar to the manner in which the space-charge layer supports the reverse voltage applied to a simple p-n junction diode. Again space-charge widening occurs as a result of increasing the applied reverse voltage. In the case of the MOSFET, this increases l_d at the expense of decreased l_s, thereby reducing the effective conducting channel length. The output conductance g_{os} for some gate bias V_{GS} can then be computed by calculating the slope of the drain-source output characteristics beyond $(V_{DS})_{sat}$ as

$$
\begin{aligned}
g_{os} &\equiv \left[\frac{\partial (I_D)_{sat}}{\partial V_{DS}}\right]_{V_{GS}} = \left[\frac{\partial (I_D)_{sat}}{\partial l_s}\right]_{V_{GS}} \left[\frac{\partial l_s}{\partial V_{DS}}\right]_{V_{GS}} \\
&= - \left[\frac{\partial (I_D)_{sat}}{\partial l_s}\right]_{V_{GS}} \left[\frac{\partial l_d}{\partial V_{DS}}\right]_{V_{GS}}
\end{aligned}
\tag{10.14}
$$

since $l_d = l - l_s$ (see Fig. 10.11). Using Eq. (10.10a), we have

$$
\left[\frac{\partial (I_D)_{sat}}{\partial l_s}\right]_{V_{GS}} \approx \left[\frac{\partial (I_D)_{sat}}{\partial l}\right]_{V_{GS}} = -\frac{2(I_D)_{sat}}{l},
\tag{10.15}
$$

which together with Eq. (10.14) gives the output conductance

$$
g_{os} = 2\left[\frac{\partial l_d}{\partial V_{DS}}\right]_{V_{GS}} \frac{(I_D)_{sat}}{l} \propto \frac{(I_D)_{sat}}{l}.
\tag{10.16}
$$

Hence the output conductance is inversely proportional to the channel length and increases with increasing drain current. This is found to be approximately true in practice (see Fig. 10.7).

A set of specification sheets for a typical commercial silicon n-channel enhancement-type MOSFET is included in Appendix F.1; Appendix F.2 gives specifications for a p-channel device.

10.3   CIRCUIT APPLICATIONS OF THE MOSFET

The high input impedance of the MOSFET accounts for a number of its circuit applications, such as in the input circuit of FM tuners and electrometers, since hardly any current is drawn from the signal source. Dual-gate depletion devices are used to provide AFC (automatic frequency control); one of the gates is used to provide this function in a feedback arrangement. However, the primary circuit application of the enhancement-type MOSFET is as an electronic switch, and this use will next be described.

A. The MOSFET Switch with Resistive Load

Consider the enhancement-type n-channel MOSFET used as a switch in the circuit sketched in Fig. 10.12a, with a load resistor R_L. A common symbol

FIGURE 10.12

a) An n-channel enhance-
ment-type MOSFET used as
a switch in an inverter
circuit with a linear ohmic
load resistor.
b) Output characteristics of
the MOSFET in Fig. 10.12a,
with a linear load line whose
slope is $-1/R_L$ also shown.

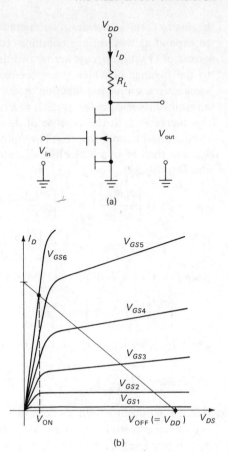

(a)

(b)

for the MOSFET is indicated in the figure. The output I_D-V_{DS} characteristics
of this transistor are indicated in Fig. 10.12b, as is a load line corresponding to
R_L. When the input voltage to this device is below the threshold voltage,
$V_{in} < V_{GST}$, the circuit current through R_L is extremely small and essentially all
the battery voltage V_{DD} appears at the output. However, if a positive gate
voltage such as V_{GS6} is applied to the input, the device is driven into bottoming
and the output voltage drops to a small voltage V_{ON}. Hence this transistor
switch is either in the high OFF state or low ON state depending on the magnitude
of the input gate voltage to the device. Since a "high" input voltage causes the
output voltage to go "low," the circuit of Fig. 10.12a is referred to as an
inverter circuit.

B. The MOSFET Switch with MOS Resistor Load

In practical MOS digital switching circuits, the ohmic or linear load resistor of
Fig. 10.12a is seldom used. Instead, the MOS resistor (MOSR), which has a

nonlinear *I-V* characteristic, is generally utilized. This device has a structure quite similar to that of the MOSFET except that its channel length is much greater. Also the MOSR gate and drain electrodes are usually connected together electrically to form a common terminal. This device is used as a load because it can be made extremely small, like the MOSFET, for high-density circuit applications. (This is not true for the integrated form of a load resistor — see Chapter 11.) Also exactly the same technology used to fabricate the MOSFET can be used to make the MOSR. The advantage of this will become more apparent when the technology of integrated circuit fabrication is described in the next chapter.

A sketch of the MOSR structure is shown in Fig. 10.13a. The relation between the MOSR current and voltage can be derived by starting with Eq. (10.9) and setting $V_{GS} = V_{DS}$. Then the MOS dc resistance can be written as

$$R_{dc} = \frac{V_{DS}}{I_D} = \frac{1}{(\mu\epsilon_{ox}L/t_{ox}l)(V_{GS} - V_{GST} - V_{DS}/2)} = \frac{t_{ox}(V_{DS}/2 - V_{GST})}{\mu\epsilon_{ox}} \frac{l}{L}.$$

$$(10.17)$$

FIGURE 10.13

a) Sketch of an MOS resistor structure showing the gate shorted to the drain. Note the relatively long channel.
b) Schematic diagram of a MOSFET inverter circuit utilizing an MOS resistor as a load.

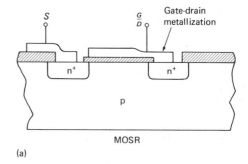

(a)

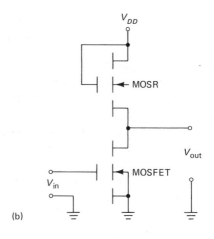

(b)

FIGURE 10.14

a) *I-V* characteristic of an
MOS resistor.
b) Output characteristic of
the MOSFET of Fig. 10.13b
along with an MOS resistor
nonlinear load line.

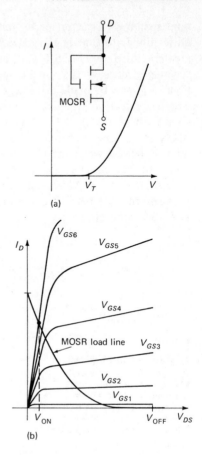

(a)

(b)

This expression indicates why the MOSR channel length l is usually made greater than the channel width L to provide a large resistance to limit the current to a small value. This is in contrast to the small value of channel length and large width required in the MOS transistor to yield high g_m (see Eq. 10.11) and large gain-bandwidth product (Eq. 10.19). The nonlinear nature[8] of the *I-V* characteristic of the MOSR will be seen to be of no consequence when the device serves as a load to limit the current in a digital circuit of the type shown in Fig. 10.13b.

The *I-V* characteristic of this MOSR is shown in Fig. 10.14a. For small applied voltages, there is insufficient induced channel charge to conduct appreciable current, and the device has a very high impedance (the curve is nearly horizontal). After the gate threshold voltage is achieved, the device current begins to increase rapidly as indicated in Fig. 10.14a. This MOSR characteristic can be used in a load-line analysis of the inverter circuit of

---

[8] This nonlinearity results from the dependence of the MOS resistance on V_{DS}.

Fig. 10.13b in a manner analogous to Fig. 10.12b, except that now the load line is no longer linear. This load-line diagram is shown in Fig. 10.14b.

In this case the OFF voltage of the circuit is nearly V_{DD} as is the situation with a linear load resistor. The small ON voltage in this circuit may be shown to be dependent on the relative geometries of the MOSR and the MOSFET. It can be calculated by noting that these devices carry a common current and employing Eqs. (10.17) and (10.9). This is the subject of a problem at the end of this chapter.

C. A Complementary-Type MOSFET Inverter Circuit

For switching-circuit applications such as the electronic wristwatch, in which low battery drain is a primary requisite, complementary MOS inverter circuits are used. This type of circuit contains both a p-channel and an n-channel enhancement device as shown in Fig. 10.15. This circuit has the property that there is negligible battery current drain in the quiescent condition in either of the two logic states, ON and OFF. Power is consumed only during the switching transient as the inverter changes its state, and this is quite small if the switching rate is not too large. This is in contrast to the n-channel inverter of Fig. 10.13b, which draws appreciable battery power in the switched ON state.

FIGURE 10.15

a) Schematic diagram of a complementary enhancement-mode inverter circuit.
b) Sketch of the cross section of a typical complementary enhancement-mode inverter of the type indicated in Figure 10.15a. The p-pocket of the n-channel device requires an extra diffusion step.

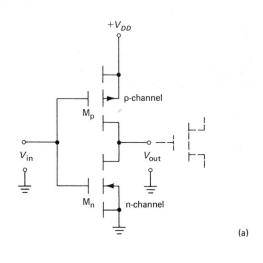

(a)

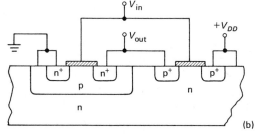

(b)

A single low-voltage COS/MOS (complementary MOS) integrated circuit, which can perform a variety of mathematical functions. (Courtesy of RCA Solid State Division, Somerville, N.J.)

To follow the operation of this complementary circuit, assume that the input terminal is at a potential $V_{in} = +V_{DD}$. This induces an n-channel in MOSFET M_n, causing it to turn ON and causing its output voltage V_{out} to drop to a small value (the MOSFET bottoming voltage) above ground potential. However, the MOSFET M_p is kept OFF by this positive input voltage, preventing anything other than a small (picoampere) leakage current to be drawn from the battery. On the other hand, if V_{in} is put at ground potential, MOSFET M_n is kept OFF while MOSFET M_p turns ON, bringing the output V_{out} to a small voltage below $+V_{DD}$. Hence the inverter can exist in either the ON or OFF state without drawing any appreciable power from the battery.

Inverters of the complementary type or of the n-channel type may be cross-coupled in a flip-flop configuration to produce a memory-type circuit function. This has already been described for bipolar transistors in Section 9.1C (The flip-flop). MOSFET memory circuits are of particular interest since the simplicity of fabrication and the small size of these devices permit large memory arrays containing 10,000 devices or more to be successfully manufactured on a small silicon chip. A typical memory element for such a device is discussed next.

D. MOSFET Memory-Circuit Element

A typical MOSFET flip-flop is shown in Fig. 10.16. It contains two cross-coupled n-channel enhancement-type MOSFETs M_1 and M_2, plus two MOS resistors MR_1 and MR_2. The description of the two possible states of this circuit is similar to that for the bipolar flip-flop previously described. If V_2, the gate potential of M_1, is positive with respect to ground, it will induce an n-channel in M_1 causing the potential V_1 to be only a small voltage above ground potential. MOSFET M_1 is ON and MR_1 conducts. If the voltage V_1 is lower than the threshold potential of MOSFET M_2, then that device will not conduct, nor will MR_2, so that potential V_2 must be up toward V_{DD} in value. The voltage V_2 can be made high enough, by appropriate design of MR_2, so that the threshold voltage of M_1 is exceeded, keeping it ON as originally assumed. This constitutes one of the two possible stable states of the flip-flop. The other state exists when M_2 is ON and M_1 is OFF. Again the state of the device may be ascertained by noting the potential of node V_1 or V_2.

MOSFETs M_3 and M_4, also shown in Fig. 10.16, are inputs used to change the state of the flip-flop. For example, if MOSFET M_1 is ON and M_2 is OFF, a positive potential applied to the gate of MOSFET M_4 will tend to force the potential V_2 to a small voltage above ground potential. Since V_2 is the gate potential of MOSFET M_1, this device will turn OFF, raising the potential V_1 and hence turning M_2 ON. Hence the flip-flop may be set in either of its two stable states by impressing a positive potential to the gates of MOSFET M_3 or M_4. This may take the form of a positive-going input pulse. The flip-flop may even be caused to change its state with a short input pulse as long as the pulse lasts long enough for the change of state to take place.

FIGURE 10.16

Schematic diagram of a typical MOSFET memory element including a basic flip-flop as well as input devices M_3 and M_4. Note that either all n-channel or all p-channel devices may be used. The only difference is the polarity required of the battery potential. Cross-coupled complementary inverters may be used alternately to provide a very low quiescent current battery drainage flip-flop.

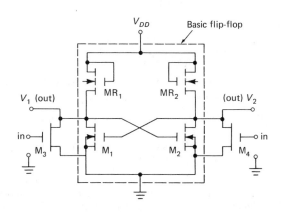

FIGURE 10.17

Schematic diagram of a simple
MOSFET amplifier. Capaci-
tors C_i and C_o represent the
input and output coupling
capacitors. R_1 and R_2
comprise a voltage divider
which sets the dc gate bias
of the MOSFET.

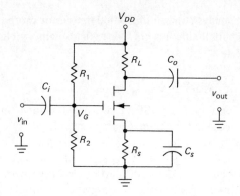

E. FET Linear Amplifier Circuit

The high input resistance and hence low input-signal-power requirement of the
field-effect transistor make it useful in the input stage of an FM tuner. In
addition, a field-effect transistor has distinctly lower noise than a corresponding bipolar transistor.[9] The JFET is somewhat better in this respect than
the MOSFET. However, the MOSFET input resistance is several orders of
magnitude greater than that of a comparable JFET. Figure 10.17 is a schematic
diagram of a simple FET amplifier. Resistors R_1 and R_2 determine the gate bias
potential $V_G = V_{DD} R_2/(R_1 + R_2)$. C_i and C_o are the input and output coupling
capacitances, and C_s is the bypass capacitor for the source resistor R_s. This
circuit is quite analogous to the bipolar transistor amplifier of Fig. 9.4. The
obvious difference is that the input (gate) in the FET circuit draws essentially no
current from the signal source.

10.4   THE FREQUENCY LIMITATION OF THE MOSFET

To estimate the gain-bandwidth capability of the MOSFET, refer to the
calculation of this quantity for the bipolar transistor given in Section 9.4B. In
analogy with that discussion the gain-bandwidth product for the MOSFET can
be written as

$$G \times BW = \frac{g_m}{2\pi C_{in}} \approx \frac{g_m}{2\pi C_{ox}}, \tag{10.18}$$

where C_{in} is the input or essentially the gate capacitance of the MOSFET which
is roughly the oxide capacitance C_{ox}. Combining Eq. (10.18) with Eq. (10.11)
for g_m yields

$$G \times BW \approx \frac{\mu}{2\pi l^2} (V_{GS} - V_{GST}) = \frac{\mu}{2\pi l^2} (V_{DS})_{\text{sat}}. \tag{10.19}$$

---

[9] This follows from the fact that the field-effect transistor is a majority carrier device and
noise is generated mainly by the bipolar device's minority carriers. See A. Van der Ziel,
Electronics, Chapter 23, Allyn and Bacon, Inc., Boston (1966).

Hence the gain-bandwidth product may be increased four times by halving the channel length. However, this will increase the device's output conductance (see Eq. 10.16) and limit its voltage capability; for if the channel length is small enough, increased drain voltage will expand the drain space-charge region l_d, and the maximum drain voltage capability will be limited to that voltage which causes the drain depletion region to reach through to the source region. At that point a rapid increase of drain current will occur, somewhat like that observed when the drain junction breaks down by avalanching. This "reach-through" phenomenon also occurs in bipolar transistors, particularly the alloy type with heavily doped emitter and collector regions. (The planar-type bipolar transistor has heavy doping near the emitter junction, which reduces the rate of space-charge depletion layer increase as the emitter is approached.)

A physical interpretation of Eq. (10.19) may be obtained by rewriting it as

$$G \times BW \approx \frac{1}{2\pi}\frac{\mu}{l}\frac{(V_{DS})_{\text{sat}}}{l} = \frac{1}{2\pi}\frac{\mu(E_D)_{\text{sat}}}{l}, \tag{10.20}$$

where $(E_D)_{\text{sat}}$ is approximately the average electric field along the channel, just at saturation. Since the channel mobility μ is defined as carrier velocity per unit electric field, $\mu(E_D)_{\text{sat}}/l$ is the average channel electron carrier velocity divided by the channel length, or the reciprocal of the carrier source-to-drain *transit time*. Another way of arriving at the same result is to consider the time for charging the gate capacitance through half the channel resistance. This is the subject of a problem at the end of this chapter. The gain-bandwidth product also supplies an estimate for the switching speed of a MOSFET, since this is limited by the gate charging time and the carrier transit time.

EXAMPLE 10.3

Determine approximately (a) the gain-bandwidth product for a MOSFET with a gate capacitance $C_{ox} = 2.0$ pF and a transconductance $g_m = 1.25 \times 10^{-3}$ ℧ and (b) the carrier transit time for this device.

SOLUTION a) From Eq. (10.18),

$$G \times BW \approx \frac{g_m}{2\pi C_{ox}} = \frac{1.25 \times 10^{-3}\,\text{℧}}{2\pi\,(2.0 \times 10^{-12}\,\text{F})}$$

$$= 99 \times 10^6 \text{ Hz} \quad \text{or} \quad 99 \text{ MHz}.$$

b) The carrier transit time is

$$\tau_t = \frac{1}{2\pi(G \times BW)} = \frac{1}{2\pi(99 \times 10^6 \text{ Hz})}$$

$$= 1.6 \times 10^{-9} \text{ sec} \quad \text{or} \quad 1.6 \text{ nsec}.$$

A. Comparison of the Bipolar and Unipolar Transistor Gain-Bandwidth Products

The gain-bandwidth products for bipolar and unipolar transistors of similar geometry may be compared by estimating the expression given in Eq. (10.18) for each type of device. The g_m of the bipolar transistor is given by Eq. (9.19a). The g_m of a MOSFET may be estimated using Eq. (10.11). If these devices are similar in size, then their input capacitances are comparable. Hence the G × BW products of these devices may be compared by taking the ratio of their transconductances.

EXAMPLE 10.4

Determine approximately the ratio of the gain-bandwidth product of a bipolar transistor operating at a collector current of 10 mA to that of a MOSFET of comparable size. The MOSFET has the following design and operating parameters: $C_{ox} = 2.0$ pF, $\mu = 0.075$ m^2/V-sec, $l = 1.0 \times 10^{-5}$ m, and $V_{GS} - V_{GST} = 5.0$ V.

SOLUTION   Using Eq. (9.19a), we have

$$(g_m)_{\text{bipolar}} = \frac{I_c}{kT/q} = \frac{1.0 \times 10^{-2} \text{ A}}{0.026 \text{ V}} = 0.38 \text{ } \mho.$$

From Eq. (10.11),

$$(g_m)_{\text{MOSFET}} = \frac{C_{ox}\mu}{l^2}(V_{GS} - V_{GST})^2$$

$$= \frac{2.0 \times 10^{-12} \text{ F } (0.075 \text{ m}^2/\text{V-sec})(5.0^2 \text{ V}^2)}{(1.0 \times 10^{-5})^2 \text{ m}^2}$$

$$= 0.0075 \text{ } \mho.$$

Using Eq. (10.18), we have

$$\frac{(G \times BW)_{\text{bipolar}}}{(G \times BW)_{\text{MOSFET}}} = \frac{(g_m/C_{in})_{\text{bipolar}}}{(g_m/C_{in})_{\text{MOSFET}}} \approx \frac{(g_m)_{\text{bipolar}}}{(g_m)_{\text{MOSFET}}}$$

if the input capacitances of these devices are comparable. Hence

$$\frac{(G \times BW)_{\text{bipolar}}}{(G \times BW)_{\text{MOSFET}}} \approx \frac{0.38}{0.0075} \approx 50.$$

This example indicates that the bipolar transistor has a higher inherent frequency capability than the MOSFET. However, because of the simplicity of the MOSFET structure, its input (gate) capacitance can be made five times smaller than the bipolar-device (emitter) capacitance, by reducing the area of the MOSFET gate. Example 10.4 then predicts that the bipolar transistor will still have a gain-bandwidth product 10 times greater than that of a comparable

MOSFET unipolar device. Hence, intrinsically, the bipolar transistor has the higher frequency and hence the faster switching capability, for devices of comparable geometry made of the same material. However, very small MOSFETs have been developed recently, using gallium arsenide as starting material, with frequency capability comparable to the best bipolar transistors of silicon. Of course, the four-times-higher electron mobility in GaAs compared with Si accounts for this result (see Eq. 10.11).

It has already been mentioned that a primary application of the MOSFET is in high-density electronic memories, where the small size and simplicity of the device are its primary advantages. High-speed capability is not the most significant requisite here; rather it is device manufacturing yield.

B. Small-Signal Equivalent Circuit for the Field-Effect Transistor

A small-signal model of the FET useful for analyzing linear amplifier circuits in the active mode will now be presented. This will be developed along the lines already taken in the derivation of the two-port hybrid circuit model of the bipolar transistor given in Section 9.2A. Since the drain current i_D is dependent on the gate-source voltage v_{GS} as well as the drain-source voltage v_{DS}, an approximate (first-order) Taylor expansion for an incremental change in drain current Δi_D about a dc operating point in the saturation (active) region set by V_{DS} and V_{GS} is

$$\Delta i_D = \frac{\partial i_D}{\partial v_{GS}}\bigg]_{V_{DS}} \Delta v_{GS} + \frac{\partial i_D}{\partial v_{DS}}\bigg]_{V_{GS}} \Delta v_{DS}. \qquad (10.21)$$

In small-signal notation this can be written as

$$i_d = g_m v_{gs} + g_{os} v_{ds}, \qquad (10.22)$$

where the lower-case subscripts refer to incremental quantities and the following definitions are used:

$$g_m \equiv \frac{\partial i_D}{\partial v_{GS}}\bigg]_{V_{DS}} \qquad (10.23)$$

and

$$g_{os} \equiv \frac{\partial i_D}{\partial v_{DS}}\bigg]_{V_{GS}}. \qquad (10.24)$$

A circuit which represents Eq. (10.22) is shown in Fig. 10.18a. This common-source circuit is useful both for JFETs and MOSFETs with substrate joined to source. Note that the open circuit between gate and source and between gate and drain indicates the negligibly small current leakage through the gate electrode. However, the finite conductance between drain and source is indicated by g_{os}. This quantity may be determined by the slope of the FET I_D-V_{DS} characteristic in the saturation or active region where the device is

FIGURE 10.18

a) Low-frequency small-signal common-source equivalent circuit for a field-effect transistor. $g_m v_{gs}$ indicates a current-source, and g_{os} is the dynamic drain-source conductance of the device biased in the saturation or active region.
b) High-frequency small-signal equivalent circuit for a field-effect transistor. C_{gs} and C_{ds} represent the gate-source and drain-source interelectrode capacitances. C_{gd} is the gate-drain capacitance.

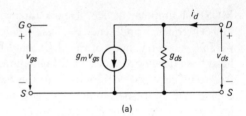

(a)

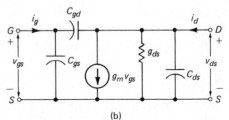

(b)

biased for amplifier operation, and is generally greater for JFETs than for MOSFETs.

The model just described is intended for low-frequency circuit analysis. A higher-frequency version of this equivalent circuit is given in Fig. 10.18b. Here the capacitor C_{gs} represents the capacitance between gate and source, which can be shown to be about two-thirds of the gate oxide capacitance C_{ox} in the case of the MOSFET.[10] It is less than the oxide capacitance since the charge induced in the semiconductor is nonuniformly distributed in the conducting channel just under the oxide, in the current saturation region. The gate-to-drain capacitance C_{gd} is normally less than C_{gs}. A small drain-to-source capacitance C_{ds} completes the simple high-frequency FET model of Fig. 10.18b.

Since the displacement current through a capacitor is given by $i = C\, dv/dt$, the gate and drain currents of the FET in this simple circuit model are given by

$$i_g = C_{gs}\frac{dv_{gs}}{dt} + C_{gd}\frac{dv_{gd}}{dt}, \qquad (10.25a)$$

and

$$i_d = g_m v_{gs} + g_{os} v_{ds} + C_{ds}\frac{dv_{ds}}{dt} - C_{gd}\frac{dv_{gd}}{dt}. \qquad (10.25b)$$

Here an incremental gate signal current i_g must be supplied to charge C_{gs} and C_{gd}. This equation for i_g is consistent with the summing of the currents at the gate node G of Fig. 10.18b. Similarly, Eq. (10.25b) can be obtained by summing the currents at the drain node, marked D.

These interelectrode capacitances can be obtained by measurements which are next described. First the output drain-to-source reverse transfer capacitance C_{rss} is measured with the gate-to-source input terminals short-circuited. This

[10] P. E. Gray and C. L. Searle, *Electronic Principles—Physics, Models and Circuits*, p. 339, John Wiley & Sons, Inc., New York (1969).

FIGURE 10.19

FIGURE 10.19

Common-source small-signal y-parameter equivalent circuit of a field-effect transistor. In general, the y parameters are frequency dependent.

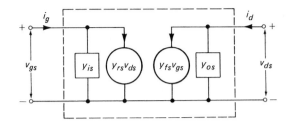

yields a value for C_{gd} in parallel with C_{ds}, which is essentially C_{gd} since normally $C_{gd} \gg C_{ds}$. Now with the output drain-to-source terminals short-circuited, a measurement of the input gate-to-source capacitance C_{iss} gives a value for C_{gd} and C_{gs} in parallel or $C_{gd} + C_{gs}$. When C_{gd} from the first measurement is subtracted from this latter measurement, C_{gs} is obtained. Since these capacitances are voltage-dependent in the case of the JFET, care must be taken to make all the measurements at the same value of voltage.

It is sometimes convenient to model the FET for small-signal high-frequency calculations with a two-port black-box equivalent circuit similar to that introduced for the bipolar transistor in Section 9.2A and in Fig. 9.8. In the present case, however, y-parameter admittances are commonly used. This circuit model is described by the following equations, where the subscripts 1 and 2 refer respectively to the input and output:

$$i_1 = y_{is} v_1 + y_{rs} v_2 \qquad (10.26a)$$

and

$$i_2 = y_{fs} v_1 + y_{os} v_2. \qquad (10.26b)$$

The values of the y parameters can be determined by measurements at the device leads. Here y_{is} is the input admittance, and y_{fs} is the forward transadmittance with the output short-circuited; y_{rs} is the reverse transadmittance, and y_{os} is the output admittance with the input short-circuited. Note that the real part of y_{fs} is g_{fs}, the forward transconductance, equal to g_m when the FET output is short-circuited. (See the device parameter specification sheets in Appendixes E.2 and F.) These y parameters are frequency dependent.

In the common-source circuit configuration, i_1 and i_2 in Eqs. (10.26) are respectively i_g and i_d; the voltages v_1 and v_2 are respectively v_{gs} and v_{ds}. This common-source small-signal y-parameter equivalent circuit is sketched in Fig. 10.19.

PROBLEMS

10.1 Explain the output characteristic of the p-channel depletion-type MOSFET for negative values of gate potential relative to the substrate.

10.2 Discuss the effect of increased substrate doping on the gate threshold voltage of an n-channel enhancement-type MOSFET.

10.3 Given for a p-channel enhancement-type MOSFET made of silicon:

$$\epsilon_{ox} = 3.8(8.85 \times 10^{-12} \text{ F/m}),$$
$$\mu = 0.020 \text{ m}^2/\text{V-sec},$$
$$l = 1.0 \times 10^{-5} \text{ m},$$
$$t_{ox} = 1.0 \times 10^{-7} \text{ m},$$
$$L = 7.5 \times 10^{-5} \text{ m},$$
$$V_{GST} = -1.5 \text{ V},$$
$$\text{substrate resistivity} = 0.01 \text{ }\Omega\text{-m, n-type.}$$

a) Compute the gate oxide capacitance.
b) Calculate $(I_D)_{sat}$ for values of V_{GS} ranging from 0 to -10 V, in 2 V intervals.
c) Determine the bottoming conductance g_{ds} of this device near $V_{DS} = 0$ for $V_{GS} = -2, -4, -6, -8,$ and -10 V.

10.4 Starting with Eq. (10.9), derive Eq. (10.10a).

10.5 For the device described in Problem 10.3:

a) Find the device g_m at $V_{GS} = -2, -5,$ and -10 V.
b) Repeat part a with the channel reduced in length by a factor of 2.

10.6 Assume that the drain space-charge width of the MOSFET of Problem 10.3 increases by 0.10 μm/V of drain-source voltage.

a) Calculate the active-region output conductance g_{os} for a gate voltage of -2 and -10 V.
b) Repeat part a with the channel reduced in length by a factor of 2.

10.7 Determine approximately the maximum drain-voltage capability of the device of Problem 10.3 if this is limited by reach-through of the drain space-charge region to the source region. How is this value affected if the channel length is halved? How is this affected by the substrate doping?

10.8 With the device of Problem 10.3 used as an MOS resistor:

a) Calculate the dc resistance at $V_{DS} = -2$ and -10 V.
b) Determine the value of L/l required to provide a 1 MΩ resistance for $V_{DS} = -10$ V.

10.9 In the inverter circuit of Fig. 10.13b, assume the MOSFET design of Problem 10.3 and an identical MOSR structure. Let $V_{DD} = -10$ V, and calculate V_{out} for $V_{in} = -10$ V.

10.10 In the inverter circuit of Fig. 10.13b, assume the MOSFET design of Problem 10.3 and a similar MOS resistor structure except that the values of l and L are interchanged. Let $V_{DD} = -10$ V, and calculate V_{out} for

a) $V_{in} = -10$ V,
b) $V_{in} = 0$ V, if the leakage current of the MOSFET with zero gate voltage is 10 nA.

10.11   For the device of Problem 10.3 at $V_{GS} = -2$ and -10 V:

a) Determine the gain-bandwidth product of the device.
b) Repeat part a with the channel length halved.

10.12   a) Derive an expression for the gain-bandwidth product of a MOSFET by calculating the time necessary to charge the gate capacitance through one-half the channel resistance. Compare this result with Eq. (10.19).
b) Compare the gain-bandwidth products for a p-channel and an n-channel silicon MOSFET of identical design.

11 Modern Solid-State Devices

11.1 INTEGRATED CIRCUITS: PHILOSOPHY AND TECHNOLOGY

The basic building block of most electronic equipment today is the integrated circuit. This was described in Section 1.2D where the integrated circuit was defined as "one piece of solid in which have been included several components, passive as well as active, without external connection between these devices." Discrete components such as planar transistors (see Fig. 1.2) and MOS transistors (see Fig. 10.1) are a primary element of these devices which "perform a complete circuit function." Vast families of these circuit blocks have been developed, and the incentive for doing so has been fourfold:

1) cost,
2) size,
3) reliability, and
4) performance.

The development of sophisticated semiconductor technology resulting in the fabrication of tens, hundreds, and even thousands of tranistors onto a chip of single-crystal silicon, less than 5 mm on each side and only 0.25 mm thick, has had low cost as a primary impetus. The planar technology of integrated circuits utilizes *batch* processing which permits hundreds of discrete components to be incorporated into a single chip of silicon at one time at about the cost of producing a single discrete transistor chip. Hence the unit cost of a transistor has been reduced to a fraction of a cent in the integrated circuit.

The use of photolithographic processing to delineate the many devices on the silicon chip not only provides a cheap process for simultaneously producing these components, but also permits them to be made extremely minute in size.

Three stages in the manu-
facture of integrated circuits.
The photograph on this page
shows the dust-off pro-
cedure, involved in the
transfer of an integrated
circuit from a negative mask
to the chemical-coated
wafers. In this operation, a
compressed-air hose dusts a
wafer before alignment and
photography (shown in the
photos on the facing page).
(Courtesy of RCA Solid
State Division, Somerville,
N.J.)

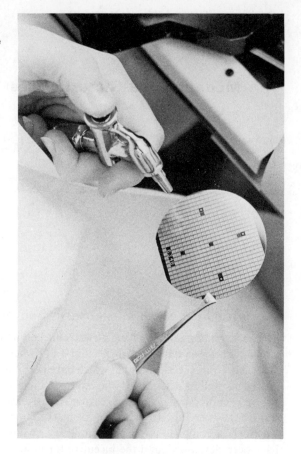

The theory of optics dictates that the minimum resolvable size of these structures is limited only by the wavelength of the radiation used to expose the
photographic emulsion utilized in the processing. This ranges from visible
light with a wavelength of several thousand angstroms to electrons with a wavelength of a few angstroms.

The technique utilized for interconnecting the various components on the
single-crystal chip involves the deposition of a thin film of metalization in a
pattern delineated again by a photolithographic process. Since the joining
together of different metals in interconnecting the various devices in an
electronic circuit is a primary source of circuit failure, the use of a *single* thin
metal film provides a highly reliable interconnection scheme. Because each
integrated circuit contains so many internally interconnected components, the
number of less reliable external interconnections made by soldering or welding
is vastly minimized. This of course also significantly reduces the cost of circuit
assembly and even permits some circuit redundancy to be practiced.

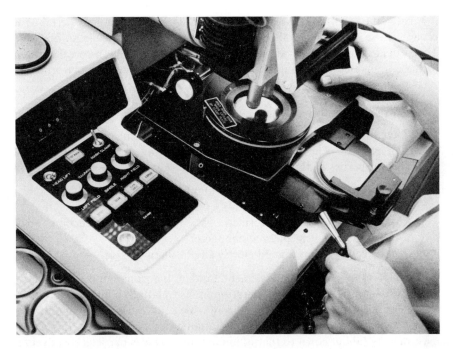

Vertical and horizontal alignment of integrated circuit wafers before final photographing. (Courtesy of RCA Solid State Division, Somerville, N.J.)

These silicon wafers, each containing hundreds of integrated circuits in a grid pattern, are ready to be broken into chips and mounted on lead frames. (Courtesy of RCA Solid State Division, Somerville, N.J.)

Finally, the batch-process aspect of integrated circuit technology permits the economic use of a large number of devices to perform a single circuit function. This provides for a sophistication of circuit design not possible before the advent of these integrated devices and hence an improvement in circuit performance.

A. Materials for Integrated Circuits

Most integrated circuit blocks in use today employ the semiconductor silicon as a starting material. The reason for this is the following unique combination of favorable properties of this semiconductor material:

1) It is an element of low atomic number.
2) Pure, single crystals are readily grown.
3) P-N junctions can be formed easily by impurity introduction.
4) It is an excellent thermal conductor.
5) It readily forms a good electrically insulating oxide.

The fact that silicon is an element (and not a compound), abundantly available in nature,[1] makes its preparation for use relatively simple and economical. The elementary nature of silicon makes its purification and single-crystal growth straightforward. Its low atomic number means it has a low mass density (about equal to that of aluminum). This is particularly advantageous in space applications where the lightweight nature of the integrated circuit is of extreme importance.

Techniques for purifying silicon are readily available, and the material is obtainable economically at a maximum impurity level of less than one part per billion. Single crystals can be grown from the material without much difficulty in a manner shown in Fig. 2.2. The need for single crystals of silicon as the starting material for the fabrication of integrated circuits was noted in Section 2.1B; it stems from the requirement for perfection and controllability of the material. This is particularly true in the case of relatively large-area circuits in which a minute imperfection can destroy the proper functioning of the entire circuit chip.

P-N junctions are readily produced in single-crystal silicon by the selective introduction of donor and acceptor impurities via sequential exposure of the material to a vapor of these impurities at an elevated temperature in excess of 1000°C. This technique of *solid-state diffusion* is not always successful in materials which are compounds but is easily accomplished, very controllably, in elements such as silicon and germanium.

All electronic devices are limited in their performance by internally generated heat. Semiconductor devices in particular are temperature sensitive, as has already been discussed in some detail. An important requisite for an integrated circuit starting material is that it have good thermal conductivity, particularly

---

[1] About 20 percent of the earth's crust is made up of silicon.

because of the minute size of these devices, so that the high heat density generated therein can be efficiently conducted to a heat radiator. Proper circuit functioning is often possible using appropriate circuit design at somewhat elevated temperatures as long as all the elements of the circuit are at the same temperature. Operation at high power density is possible using silicon, which has a thermal conductivity at room temperature about one-third that of the metal copper. This is particularly high for a semiconductor substance. For example, germanium has only about one-half the thermal conductivity of silicon. In addition the maximum operating temperature of a germanium device is about 100°C, compared to 200°C for silicon devices, because of germanium's smaller energy gap. The p-n junction leakage current due to thermal electron-hole generation at 100°C is already excessive in the case of germanium but not so in the case of silicon.

In a final comparison of silicon with other semiconductors such as germanium it is important to point out that silicon forms an excellent, stable, insulating oxide which plays an important role in integrated circuit technology. This is in contrast to germanium, which forms two types of oxides: one is water soluble and the other is unstable at high temperatures. Silicon dioxide (SiO_2) is an electric insulator with a resistivity in excess of 10^{12} Ω-m. The common name for this material is quartz, known to be very stable at high temperatures and having a melting point in excess of 2000°C. Silicon dioxide is used for two separate purposes in the fabrication of integrated circuits. Its electric insulating property is utilized in coating and protecting p-n junctions from contamination by unwanted impurities in the surrounding ambient. In addition, it is used as a stable high-temperature masking material which selectively blocks donor or acceptor impurities from diffusing into the silicon where they are not desired. The protective property of silicon dioxide has been discussed in Section 5.6; it may often be enhanced by the addition of a layer of silicon nitride over the oxide.

Material preparation for integrated circuits   The basic form of the starting material for the silicon monolithic integrated circuit[2] is a slice of silicon cut from a single crystal of the type pictured in Fig. 2.2a. The diameter of the crystals presently in use is 2–3 in. or about 6 cm. These are then cut with a diamond-primed saw,[3] like baloney, into slices less than half a millimeter thick. Thousands of individual circuit chips 5 mm² or more in area are eventually produced by dicing this slice with a tungsten carbide scribe which is used to cut the silicon just as glass is cut. This is shown in Fig. 11.1. However, the silicon slice is kept intact during most of the fabrication so that, by batch

---

[2] "Monolithic" means all of a uniform nature. Hence the monolithic circuit is one in which all devices, both active and passive, are fabricated basically from one material and in one block of the substance. The material preparation described here is primarily for an integrated circuit containing bipolar transistors.
[3] Diamond must be used since silicon is a very hard material.

FIGURE 11.1

a) Single crystal of silicon
about 6 cm in diameter cut
into slices about 0.5 mm
thick.
b) Single slice of silicon
diced into chips about 2.5
mm on each side.

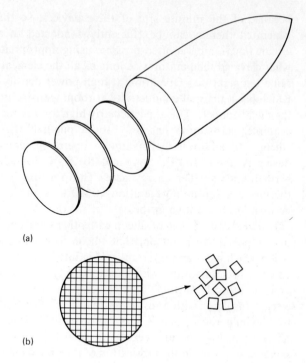

(a)

(b)

processing, thousands of identical circuits may be made at once. The dicing
operation, one of the final steps in integrated circuit fabrication, occurs just
prior to circuit packaging.

After sawing, the silicon slices are lapped and polished to a mirrorlike
finish. The final step in the polishing is chemical treatment in a mixture of
nitric and hydrofluoric acids to dissolve away any roughness in the silicon
crystal surface. The result is an extremely flat and smooth finish.

The slices so treated are generally cut from 0.001–0.01 Ω-m p-type silicon
single crystal. These are then placed in a reactor for the deposition of a thin
(about 5–20 μm) layer of 0.001–0.01 Ω-m n-type silicon single crystal on one
side of the silicon slice by *epitaxial growth*. Epitaxial growth refers to the
deposit of a thin single-crystal layer onto a single-crystal substrate of a basically
similar material, from the vapor phase. This vapor-phase growth is in contrast
to the growth from the liquid phase (illustrated in Fig. 2.2c) in that the seed in
epitaxial growth is a flat silicon substrate slice which is exposed to a silicon-
containing vapor at an elevated temperature *below* the silicon melting point.
In this way a thin n-layer of extremely uniform thickness may be prepared.
The apparatus for accomplishing this deposition is shown in Fig. 11.2a. The
silicon source material is a few percent of silicon tetrachloride in hydrogen gas,

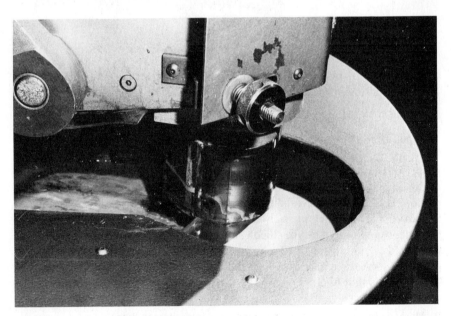

Diamond saw for slicing silicon crystals. The closeup shows a crystal being sliced. (Photos courtesy of Silicon Technology Corporation.)

This employee is operating a
laser scriber that cuts
through processed wafers
so that each can be separated
into hundreds (or thousands)
of chips. (Courtesy of RCA
Solid State Division, Somerville, N.J.)

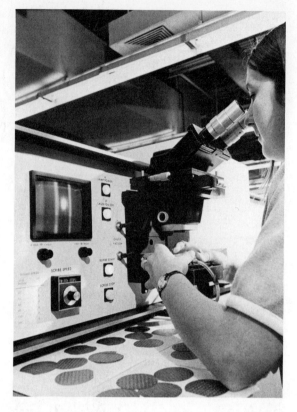

and the chemical reaction that takes place at temperatures between 1000 and
1200°C, resulting in the silicon deposition, is given by

$$SiCl_4 + 2H_2 \leftrightarrows Si\downarrow + 4HCl. \qquad (11.1)$$

In addition to accurate control of the thickness of this deposited n-silicon
layer, its impurity content must also be controlled to yield material of the
proper resistivity. For this purpose an extremely small, but accurate, quantity
of donor material is added to the $SiCl_4$. This is usually in the form of phosphine, PH_3. Figure 11.2b shows in cross section a typical starting slice for the
fabrication of the integrated circuit, including a thin n-type silicon epitaxial
layer deposited onto a p-type silicon substrate. All the electronic devices in the
integrated circuit are included in the n-type epitaxial film. The p-type substrate
primarily serves as a base on which the epitaxial layer is deposited, for ease in
handling this thin film.

B. Methods for the Controlled Introduction of Impurities in Silicon

A primary method for the introduction of donor and acceptor impurities into
silicon is the technique of solid-state diffusion. This involves exposing the

silicon to the vapor of an appropriate impurity at a temperature in excess of 1000°C. Another technique in use is the ion implantation of donors and acceptors into the semiconductor crystal. Here ions of doping atoms are accelerated to an energy of a few hundred thousand electron-volts and "shot" and implanted into the silicon. Its simplicity makes diffusion the most-used technique for impurity introduction at the moment. All that is required in the way of equipment is a furnace capable of raising the semiconductor crystal to temperatures in excess of 1000°C. However, oxide-masking techniques must be employed to ensure that the impurity is introduced into the crystal slice in only certain areas selected by design. In contrast, ion implantation requires a rather complex and expensive ion accelerator. However, in principle, doping atoms can be electronically programmed to be introduced in only certain selected areas, automatically and without oxide masking. Thermal diffusion is a more gentle process; it causes much less damage to the semiconductor crystal than the bombardment required for ion implantation. Thermal treatment is required after implantation to anneal out structural defects in the crystal.

The solid-state diffusion technique   The process of solid-state diffusion is comparable in a sense to the diffusion of minority electrical carriers in the silicon lattice described in Section 5.4A. That is, the atoms in the vapor undergo

FIGURE 11.2

a) Apparatus for depositing thin single-crystal layers of silicon onto single-crystal silicon substrates; phosphine gas, PH_3, is introduced to yield n-type layers, while diborane gas, B_2H_6, is used for p-type layers.
b) Cross section of a typical silicon epitaxial starting slice for the fabrication of integrated circuits.

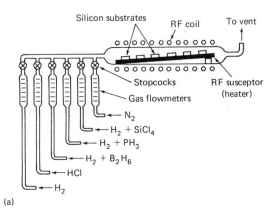

(a)

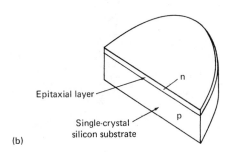

(b)

random motion, but there is a statistical probability that the impurities will tend to enter the pure silicon crystal lattice, where few such doping atoms are present, from the impurity-rich vapor. The formulation of the problem of the flow of these impurity atoms from the vapor into the solid results in a diffusion equation similar to Eq. (7.5) for the diffusion of minority carriers in a crystal. However, no recombination term is needed for the case of the diffusion of impurity atoms, so that the relevant equation can be written as

$$D\frac{\partial^2 N}{\partial x^2} = \frac{\partial N}{\partial t},$$ (11.2)

where $N(x, t)$ represents the density of impurity atoms at a distance x from the surface of the crystal after a diffusion time t at an elevated temperature. Here D is the diffusion constant which represents the rate of penetration of the doping impurity into the crystal. This parameter may be expected to be temperature dependent, the penetration rate increasing with increasing temperature as the random motion of the impurity atoms becomes more vigorous. The variation of the diffusion constant with absolute temperature T is given by

$$D = D_0 e^{-E_D/kT},$$ (11.3)

where E_D is the *activation energy for diffusion* which for silicon has the value of about 4.3 eV for nearly all acceptor and donor impurities, and D_0 is a temperature-independent constant. A semilogarithmic plot of D versus $1/T$ results in a straight line whose slope is determined by E_D. A graph of the diffusion constant of the primary donor and acceptor impurities in silicon versus temperature is given in Fig. 11.3.

Diffusion furnace for fabrication of integrated circuits. (Courtesy of Lindberg, Division of Sola Basic Industries.)

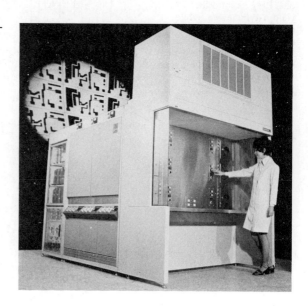

FIGURE 11.3

Diffusion coefficients of the primary donor and acceptor impurities in silicon versus temperature. Data are shown for the p-type impurities aluminum, gallium, boron, and indium and the n-type impurities phosphorus, arsenic, and antimony.

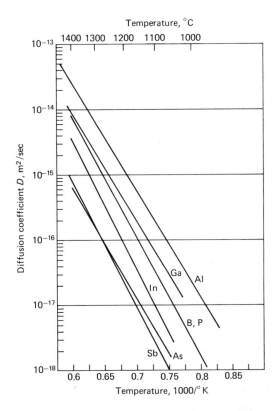

If the silicon crystal is in contact at all times at an elevated temperature with a constant concentration of donor or acceptor atoms in a vapor, the impurity-atom distribution in the crystal is given by

$$N(x, t) = N_0\left[1 - \text{erf}\left(\frac{x}{2\sqrt{Dt}}\right)\right] \equiv N_0 \text{ erfc}\left(\frac{x}{2\sqrt{Dt}}\right). \qquad (11.4)$$

This equation results from the solution of Eq. (11.2) subject to appropriate boundary conditions. Here N_0 represents the concentration of the impurity at the surface ($x = 0$) of the semiconductor, assumed independent of time, and erf is the mathematical error function which previously appeared in Eq. (7.24). Erfc is by definition the complementary error function (erfc $Z \equiv 1 - \text{erf } Z$) which is a tabulated function that is graphed in Fig. 11.4a. With Eq. (11.4) and Fig. 11.4a, the distribution of impurity atoms which have penetrated into the surface of a semiconductor crystal may be predicted. This is represented in the graph of Fig. 11.4b, which shows the increasing penetration of the impurity atoms beyond the surface as time progresses with the crystal held at some temperature T. Since the crystal is normally exposed to a dense vapor of donors or acceptors, the crystal surface is saturated with this impurity and

FIGURE 11.4

a) Complementary error function of Z, erfc Z, versus Z (erfc $Z = 1 -$ erf Z).
b) Graph of increasing penetration at $T = T_1$ (°K) of impurity atoms below the silicon surface, which is located at $x = 0$.

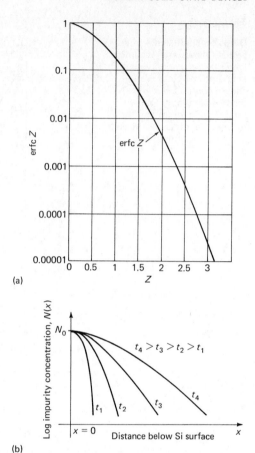

(a)

(b)

hence N_0 in fact represents the *maximum solubility* of the impurity atom in the semiconductor. Values for the maximum solid solubility of various donor and acceptor atoms in silicon in the normal diffusion-temperature range of 1100–1300°C are given in Table 11.1.

To form n-p-n-type structures as for planar bipolar transistors (see Fig. 1.2), p-type and then n-type impurities must be sequentially diffused into a crystal

TABLE 11.1 Maximum solid solubility of donor and acceptor impurities in silicon

	Donors			Acceptors		
Impurity	P	As	Sb	B	Ga	Au
Solid solubility, atoms/m³ (1100–1300°C)	4×10^{26}	2×10^{27}	5×10^{25}	5×10^{26}	4×10^{25}	1×10^{23}

TABLE 11.2

Ratio of the diffusion coefficients in SiO_2 compared with Si for donors and acceptors in Si between 1100 and 1300°C

	Donors		Acceptors	
Impurity	P	Sb	B	Ga
Ratio of diffusion coefficients SiO_2/Si (1100–1300°C)	1×10^{-4}	3	3×10^{-3}	50

containing n-type impurities. In addition, these diffusions must selectively introduce the doping impurities only into certain areas of the surface of the crystal. The presence of silicon dioxide in an area of the silicon surface tends to retard the introduction of these impurities there. This results from the fact that certain impurities like boron and phosphorus diffuse more slowly in silicon dioxide than in silicon. The ratios of diffusion constant of the donor and acceptor impurities in SiO_2 relative to silicon in the diffusion-temperature range for some common doping impurities are given in Table 11.2. The several impurities with low ratios are said to be effectively masked by this thermally stable oxide which is used to delineate device structures on the silicon crystal surface.

Gold is normally introduced into silicon by diffusion for control of minority carrier lifetime, as discussed in Section 7.3A. The diffusion coefficient of Au in Si is quite rapid, about 10^{-9} m^2/sec, so that this diffusion is normally accomplished at temperatures as low as 800°C.

Ion implantation The process of ion implantation of boron atoms in silicon requires the use of a boron-ion source as well as an accelerator to raise the kinetic energy of the ions to a few hundred kilovolts. Positively charged boron ions can be produced by exposing boron gas to a radiation source. These ions are then caused to fall through a large dc potential difference between a capacitorlike metal plate and the silicon slice. These high-energy boron ions penetrate a distance of micrometers into the negatively charged silicon crystal, capturing electrons there and becoming p-type doping atoms. An apparatus for performing this implantation is shown in Fig. 11.5a. Masking of these ions can be accomplished with a thin metal film or an oxide layer. In addition, TV-raster type electric deflection may be used to scan the silicon target. This is shown in Fig. 11.5b. The bombardment damage in the crystal is later removed by moderate-temperature ($\sim 600°C$) thermal annealing. This technique is very useful in sensitively adjusting the doping of an MOS transistor surface channel which permits accurate control of the MOSFET threshold voltage. Also, by modulating the ion beam, a predesigned pattern of impurity atoms may be introduced into the crystal in this manner.

FIGURE 11.5

a) Drawing of a 300 kV ion-implantation system with three arms for separate applications of the ion beam.

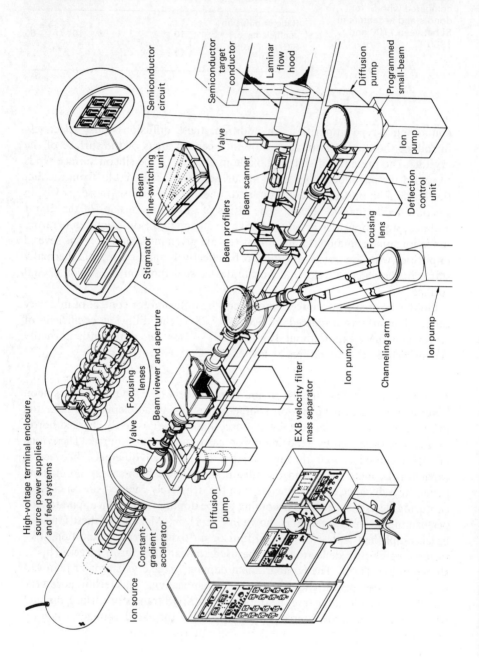

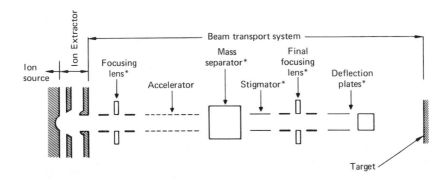

*Not required in all systems; may be arrayed in
different sequence depending on system use

b) Sketch of an ion-implantation system including an ion source, focusing electrodes, acceleration tube, mass analyzer, neutral trap, beam deflector, and wafer chamber. Since some ions are neutralized in transit, these must be eliminated before reaching the target. (From R. G. Wilson and G. R. Brewer, *Ion Beams*, John Wiley & Sons, New York, 1973.)

Commercial ion-implantation equipment (a 150 kV production unit). (Courtesy of Danfysik/High Voltage Engineering Corporation.)

FIGURE 11.6

a) Epitaxial wafer after p-type impurity diffusion resulting in isolated "n-pockets" in p-type silicon. b) Silicon dioxide thin-film isolation of single-crystal n-type pockets of silicon embedded in an amorphous silicon substrate.

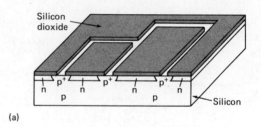

(a)

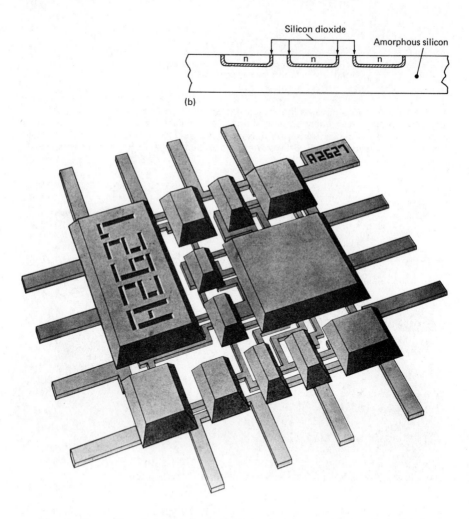

(b)

c) Sketch of rear view of a simple beam-lead circuit structure showing the isolated segments of the silicon and gold-plated beam leads. (Courtesy of Bell Laboratories.)

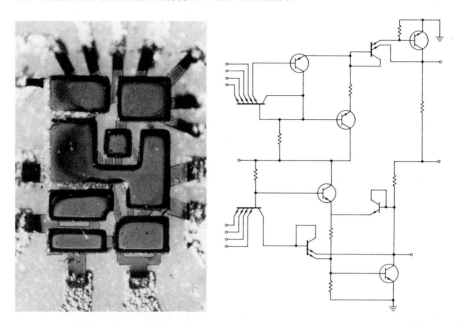

Actual photograph taken from the rear of a simple beam-lead structure. This shows the beam-leaded device joined to the metalized areas of a ceramic circuit board. (Courtesy of Bell Laboratories.) At the right is a schematic diagram of a simple beam-lead circuit.

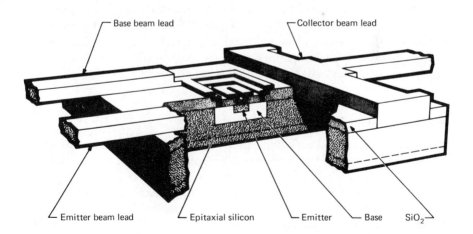

Detailed sketch of a beam-leaded bipolar transistor structure. (Courtesy of Bell Laboratories.)

C. Device Isolation Techniques

A monolithic integrated circuit may include many transistors in a single block of silicon. If a number of n-p-n planar devices of the type illustrated in Fig. 1.2 are included in the epitaxial n-silicon layer, it is apparent that their collector regions will all be electrically connected together by the silicon. This in general does not satisfy circuit connection requirements. Hence some method must be available to electrically isolate bipolar devices on the silicon chip from each other. There are two basic techniques in general use for accomplishing this task:

1) isolation by introducing a high-impedance p-n junction between each pair of components, and
2) isolation by producing an electric nonconducting insulating material between the various components.

The first method involves diffusing p-type regions through openings in an oxide mask on the n-type epitaxial layer so as to produce p-type isolated " n-pockets," as shown in Fig. 11.6a. Other somewhat different diffusion procedures have been proposed for accomplishing the same purpose as well as achieving more economy in the surface area devoted to this isolation step — and hence permitting higher-component-density circuits to be integrated. In all these methods, isolation is accomplished by a very-high-resistance, low-leakage-current reverse biased p-n junction. The second method involves the use of a thin (1 μm) dielectric film of silicon dioxide to surround each n-pocket and isolate it from the amorphous silicon substrate. This is shown in Fig. 11.6b. Another implementation of this dielectric isolation technique uses air as an insulator to isolate the different devices electrically. This is illustrated in Fig. 11.6c, which is the *beam-lead* structure. This is produced by chemically etching away the semiconductor material between components. The device is usually joined to a package upside-down by welding the many thick (25 μm) gold-plated extended leads to metallized areas on the package.

D. Devices for Integrated Circuits

The minute device structures of integrated circuits are delineated on the silicon surface with the aid of photolithographic techniques. The first step in the fabrication procedure is to coat an oxidized epitaxial slice of silicon with a liquid photosensitive material called *photoresist*, which is basically a polyvinyl alcohol in a solvent such as xylene. The solvent is then baked out to harden the coating. Then the whole coated slice is placed under a photographic negative glass plate that contains opaque and transparent areas which comprise device patterns that are to be transferred to the oxide. Now the oxidized silicon slice, covered by the masking plate, is exposed to ultraviolet radiation. The photoresist becomes polymerized and hence solvent-insoluble

Mask-making facility, the totally integrated GCA/David W. Mann "Turnkey System." This includes (clockwise, starting at the lower left) : the Pattern Compiler, the Pattern Generator, the Photorepeater ®, and the Contact Printer or Mask Replicator. (Courtesy of GCA Corporation.)

in the areas exposed to ultraviolet light; it is soluble in a solvent such as xylene where it has been protected from light by the mask. Now the coated slice is washed with a solvent (developed) which leaves a desired photoresist pattern on the oxide. The oxide is then selectively removed in certain areas and allowed to remain in others; this is accomplished by subjecting the slice to a buffered solution of hydrofluoric acid which dissolves silicon dioxide where it is not protected by the photoresist. In this way an oxide mask is produced which will selectively inhibit the diffusion of doping impurities such as boron and phosphorus when the slice is later subjected to a vapor of these impurities at a temperature in excess of 1000°C. The aluminum interconnection pattern is also delineated like the oxide pattern by depositing the metal everywhere on the slice and using photoresist for protection and phosphoric acid to dissolve the metal where it is unwanted.

The basic procedures outlined above are repeated in sequence to produce the final integrated circuit device structure. A typical set of steps for fabricating a monolithic integrated device slice is outlined in Fig. 11.7.

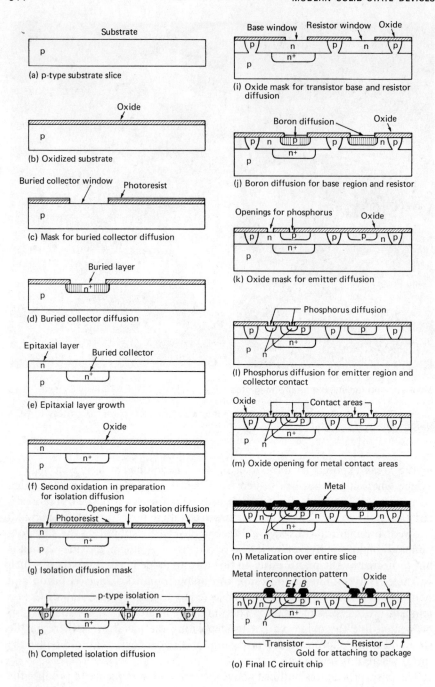

(a) p-type substrate slice

(b) Oxidized substrate

(c) Mask for buried collector diffusion

(d) Buried collector diffusion

(e) Epitaxial layer growth

(f) Second oxidation in preparation for isolation diffusion

(g) Isolation diffusion mask

(h) Completed isolation diffusion

(i) Oxide mask for transistor base and resistor diffusion

(j) Boron diffusion for base region and resistor

(k) Oxide mask for emitter diffusion

(l) Phosphorus diffusion for emitter region and collector contact

(m) Oxide opening for metal contact areas

(n) Metalization over entire slice

(o) Final IC circuit chip

FIGURE 11.7

Typical set of steps in the fabrication of a monolithic integrated circuit device.

This process inherently permits the basic device pattern to be repeated on the silicon slice thousands of times and more; it should be clear that all the fabrication steps outlined are of the batch fabrication type. After the slice is completed, it is scribed and broken into separate silicon chips a few millimeters on a side. From that point on, the chips are mounted into packages and leads are bonded to the devices, individually, no longer in a batch procedure. Hence these latter steps are the most expensive ones in the fabrication; some successful attempts have even been made to attach the leads all at once in a batch procedure. The beam-lead technique is a process which involves the simultaneous electroplating of all the leads, between 14 and 40 in number, at once (see Fig. 11.6c).

Having described the procedures for producing integrated circuit devices, we shall next discuss the various components of these circuits. The differences between the components included in these monolithic integrated circuits and the comparable discrete components will be emphasized.

Integrated transistors and diodes Figure 11.8a shows a cross section of a typical integrated version of an n-p-n bipolar transistor. Note how this device is electrically isolated from the p-type substrate by a p-n isolation junction. Note too that the collector contact and the emitter and base metal contacts are made on the upper surface of the silicon slice, since all interconnections must be made on the top surface of this monolithic integrated circuit chip (see Fig. 11.9). This is in contrast to the discrete planar transistor structure of Fig. 1.2 where the collector contact is made at the bottom of the chip. The relatively long path through the collector region to the collector contact in the integrated transistor as compared to that in the discrete transistor accounts for the relatively high inherent series collector resistance of the integrated component. The n^+-*buried-collector* low-resistivity region shown in the figure is introduced to effectively "short out" this high-resistance collector current path.

The n-p-n transistor structure is generally used in preference to the p-n-p design, owing to the fact that high-mobility electrons transport the injected current from emitter to collector. This yields higher frequency capability and faster switching speed for the component. In certain types of circuits, however, there is a need for p-n-p bipolar transistors to complement the n-p-n components. The production technology is not compatible with producing both n-p-n and p-n-p devices of comparable structure at once. However, the p-type diffusion for the n-p-n base region may be used at the same time to produce the emitter and collector regions of a "lateral" p-n-p transistor, as shown in Fig. 11.8b. This p-n-p transistor normally has rather wide spacing between emitter and collector, and hence the component has low current gain, low frequency, and slow switching-speed capability. A number of linear-amplifier integrated circuits use this device for polarity reversal and level shifting where gain and frequency capability are of no consequence.

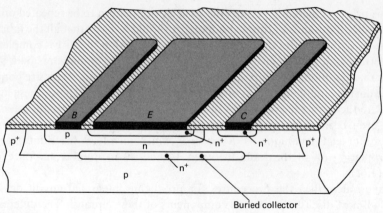

Buried collector

FIGURE 11.8

a) Cross section and top view of a typical integrated version of an n-p-n bipolar silicon transistor. Note the utilization of a "buried" n⁺-layer in the collector region to reduce series collector resistance. The emitter, base, and collector metalizations are marked E, B, and C. The crosshatched regions represent the insulating oxide.

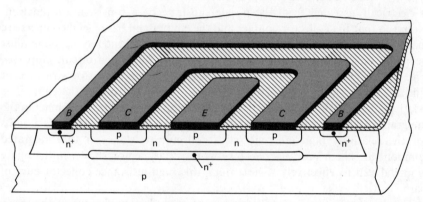

b) Cross section of a typical "lateral" p-n-p bipolar transistor often utilized in integrated circuits. The "buried" n⁺-collector layer is used here to improve the current gain. This gain is normally low relative to the gain of "vertical" transistor structures (as in Fig. 11.8a) owing to typically 5–10 times larger emitter-to-collector spacings. The base and collector metalizations are usually in the form of a U surrounding the emitter.

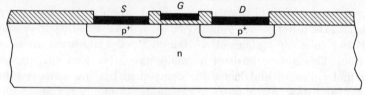

c) Cross section of a p-channel MOS transistor. The source, gate, and drain metalizations are marked S, G, and D. The crosshatched regions represent the insulating oxide, which is thinnest under the gate electrode to provide high transconductance.

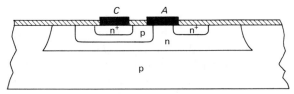

d) Cross section of an n$^+$-p integrated form of a diode. In this version the p- and n-regions (base and collector) are electrically shorted to form the anode A. An emitter region is the cathode C. This results in a fast-switching but low avalanche-breakdown voltage (6–8 V) diode.

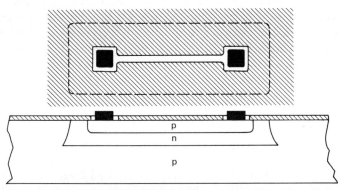

e) Cross section of a diffused silicon resistor used in monolithic integrated circuits. The current flows from one metalized area (darkened area) through the thin p-region to the other metalized area. The high-impedance p-n junction isolates the current in the upper p-region and prevents it from entering the n-region or the substrate.

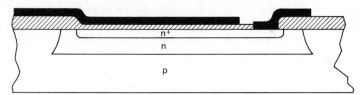

f) Cross section of a monolithic integrated capacitor. Here the thin insulating layer (shown crosshatched) is usually silicon dioxide thermally grown onto the silicon substrate by exposing it to oxygen or water vapor at a temperature exceeding 1000°C.

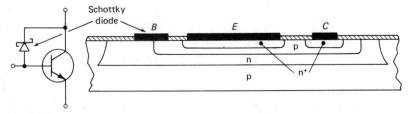

g) Cross section of an integrated Schottky diode connected in parallel with the collector-base junction of an integrated n-p-n transistor. The metalization marked B is both the anode of the Schottky diode and the transistor base contact. The cathode of the Schottky diode is the n-type collector region of the transistor. A schematic diagram of this device is shown to the left.

A single p-type diffusion will produce the p-channel MOSFET component shown in Fig. 11.8c. Comparing this structure with the bipolar structure of Fig. 11.8a confirms the earlier claim of simplicity of MOSFET fabrication technology. This is particularly evident since no isolation pockets are needed for MOSFET circuits. They are "self-isolated" since the current is automatically confined to the small channel region at the silicon surface, between source and drain, as discussed in Chapter 10.

The diode in integrated circuits is actually a transistor structure with two regions shorted together by aluminum metalization. For example, the base and collector regions may be connected together to form one electrode of the diode while the emitter electrode forms the other connection. Any other two regions may be connected together to form a different diode structure; the electrical behavior of the component is determined by which two are interconnected.[4] A form of integrated diode is shown in Fig. 11.8d.

Integrated diffused resistors The form that discrete resistors often take is a thin metallic film composed of a high-resistivity material such as nichrome. Such a film could be deposited onto the oxide of the integrated circuit chip and delineated into the form of thin-film resistors using photoresist. This would require the introduction of still another material and step in the integrated circuit process. Instead, silicon diffused resistors are normally used for the purpose of limiting current and establishing biases in integrated circuits. The integrated resistor in its usual form is shown in Fig. 11.8e. It consists of a long narrow diffused p-type region embedded in an n-type region and hence electrically isolated from it by a p-n junction. This p-type region is diffused into the silicon at the same time as the integrated transistor base region. It is this compatibility with the standard diffusion technology which accounts for its general use in bipolar circuits.

Since the basic material of the diffused resistor is the semiconductor silicon, its resistance value is subject to rather large variations with temperature. In fact, the integrated resistor increases in value by about 3000 ppm/°C temperature increase, in contrast to only 50 ppm/°C for a thin-film nichrome resistor. However, this is of little consequence in digital switching circuits which do not require accurate resistor values, since it is only necessary to maintain two separate circuit states, ON and OFF. Linear-amplifier circuits require much more strict control of resistor values, however. In the commonly used differential-amplifier integrated circuits, it is only the matching and tracking with temperature of a pair of resistors which is required. This is automatically accomplished in integrated circuit technology since both resistors in the pair are produced at the same time and in close proximity to each other on the

---

[4] For a more complete discussion of integrated circuits see, for example, J. Millman and C. C. Halkias, *Integrated Electronics*, Chapter 7, McGraw-Hill Book Company, New York (1972).

silicon chip. Resistor values ranging from about 50 to 30,000 Ω are conveniently attained by this method.

Integrated capacitors   It is quite natural to expect the silicon dioxide layer on the integrated circuit chip to be used as the dielectric material in a parallel-plate type of capacitor structure. This is in fact the form of the capacitor commonly used in integrated circuit chips, as is illustrated in Fig. 11.8f. The upper plate of this sandwichlike structure is aluminum metalization; the middle section is the insulator silicon dioxide. The lower plate is basically a heavily doped n-type silicon layer which is contacted with aluminum metalization. The relative dielectric constant of silicon dioxide is unfortunately only 3.8, but in other respects it is an excellent insulating material. Since the thinnest pinhole-free oxide that can be produced with good yield at this time is 1000 Å, it can be shown that the capacitance per unit area achieved in this manner is only about 350 pF/mm^2. However, a square millimeter constitutes a substantial part of an integrated circuit chip. In fact, about 50 transistors or more can be integrated per square millimeter. Hence it is clear that the use of integrated capacitors in integrated circuits must be restricted. This accounts for the general use of directly coupled amplifier circuits, which do not require isolating or bypass capacitors, in the integrated form.

Integrated Schottky diodes   The important use of the Schottky diode for speeding up digital switching circuits was mentioned in Section 6.5. This majority carrier device is shown in Fig. 11.8g in its integrated form, shunting the collector junction of an n-p-n transistor. The simplicity of structure of this metal-semiconductor diode accounts for its frequent use in transistor logic integrated computer circuits — to reduce carrier storage time in this saturating form of logic circuit (see Section 9.5B). For this purpose the anode of the diode is connected to the transistor base, and the cathode to the collector region. The Schottky diode metalization is done at the same time as the interconnection metalization for the integrated circuit and hence is compatible with standard integrated circuit technology. The etching of a hole in the silicon dioxide to expose bare semiconductor for the aluminum-silicon Schottky contact is accomplished in the same way in which openings are made in the oxide for impurity diffusion.

Figure 11.9a shows an integrated inverter circuit illustrating a very simple version of a monolithic silicon integrated circuit. The interconnection metalization, which is confined to the silicon wafer upper surface, is shown. The aluminum metalization paths eventually terminate at square bonding "pads" at the edges of the integrated circuit chip of silicon so that fine (20 μm diameter) lead wires may be attached. These wires are then connected to the various metal terminals which are part of the device package. The common methods of lead attachment are thermal compression bonding and ultrasonic bonding as shown in Figs. 11.9b and 11.9c.

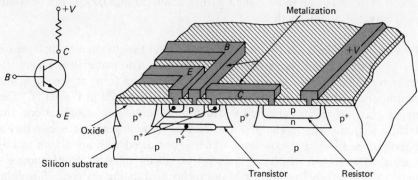

FIGURE 11.9

a) Simple silicon monolithic integrated inverter circuit showing the surface-interconnection metalization. A schematic circuit diagram of this structure is shown to the left.

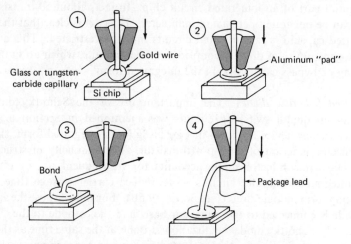

b) Sequence of steps in the "thermocompression" bonding of gold wire to an aluminum pad on a silicon circuit chip. The chip is heated to about 300°C in this operation.

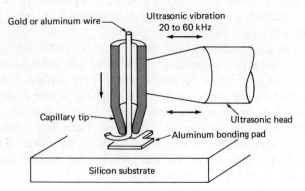

c) Ultrasonic bonding of a fine gold or aluminum wire to an aluminum pad on a chip. The ultrasonic vibrations break through the aluminum oxide. No heating of the wafer is required in this process.

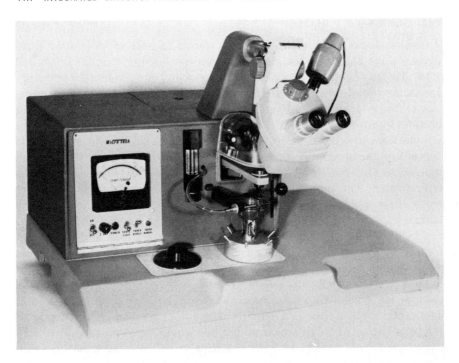

Thermocompression nailhead bonder. (Courtesy of Micro Tech Mfg., Inc., Worcester. Mass.)

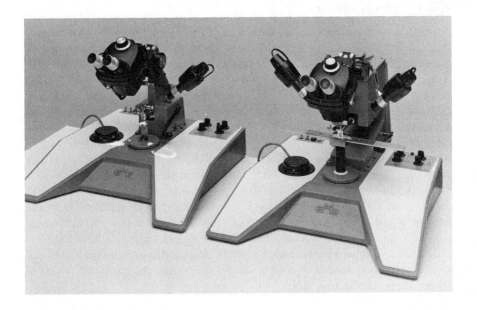

Two ultrasonic wire bonders. (Courtesy of Tempress, Division of Sola Basic Industries.)

11.2 GUNN-EFFECT DEVICES

A semiconductor device, the Gunn oscillator, capable of generating high-frequency oscillations in the gigahertz range was briefly referred to in Section 1.3B. This two-terminal device offers the possibility of extending the frequency spectrum of solid-state electronic components beyond the few-gigahertz limitation of transistors, into the 10 and 100 gHz range. This microwave component operates by a *transferred-electron* mechanism, is a bulk device, and requires no p-n junctions in order to operate. The fundamental mechanism of operation depends on the transfer of conduction electrons from the normal state of high mobility to a low mobility, as the voltage across this diode is increased beyond a critical value. This mobility decrease produces a decreasing conductivity with electric field or voltage/length of the sample (see Eq. 1.3) and hence a decreasing current with voltage as shown in Fig. 1.11. This negative resistance offers the possibility of the generation of high-frequency oscillations and even high-frequency, high-power amplification. The device is an example of a component which requires certain intermetallic compounds as starting materials. For example, the semiconductors gallium arsenide, indium phosphide, and cadmium telluride make good Gunn oscillators, whereas silicon and germanium do not. The explanation of this fact requires a quantum mechanical discussion of the band theory of solids which we are now equipped to pursue. The semiconductor laser described in Section 1.3C also is fabricated from the semiconductor gallium arsenide as well as other semiconductor materials, but not silicon or germanium. The exploration of this fact must also be made along quantum mechanical lines and will be pursued in Section 11.3.

A. GUNN-EFFECT PHYSICS

An explanation of the negative-resistance region in the *I-V* characteristic (Gunn effect) of the device pictured in Fig. 1.11 will now be presented, based on the band theory of solids. Figure 11.10 illustrates schematically the band structure of GaAs, a direct-gap semiconductor material, as discussed in Section 5.5D and shown in Fig. 5.15a. An energy gap of 1.43 eV is indicated here between the top of the valence band and the bottom of the conduction band at $k = 0$. However, in addition to these extreme values of energy, calculations predict a set of subsidiary or *satellite* conduction band minima, two of which are shown in the figure at values of $k = \pm 2\pi/a$, a being the lattice constant. In n-type GaAs at room temperature the valence band is nearly filled and the conduction band contains some electrons near $k = 0$. Energetically, the probability is extremely low for the excitation of electrons from the bottom of the conduction band to these satellite *valleys*, since an additional 0.35 eV is necessary for this transition and the thermal energy available, kT, is only 0.026 eV. Also a participating phonon is needed to provide the additional k (or momentum) needed to bring a valence electron to the minimum of one of these valleys (see Section 5.5E).

FIGURE 11.10

Band structure of gallium
arsenide showing a direct
energy band gap of 1.43 eV
as well as two satellite
(indirect) conduction band
minima at $k = \pm 2\pi/a$
separated from the minimum
of the central valley at $k = 0$
by 0.35 eV. Note the larger
E-versus-k curvature at $k = 0$
compared with $k = 2\pi/a$,
indicating the larger effective
mass of electrons in the
satellite valleys relative to the
central valley.

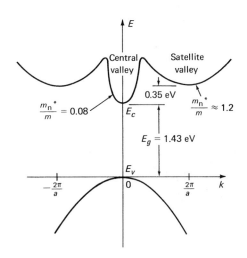

However, when a GaAs crystal is subjected to a high electric field (300 kV/m), some electrons in the normal conduction band gain sufficient energy from this field to be excited into one of the satellite valleys. Now, the effective mass of electrons in this subsidiary band is considerably larger than that of electrons in the normal conduction band centered about $k = 0$, as indicated by the small curvature of E versus k for $k = 2\pi/a$ (see Eq. 4.18). Hence these excited electrons will be less mobile than those in normal conduction. As the electric field is raised beyond the threshold value for excitation, more electrons appear in the satellite conduction band and so the average electron mobility, and hence the electric conductivity of the crystal, decreases. This results in a reduction in current as the voltage is raised above the threshold value and accounts for the negative-resistance region in the Gunn diode I-V characteristic of Fig. 1.11. This of course offers the possibility of using the device to generate oscillation energy; the physical mechanism by which this occurs will be described soon.

The tendency for the occurrence by this mechanism of a negative-resistance region in the I-V characteristic of a diode fabricated from a particular semiconductor depends on the following material characteristics:

1) The material must have a satellite band with an energy minimum which is above the normal conduction band minimum.
2) The E-versus-k curvature for this satellite band must be less than for the normal conduction band minimum. (The effective mass must be greater in this satellite band.)
3) The satellite band must have an appreciable density of energy states.
4) The energy separation between the minima in the normal and satellite conduction bands must be greater than kT to ensure that the

excitation to the low-mobility band is caused by the electric field and not by thermal energy.

5) The energy separation between the minima in the normal and satellite conduction bands must be much less than the product of the avalanche-breakdown voltage and the electronic charge, for then avalanching would preclude the occurrence of a negative-resistance region.

6) Some mechanism must be available to supply the k-value (momentum) change required for transition to the satellite band.

That gallium arsenide is a material which meets all these requirements should be clear from the previous discussion. Note that a large density of available states in the satellite band is ensured by the large effective mass there compared to the normal conduction band, according to Eq. (4.9). It is believed that a GaAs lattice phonon supplies the momentum required for the k-value change necessary for transition into the satellite band.

B. The Growth of Space-Charge Domains and Electrical Oscillations

Biasing of the Gunn diode in the negative-resistance region results in an unstable situation. It was previously stated that in a uniformly doped conducting semiconductor material any localized space charge will tend to be eliminated rapidly with a time constant which for gallium arsenide is about 10^{-12} sec. This so-called *dielectric relaxation time* τ_d is given by the theory of electromagnetism as

$$\tau_d = \frac{\epsilon}{\sigma}, \tag{11.5}$$

where ϵ is the dielectric constant, and σ the electric conductivity of the material. The rate of charge dispersement is exponential in time according to a factor $\exp(-t/\tau_d)$. In the negative-resistance region of GaAs the conductivity is negative, making τ_d negative according to Eq. (11.5). This predicts that the space charge will rise exponentially in time, or an instability will result. That is, random fluctuations in space charge will build up suddenly rather than be neutralized after a short time, for the Gunn device biased in the negative-conductivity region. If a small nonuniformity in the electron density occurs at some place in the device, a charge dipole can occur locally as shown in Fig. 11.11a. If the Gunn diode is biased in the negative-conductivity region, this unneutralized charge will build up according to Poisson's equation and so will the field there, as shown in Fig. 11.11b. This charged region will grow in strength as it drifts along in the applied electric field. When the drifting charge region or domain reaches the ohmic contact at the anode, it sends a current pulse into the connected wire and hence into the external circuit. Now another space-charge domain forms at the cathode and the process repeats.

FIGURE 11.11

a) Charge dipole due to local nonuniformity in the electron density in a crystal of gallium arsenide.
b) Field buildup due to an initial local space charge with the device biased in the negative mobility region, at time t_1 and a later time t_2. $\mathscr{E}_{th}$ represents the minimum value of electric field to make charge buildup possible.

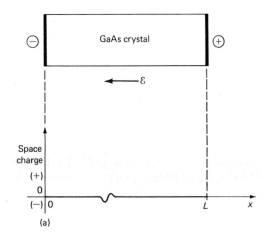

(a)

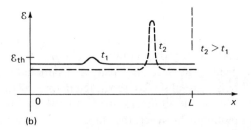

(b)

FIGURE 11.12

Electron drift velocity versus electric field in a gallium arsenide crystal; $v_d = 2 \times 10^5$ m/sec at the critical field of 3×10^5 V/m and the saturation velocity $v_s = 1 \times 10^5$ m/sec at fields about 2×10^6 V/m.

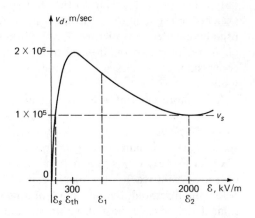

The movement of the high-density electron region may be described in more detail as follows: Figure 11.12 shows a plot of the electron drift velocity versus electric field for GaAs, indicating the drop in velocity above a critical electrical field caused by electron scattering into the high-effective-mass or low-mobility state. Consider the device biased so that the average field is $\mathscr{E}_1$, directed to the left. Since the field in the space-charge region is somewhat higher, the electrons there will move more slowly than those in the neutral regions. Hence the electron density will increase, causing augmented accumulation of electrons on the left side and depletion on the right side of this space-charge layer. This will cause the field in the space-charge region to increase until it reaches $\mathcal{E}_2$ (see Fig. 11.12), when a steady state is achieved. Then the charged domain drifts toward the right without growth and with a field inside this region of $\mathcal{E}_2$, and outside of $\mathcal{E}_s$, at a constant velocity v_s.

In this mode of operation, originally observed by J. B. Gunn, the space-charge domain grows to its stable configuration before the ohmic contact at the end of the device is reached. For this to happen the time for domain formation, τ_d, must be less than the transit time through a device of length L. That is,

$$\frac{\epsilon}{\sigma} = \frac{\epsilon_0 \epsilon_r}{q|\mu|n} < \frac{L}{v_s}, \tag{11.6a}$$

or

$$nL > \frac{\epsilon_0 \epsilon_r v_s}{q|\mu|} \qquad (\approx 10^{16} \text{ m}^{-2}) \tag{11.6b}$$

for n-type GaAs containing n electrons/m^3 where $v_s \approx 10^5$ m/sec, $\epsilon_r = 13$, and the negative mobility is about -0.01 m^2/V-sec. Hence when this critical product of electron density and sample length is present, short output current pulses are observed, separated in time by the space-charge domain transit time. This mode of operation is simple to implement since only a dc source is required; however, the efficiency of conversion to ac high-frequency energy is only a few percent, owing to the short pulses. Since for a typical device as sketched in Fig. 1.11 the pulse rate is in the tens-of-gigahertz range, the diode is usually operated in a microwave cavity whose resonant frequency is about that of the reciprocal of the domain transit time in the device. DC-to-microwave conversion efficiencies approaching 10 percent have been achieved in this mode.

A far more efficient mode of operation of this Gunn device, recently discovered, does not involve charge-domain formation. In fact, the LSA (*limited space-charge accumulation*) mode of operation requires the suppression of these space-charge domains during operation in the negative-resistance region.[5] If a microwave cavity is caused to interact with a Gunn device so that, when the charge domain reaches the ohmic contact, the total field across the device is below the threshold, the reforming of a new domain will be delayed. In addition, if the ac field created within the diode by this external circuit is large enough so that the net field goes below the minimum sustaining value while the domain is in transit, then any charge domain will be dispersed. Both these effects can be used to cause the oscillation frequency to be controlled by the microwave circuit over a range of frequencies exceeding the frequency capa-

[5] J. A. Copeland, "A New Mode of Operation for Bulk Negative Conductance Oscillators," *Proc. IEEE*, Vol. 54, p. 1479 (1966).

bility limited by domain transit time. If the transit time is less than the time for domain formation, the LSA mode can arise. Then the domains do not form, and the electrons drift through a device which exhibits negative conductance in an essentially uniform field through most of the cycle. In this mode the maximum oscillation frequency can be significantly greater than the pulse frequency rate of the normal Gunn-effect device. The useful range of LSA frequency of operation f is given by[6]

$$2 \times 10^{11} > \frac{n}{f} > 2 \times 10^{10} \; \text{sec/m}^3. \tag{11.7}$$

Gunn-effect mode operated devices must be very thin (L small) to yield high-frequency operation. Since the operation frequency of the LSA mode device is circuit determined, this device chip need not be thinned down for increased frequency requirements. Thinning down means reduced voltage for the required threshold field, so that LSA devices have higher power and frequency capability than Gunn-effect mode operated devices. The efficiency of conversion of dc to microwave energy for the Gunn device operating in the LSA mode can approach 20 percent.

11.3 THE SEMICONDUCTOR LASER

The name laser is drawn from the words which describe its operation: *light amplification* by *stimulated emission* of *radiation*. As already indicated in Section 1.3C and pictured in Fig. 1.12, the semiconductor laser is a source of highly coherent, nearly monochromatic and very directional light.

The laser is fundamentally an electronic device which provides coherent light emission when electrons in upper, excited energy states in a material fall to lower energy levels, giving off this energy difference as photons according to Planck's law. When two specific energy levels are involved, Planck's law provides for the monochromatic nature of the radiation. In a laser, stimulated emission is made possible by means of a *population inversion* which results when an upper energy level has a greater probability of being occupied by an electron than a lower energy level. (This is energetically an unstable situation and must be provided for by *pump* energy.) In the presence of a field of photons whose energy is equal to that between the two energy levels under consideration, an inverted electron population in a semiconductor will provide stimulated emission, since the probability of a downward electron transition exceeds that of an upward transition. Light amplification can occur if a proper resonant structure (Fabry-Perot[7]) is provided. This is necessary since spontaneous emission due

[6] J. A. Copeland and R. R. Spiwak, "LSA Operation of Bulk GaAs Diodes," *Dig. Int. Solid-state Circuits Conf. (IEEE)*, p. 26 (1967); J. A. Copeland, "LSA Oscillator Diode Theory," *J. Appl. Phys.*, Vol. 38, p. 3096 (1967).
[7] See, for example, F. A. Jenkins and H. E. White, *Fundamentals of Physical Optics*, 3rd edition, McGraw-Hill Book Company, New York (1957).

FIGURE 11.13

Density of states, $\rho(E)$, in the conduction and valence band of an intrinsic direct-gap semiconductor versus energy E.
a) At $0°$K, in thermal equilibrium, with electron-filled valence band and empty conduction band.
b) At $0°$K, showing an inverted electron population caused by external excitation, with filled electron states in the conduction band at higher energies than empty states in the valence band. Incident radiation with energy $h\nu$ will cause stimulated emission.
c) The same situation as in (b) but at $T > 0°$K. The quasi-Fermi levels E_{Fn} and E_{Fp} are indicated under nonequilibrium conditions.

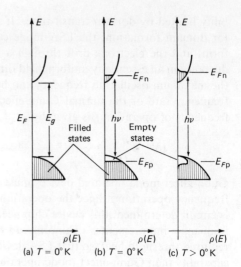

(a) $T = 0°$ K   (b) $T = 0°$ K   (c) $T > 0°$ K

to random electron-hole recombination in the manner described in Section 5.5F must be overwhelmed by stimulated light emission in an enclosure which then provides a nearly coherent light source. Light coherence occurs when all the photons emitted are *in phase* with each other, that is, interfering in a constructive or reinforcing sense.

In this way, pulses of light power of 50 W have been produced using semiconductor lasers. When focused by a lens to a spot of diameter of the order of the light radiation wavelength, they can produce a local power density in excess of 10^{14} W/m^2; this significantly exceeds the power density of about 10^9 W/m^2 observed at the surface of the sun. The efficiency of production of coherent directional light by the semiconductor laser is greatly affected by whether the downward electron transition in the semiconductor material is direct or indirect (see Sections 5.5D and E). In fact, all successful semiconductor lasers are fabricated from semiconductor crystals in which the electron-hole recombination process is known to be of a direct type.

A. Population Inversion and Lasing Threshold in an Intrinsic Semiconductor

Figure 11.13 shows the usual graph of the parabolic density-of-states function versus energy in the conduction and valence bands of an intrinsic semiconductor. The equilibrium situation at $T = 0°$K is indicated in Fig. 11.13a. Here all the states in the valence band are filled with electrons and the conduction band is completely empty (see Fig. 4.10c), typical of an intrinsic semiconductor at $0°K$. The Fermi energy E_F is at the center of the energy gap. If a light beam containing photons with energy $hv > E_g$ (the width of the forbidden energy gap) is incident on the semiconductor crystal, electrons will be excited from the valence band into the conduction band. This *nonequilibrium* state is shown in Fig. 11.13b, which indicates an inverted population of electrons due to "pumping" by the incident light beam. The conduction band at $T = 0°K$ is now filled with electrons up to E_{Fn}; the valence band is devoid of electrons (filled with holes) down to E_{Fp}. Here E_{Fn} and E_{Fp} are the quasi-Fermi levels for electrons and holes respectively. As previously mentioned in Section 6.2, the Fermi energy is only defined under equilibrium conditions and is an expression then of the density of electrons and holes. It is convenient to express the steady-state density of *excess* minority carriers, as in this case, with the aid of these quasi-Fermi energies.[8] A high value of E_{Fn} represents a large nonequilibrium presence of excess electrons. If the crystal in this excited state is introduced into a cavity containing photons with energy hv such that $E_g < hv < E_{Fn} - E_{Fp}$, stimulated emission corresponding to the *downward* transition of the pumped-up electrons to recombine with the holes in the valence band will take place; if instead $hv > E_{Fn} - E_{Fp}$, *upward* transitions of the valence electrons into the conduction band will occur. The condition that the number of photons emitted exceed the number absorbed, resulting in the possibility of lasing, is hence expressed for a block of intrinsic semiconductor material by

$$E_{Fn} - E_{Fp} > hv. \qquad (11.8)$$

Although this discussion applies for $T = 0°K$ and a pure semiconductor, a similar description applies for a doped semiconductor at $T > 0°K$.

To ensure light amplification, stimulated emission plus a condition of sufficient gain in the cavity must exist to compensate for various loss mechanisms in the material. For a Fabry-Perot cavity the minimum condition for lasing is that a stream of photons make a traversal between the two reflecting end mirrors with no net loss, which is expressed by

$$R \exp [(G_{th} - \alpha) L] = 1. \qquad (11.9)$$

Here G_{th} is the gain in photons per meter, R is the reflectivity of the end mirrors, L is the cavity length, and α is the loss coefficient.

B. P-N Junction or Injection Lasers

Semiconductor lasers have produced coherent light with wavelengths ranging from 0.32 μm in the ultraviolet to 16.5 μm in the infrared portion of the electromagnetic spectrum, depending on the energy gap of the semiconductor.

---

[8] The quasi-Fermi levels for electrons and holes are defined respectively by $n = n_i \exp (E_{Fn} - E_i)/kT$ and $p = n_i \exp (E_i - E_{Fp})/kT$, where n_i and E_i are respectively the intrinsic density of carriers and the intrinsic Fermi energy for a semiconductor material.

FIGURE 11.14

Sketch of a gallium arsenide p-n junction laser mounted on a heat sink showing the flat reflecting front and back surfaces as well as the rough side surfaces. Coherent infrared light is emitted in a very narrow beam from the front and rear surfaces. Ohmic contacts are provided on the upper and lower surfaces of the GaAs chip for connection to a current source.

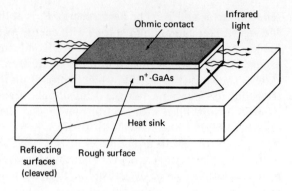

The radiation output can be obtained from the injection laser of Fig. 1.12 (which is redrawn in a practical version in Fig. 11.14) by applying an electrical input to the p-n junction, causing a population inversion of electron or hole carriers which on recombination produce light emission. The use of the semiconductor laser in communications was briefly discussed in Section 1.3C, where the vast information-transmission capability was indicated as well as the extreme ease with which the beam can be modulated for signal transmission. Semiconductor injection lasers have been produced from materials such as GaAs, InSb, InP, PbTe, PbS, and GaAsP. The "pumping" of the p-n junction laser to produce population inversion may be provided by applying forward bias to the diode, causing carrier injection. However, other methods of pumping are possible, including the use of electron-beam or optical excitation and carrier production by avalanche multiplication.

The junction laser is usually very small in size, perhaps 100×250 μm in area and 50 μm thick. Both the p- and n-sides of the laser diode must be heavily doped, each to about 10^{24} impurities/m^3, which is just short of the doping required for tunnel diodes. The light is produced in the close proximity of the p-n junction, and this radiation must undergo many internal reflections on the front and rear faces of the semiconductor chip (which must be perfectly flat and parallel to each other) to produce gain and hence light amplification. In a gallium arsenide laser chip the front and rear, plane and parallel, reflecting surfaces may be produced by cleaving the chip on two edges. Cleaving is the process by which a crystal separates on certain specific crystal planes when the material is abruptly struck in the appropriate crystallographic direction with a sharp instrument. This eliminates the necessity for accurately cutting and polishing the two parallel faces. The resulting structure is known as a *Fabry-Perot resonant cavity*. A high percentage of reflection is obtained automatically on these faces because of the high index of refraction of most semiconductors like GaAs ($n_0 = 3.6$), and as provided by Fresnel's law of reflection. (See footnote 7.) The light transmitted through one of the faces is the output of the laser. For good light amplification, the gain of photons along the cavity must exceed the transmission at the ends. With a Fabry-Perot cavity, the wavelength λ of the various possible oscillation modes is given by

$$m\lambda = 2n_0 L, \qquad (11.10)$$

where m is the mode number, and L is the distance between the front and rear reflecting faces. The value of L is not critical, since $\lambda \ll L$ and many values of integral mode numbers and wavelengths will satisfy this relationship. However, in practical situations often only one mode is favored, accounting for the monochromatic emission.

C. P-N Junction Laser Population Inversion and Lasing Threshold

The energy band diagram of the p-n junction laser will now be utilized to illustrate how the various requirements for successful lasing are met by the device illustrated in Fig. 11.14. Figure 11.15a shows the energy band diagram for a semiconductor p-n junction, in thermal equilibrium, which is very heavily doped both with acceptor impurities on the p-side and donor impurities on the n-side. This drawing may be compared with the energy band diagram for a conventional p-n diode in equilibrium in Fig 6.3b. Note that, owing to equilibrium, the Fermi energy is the same throughout the structure. However, whereas the Fermi level lies in the forbidden energy gap in the conventional diode, in the laser diode the Fermi energy lies in the conduction band on the n-side and the valence band on the p-side. This is consistent with the raising of the Fermi energy on the n-side as donor impurities are added to the semiconductor

FIGURE 11.15

a) Gallium arsenide p-n junction laser energy band diagram under thermal equilibrium conditions, showing electron-filled states in the conduction band and empty states in the valence band.
b) Energy band diagram in the steady-state situation with a potential V_A applied. An inverted-population situation occurs near the junction, and stimulated emission of radiation energy $h\nu$ is obtained. The quasi-Fermi levels in the conduction and valence bands are indicated by E_{Fn} and E_{Fp} respectively, and are separated by qV_F, where V_F is the applied forward voltage.

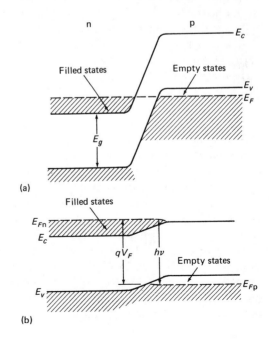

according to the description in Section 5.3D. Also, enough electrons are supplied by the large density of donor impurities on the n-side to fill the conduction band up to the Fermi level there. This large density of electrons causes the semiconductor material to begin to approach the metallic state; the presence of the Fermi energy at the top of the electron distribution in a metal was already referred to in Section 4.2A. The lowering of the Fermi level due to heavy doping of a p-type semiconductor, discussed in Section 5.3D, accounts for the Fermi level being even below the valence band edge on the p-side of the laser diode of Fig. 11.15a.

Now if a forward bias voltage V_F is applied to the structure, it will raise the energy levels on the n-side relative to the p-side; this nonequilibrium energy diagram is pictured in Fig. 11.15b. Note that in this diagram the forward bias was taken large enough so that $|qV_F| > E_g$, the energy gap. Again the Fermi level is not defined for this nonequilibrium situation and, as before, the concept of the quasi-Fermi energy level is useful, the level on the n-side being indicated by E_{Fn} and on the p-side by E_{Fp}. The forward bias reduces the barrier height, of course, causing holes to flow from the p-side to the n-side and electrons from the n-side to the p-side. The electrons flowing downhill across the barrier into the p-side cause population inversion there, and downward transitions into the many empty states available in the valence band then take place; these electron-hole recombinations tend to release energy approximately equal to E_g in the form of photons, efficiently, if this takes place in a direct-transition semiconductor. If the Fermi level originally is not in the conduction and valence bands, then the voltage required for population inversion would reduce the barrier height nearly to zero and damagingly high current would tend to flow.

Hence if the applied voltage is sufficient, a population inversion of minority electrons will appear just on the p-side of the junction and this particular requisite for lasing is satisfied. Here pumping is provided for by minority carrier injection. In fact, just as in the case of the laser action already described for an intrinsic semiconductor, the condition for lasing is given by $E_g < h\nu < E_{Fn} - E_{Fp}$. This expresses again the need for heavy doping so that the quasi-Fermi levels lie in the conduction band on the n-side of the diode and in the valence band on the p-side. A symmetric argument would apply for the holes flowing over the junction barrier and recombining on the n-side.

As the forward voltage applied to the diode is increased from zero, initially normal injection of minority electrons into the p-region and holes into the n-region takes place with corresponding electron-hole recombination, accompanied by light output. This is the case of spontaneous light emission (see Section 5.5B). When the voltage is raised still further until the condition pictured in Fig. 11.15b is reached, a sudden increase in light output will be observed if the device is in a Fabry-Perot cavity, as the condition of population inversion is achieved. Whereas the initial spontaneous radiation is of a random nature, at the lasing threshold the emission of coherent light is suddenly observed. At first, the emission may take place in a number of the possible

FIGURE 11.16

Radiation spectrum of a gallium arsenide p-n junction
laser showing the many
modes of oscillation at
$T = 10°K$.
a) Output at a current I_1.
b) Output at a current
$I_2 > I_1$. This shows how the
radiated energy can tend to
be confined to a single mode
as the current is increased.

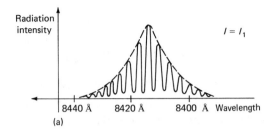

(a)

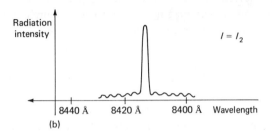

(b)

resonant modes provided by the cavity. However, when sufficient bias is
applied a few very intense modes tend to be observed, and even a single line
of monochromatic output can result as shown in Fig. 11.16. There is, however,
always a small background radiation due to spontaneous emission.

The point of lasing is usually expressed by a current density threshold, J_{th},
given by

$$J_{th} = \frac{8\pi q n_0^2 w v^2 \, \Delta v}{\eta c^2} \left(\frac{1}{L} \ln \frac{1}{R} + \alpha \right), \qquad (11.11)$$

where η is the quantum efficiency of light production, w is the width of the
lasing region, c is the velocity of light, Δv is the emitted line width, and the
other quantities already are defined in relation to Eqs. (11.8), (11.9), and
(11.10). Some of the parameters in this expression are quite temperature
sensitive. In the case of GaAs a current density of about 10^7 A/m^2 is the lasing
threshold at 77°K, but it rises to more than 10^9 A/m^2 at room temperature.
This, coupled with the approximately 1 V required to drive this current, indicates that a power density of about 10^7 W/m^2 must be extracted at 77°K by
an appropriate heat sink if the device is to be operated in the continuous, so-
called CW, mode. This accounts for the general use of the p-n junction laser
in the pulsed mode (less than 1 μsec wide pulses) to prevent excessive heating of
the diode. Of course, this fact also explains the reason for the operation of
semiconductor lasers at reduced temperatures. The pulse height at these low
temperatures is usually less than an ampere.

Pulsed operation is quite adequate for the purpose of communication
using digital signals. Increasing the height of the input current pulse increases

the light-pulse output so that pulse-height modulation may be readily used to transmit information. This method of modulation is extremely simple compared with the techniques which have been employed to modulate the output of gas lasers for communications applications.

11.4 CHARGE-COUPLED DEVICES (CCD)

As a further example of an electronic device with considerable capability whose operation may be explained by applying the principles of semiconductor physics covered in previous chapters, consider the family of charge-coupled devices (CCD).[9] These devices have unique applications in computer circuits and optical imaging. The microminiature nature of the device structure and the ease of fabrication yield the possibility of extremely high-density arrays of such devices which can economically provide memory, logic, delay-line, and optical-imaging functions.

The charge-coupled device structure bears a certain relationship to the MOS structure in that the input configuration consists of a metal-insulator-semiconductor (MIS) sandwich. The charge-coupled device can produce and store minority charge carriers in potential wells created along the surface of a semiconductor just below a series of small metallic regions which are biased with respect to the underlying crystal by an applied potential of a few volts. Just as in the case of the MOS device, surface effects of the type discussed in Section 5.6 affect the operation of the CCD.

The concept of space-charge or depletion-layer formation described in Section 6.1C will be useful in describing this device. When excess mobile charges are introduced into one of these depleted potential wells either electrically or by optical or other radiation, they can be moved along the semiconductor surface by altering the potential on the successive metalizations. The device operation involves the introduction of some charge into a potential well, the shifting of these mobile carriers to another location along the semiconductor surface, and the detection of the presence and magnitude of this charge at some other location in the semiconductor crsytal. Since a finite time elapses between the charge introduction to the first unit and the charge reaching the last unit, the device can serve as a delay line.

A. The MIS Structure

An MIS structure on an n-type semiconductor is sketched in Fig. 11.17a. A simple implementation of this CCD device uses silicon as the semiconductor, its grown oxide as the insulator, and aluminum as the metal. The technology for fabricating this type of device is entirely compatible with the integrated circuit technology described in Section 11.1D; in fact the CCD originated from the development of that technology. Figure 11.17b shows the energy band diagram

---

[9] W. S. Boyle and G. E. Smith, "Charge-Coupled Devices—A New Approach to MIS Device Structures," *IEEE Spectrum*, Vol. 8, pp. 18-27 (1971).

FIGURE 11.17

a) Basic charge-coupled device (CCD) metal-insulator-semiconductor (MIS) structure.

b) Energy band diagram for an MIS structure at $t = 0^+$, just after the voltage V_A is applied.

c) A long time later, at $t = \infty$, when the steady state is reached and holes from the bulk n-type semiconductor are collected at the oxide-semiconductor interface. Note that in the steady state the potential drop across the insulator is greater than at $t = 0^+$, and the drop in the semiconductor is less than at $t = 0^+$, owing to the presence of the stored hole charge at the surface of the semiconductor.

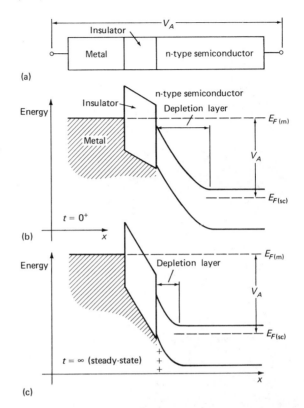

(a)

(b)

(c)

for the MIS device structure just after $(t = 0^+)$ a potential $-V_A$ is applied to the metal electrode with respect to the n-type semiconductor substrate taken as ground or zero potential. If this potential is only a few volts negative with respect to the semiconductor substrate, a depletion region will be set up under the oxide in a very short time (of the order of the semiconductor dielectric relaxation time). The bending of the conduction and valence energy band edges due to this applied voltage is indicated in the figure, as is the quasi-Fermi level $E_{F(sc)}$ in the semiconductor. After some time delay, minority holes in the n-type semiconductor tend to collect in the crystal at the oxide-semiconductor interface, where they are swept by the electric field present in the depletion layer after diffusing into this region. As the holes accumulate at the semiconductor surface, the potential there becomes more positive and the depletion layer tends to shrink in thickness. That is, since there are some additional positive hole charges in the depleted region now, less positively charged ionized donors are needed there to terminate the electric field flux lines generating from the negative charges placed on the metal electrode by the applied voltage. Eventually, the steady-state situation $(t = \infty)$ pictured in Fig. 11.17c is established. Note that in this case more of the applied voltage appears across the insulating layer relative to the voltage drop associated with the depletion region in the

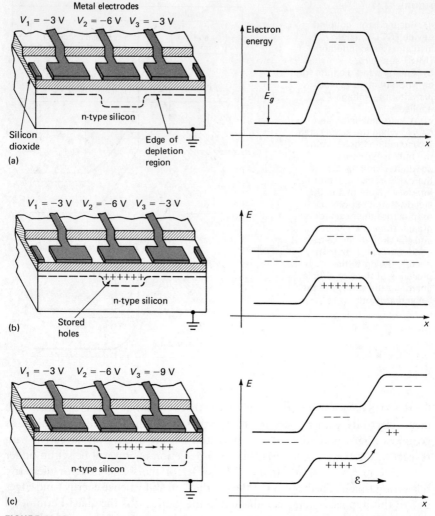

FIGURE 11.18

Schematic diagrams showing a string of three MIS devices capable of the movement of some hole charges along the surface of an n-type semiconductor, from place to place.

At the left of the figure, a sketch of the physical appearance of the device is shown. To the right, the equivalent energy band diagram for the situation just to the left is indicated. The dashed lines represent the quasi-Fermi levels in the three regions.

a) Depletion region formed in the n-type semiconductor just after the application of the potentials V_1, V_2, and V_3.

b) Same situation except that some holes are introduced in the central region and trapped there.

c) Start of transfer of the hole charge stored in the central region to the right when the right-hand electrode is raised to a potential more negative than the central electrode, producing an electric field directed to the right.

semiconductor. Holes may also be introduced and trapped at the surface in this potential well by optical excitation of hole-electron pairs by light or other radiation incident on the semiconductor.

The manner in which some hole charges trapped in the potential well of an MIS device like that sketched in Fig. 11.17a can be moved along the semiconductor surface may be explained by referring to Fig. 11.18. In Fig. 11.18a, three metal electrodes are shown; the outer two are charged quickly to −3 V with respect to the n-type substrate while the central electrode is charged to −6 V. The depletion region produced in the semiconductor by this electrode potential distribution is also indicated in the figure on the left and shows a potential well under the central electrode. The energy band diagram for this situation is shown on the right of Fig. 11.18a. This space-charge configuration will persist for the order of a second before holes diffuse into this depletion region and are swept to the silicon–silicon dioxide interface, forming an inversion layer there. Assume that in a time much shorter than the time necessary to establish this steady-state inversion layer (say, about 1 μsec) some hole charges are optically injected into the potential well at the central electrode. This situation is sketched in Fig. 11.18b. Note that this accounts for a small reduction of the space-charge-layer width under the central metalization and hence a small increase in the capacitance of this central electrode relative to ground. (This provides one technique for detecting the quantity of hole charge stored in this potential well.) If now the third electrode on the right has its potential decreased to −9 V, an electric field will be set up parallel to the semiconductor surface, drawing the holes stored in the central region to a position under the metal electrode on the right. This is shown in Fig. 11.18c, which also indicates the deepened potential well under this third electrode. If, finally, the two electrodes to the left are brought to −3 V and the one at the right is maintained at −6 V, the original charge and space-charge configuration as shown in Fig. 11.18b is reestablished, except that now the stored holes have been moved one electrode distance to the right. The successful operation of this device requires that the charge transfer time be much shorter the than time for the MIS device to achieve its steady-state condition.

This type of operation is characteristic of a device known as a *shift register*. This particular shift register has three stations, but devices with many more stations can be constructed using standard integrated circuit fabrication techniques. Each station can store a *bit* of information in the digital sense. That is, a capacitance measurement can detect whether charge is stored at any electrode or not. However, analog information can be stored as well, since the magnitude of capacitance change in each MIS structure is related to the quantity of charge stored at that station.

B. A Three-Phase Charge-Coupled Shift Register

The shift register is an extremely important device used extensively in computers to store information temporarily and then to serialize this information.

FIGURE 11.19

Schematic illustration of the operation of a three-phase shift register. A three-phase voltage supply is needed to drive the three lines shown (A, B, C) and the drive voltages are indicated by $-V_1$, $-V_2$, and $-V_3$ where $V_3 > V_2 > V_1$. After hole charges are introduced into the potential well under electrode No. 1 in (a) and (b), they are shifted sequentially to the right to the potential well under electrode No. 2, since $V_3 > V_2$. The stable location of the charge under electrode No. 2 is shown in (c). Repeating this sequence can move the charge to the end of the register, under electrode No. 9.

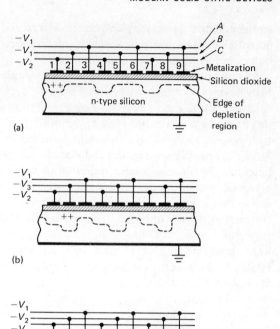

That is, the information which is shifted down to the end of the register can be "read" there; it arrives in the sequence in which it was placed in the shift register. The manner in which this shifting may be implemented is illustrated by the charge-coupled shift register driven by a three-phase clock, as shown in Fig. 11.19. Note that the electrodes are connected together in groups so that every third electrode is tied to a common line. In this way all the electrodes are contacted by using three separate lines, labeled A, B, and C. Suppose initially that the top two lines A and B are connected to a common potential $-V_1$, and the third line C is at a potential $-V_2$, where $-V_1$ and $-V_2$ are negative potentials relative to the substrate, and $V_2 > V_1$. This is illustrated in Fig. 11.19a, which shows deeper potential wells under electrodes No. 1, 4, and 7. Consider now that some positive hole charges are introduced in the depletion layer of station No. 1, also shown in Fig. 11.19a. It is now desired to transfer this hole charge along the n-type semiconductor surface, to the right. The manner in which this may be accomplished is as follows: Line B is now set at a potential $-V_3$, where $V_3 > V_2$, and lines A and C are left at their previous values, respectively $-V_1$ and $-V_2$. This is pictured in Fig. 11.19b. The more negative potential of electrode No. 2 has now succeeded in transferring the excess holes previously introduced at station No. 1 to station No. 2. To complete this transfer sequence, line C is set back to $-V_1$ and line B is set at $-V_2$. Now the situation is as shown in Fig. 11.19c, which is identical to that at the start (Fig. 11.19a) except that the stored hole charge has been shifted one station to the right. This of course assumes that this charge transfer is efficient and that the transfer is fast relative to the steady-state adjustment time required for these MIS devices to come to a steady state with respect to the minority holes in the n-type semiconductor substrate bulk. The time to establish this steady state has been found to be of the order of 1 sec. An upper-limit estimate of the time for this charge transfer to take place can be made assuming that the holes move from one station to the next by diffusion. If the distance d between electrode centers is about 10 μm and the hole diffusion coefficient in silicon is about 0.001 m^2/sec, the transfer time may be approximated by calculating the time t_{tr} for traversing a hole diffusion length (see Section 6.2A). That is,

$$t_{tr} = \frac{d^2}{D_p} = \frac{(10 \times 10^{-6})^2 \text{ m}^2}{10^{-3} \text{ m}^2/\text{sec}} \tag{11.12}$$

$$= 10^{-7} \text{ sec} \quad \text{or} \quad 0.1 \text{ } \mu\text{sec}.$$

Of course, there is some aiding field along the semiconductor surface which urges the holes to the right, so that the transfer certainly takes place in a time which is a small fraction of a microsecond. Hence the transfer not only takes place in a time short enough for proper device functioning but the shift register should be capable of operating at a several-tens-of-megahertz clock rate (the rate at which the potentials on lines A, B, and C are shifted). Measurements have also shown that charge-transfer efficiencies in excess of 99.9 percent can occur. However, this transfer loss of charges tends to limit the length of the string of CCDs in a shift-register application.

Shift registers can be constructed of typical switching devices such as bipolar or MOS transistors. However, the extreme simplicity of the CCD offers the possibility of implementing ultra-high-density electronic functions in a very small area on the surface of a semiconductor chip. No solid-state impurity diffusion is necessary to create the CCD, in contrast to several diffusions in the case of bipolar devices and at least one diffusion for the MOS transistor.[10] This makes the surface area of the chip occupied by a CCD about 10 times smaller than a comparable MOS transistor and perhaps 50 times smaller than a bipolar device, which accounts for its importance in constructing high-density memory arrays and high-resolution optical-imaging devices.

C. Input Schemes for CCD Shift Registers

It is clear that a primary application of the CCD is in some sort of shift register. Hence it is necessary to provide a method of introducing information

---

[10] This makes possible the fabrication of CCDs using semiconductor materials in which it is difficult to produce p-n junctions but where surface states are low in density.

FIGURE 11.20

Schematic representation of
three input schemes for a
shift register using charge-
coupled devices.
a) Field-effect type input
using a p$^+$-region as a
source of holes and a negative voltage relative to the
substrate, $-V_1$, applied to a
metal gate to produce a
p-channel to conduct holes
to the first CCD device
which has a potential well to
trap these holes.
b) A negative potential
$-V_2$ is applied to the device
at the left. It is high enough
to create the avalanche electric
field for silicon which provides hole-electron pairs and
hence a source of holes.
c) Hole-electron pairs are
excited by light at the back
surface of the silicon, and
they drift to the top surface
to be stored in the potential
wells there.

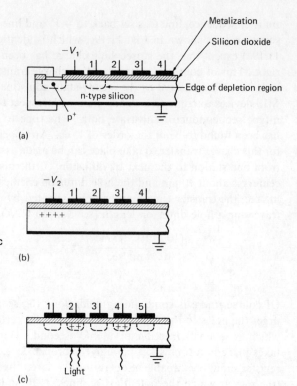

at the input of the register as well as some method of detecting the output signal
at the end of the string of charge-coupled devices.

Input schemes for a CCD shift register are sketched in Fig. 11.20. Figure
11.20a shows how a p-n junction can be utilized as an input device. A negative
potential applied to the MIS device just to the right of the p-n junction produces
an inversion layer of holes just under its metalization; this provides a conducting channel between the p-region of the p-n junction on its left and the
CCD on its right, in a manner similar to MOS transistor operation. The
heavily doped p-region is a source of holes which are conducted to the right
along the semiconductor surface to the first CCD, which is biased to attract
the holes. The potential well of the first CCD device is sketched in the figure.
Note that it is simply necessary to tie the p-region of the p-n junction to the
semiconductor substrate to provide a hole source, similar to the manner in
which the source of an MOS transistor is connected to substrate.

This method involves the diffusion of a p-type impurity into the semiconductor material to form a p-n junction. This negates one of the advantages of the
CCD in that no high-temperature diffusion step is needed to form a shift
register. Two techniques which require no additional diffusion step are discussed next. The first involves using a regular MIS structure in the avalanche

mode for an input device, as shown in Fig. 11.20b. Here a high negative voltage pulse is applied to the first device on the left; it is high enough so that the electric field in the depletion layer reaches the avalanche value. The avalanche mechanism in the depletion layer of a p-n junction, described in Section 6.3C, Avalanche breakdown, is satisfactory for describing the avalanching in this space-charge region in an MIS structure. There it was seen that impact ionization causes the creation of hole-electron pairs. These minority holes then diffuse into the first CCD of the shift register to the right. The last hole injection input technique uses photo-injected carriers as illustrated in Fig. 11.20c. This may be used as an optical imaging technique, to be described later, which uses hole-electron pair production by photons whose energy $hv > E_g$, the semiconductor energy-gap width.

D. Output Schemes for CCD Shift Registers

Schemes for detecting the output signal at the end of a CCD shift register are sketched in Fig. 11.21. Figure 11.21a shows how a p-n junction with a large reverse potential is used to detect the holes in the last CCD just to its left.

FIGURE 11.21

Schematic representation of three output schemes for a shift register using charge-coupled devices.
a) A p-n junction diode at the end of the register, with a large reverse potential V_o, acts as a bipolar transistor collector to attract holes, resulting in an output current I_o.
b) A positively pulsed MIS element yields an electric field in the silicon which drives holes from the last register element to ground, giving rise to an output current I_o and an output voltage V_o.
c) The negative voltage $-V$ is shared between the output MIS device and the output capacitor in series. As holes are collected in the potential well of the MIS device, the well depth decreases and the MIS capacitance increases, causing it to take a smaller share of $-V$. Hence V_o assumes a smaller negative potential relative to ground.

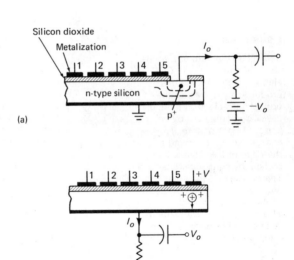

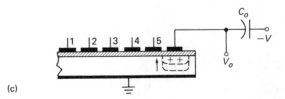

Because the reverse junction potential is even more negative than the CCD potentials, holes are collected by this diode as at the collector junction of a bipolar transistor, and output voltage is developed across the resistor R_o in proportion to the collected current and hence collected holes.

Again the creation of a p-n junction requires additional technology; hence two methods not using p-n junctions will now be described. The first uses a positively biased (relative to substrate) MIS output device, as shown in Fig. 11.21b. When holes from the last CCD in the shift register drift toward this output electrode they are repelled into the substrate and constitute a current to ground which is detected by the voltage drop across the resistor R_o. Another detection technique uses the increase in output capacitance of a conventional negatively biased MIS device when excess holes from the previous CCD device diffuse into its depletion layer, reducing its width. This capacitance increase was described in Section 11.4A. It will give rise to a shift in the voltage marked V_o, since this reflects the change in the proportion of the applied voltage $-V$ across the output MIS device capacitance compared with the fixed output capacitor C_o. In this way the relative quantity of charge on the last CCD in the shift register may be detected.

FIGURE 11.22

Schematic diagram of a solid-state TV camera using CCD elements to store charges produced by optical excitation of hole-electron pairs in the silicon substrate by an imposed light pattern. These charges are then moved to the edge of the device, line by line, by shift-register action. The charges on the edge line are then sequentially shifted to the output terminal.
This raster-type arrangement serializes the optical pattern information.

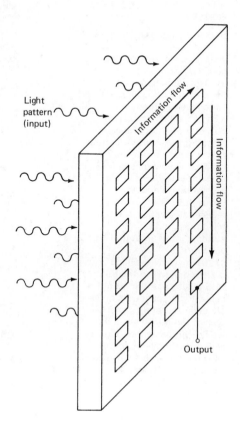

E. A Solid-State Optical-Imaging Device

The CCD array also offers other exciting application possibilities, such as a solid-state TV camera. For this a large, closely spaced, two-dimensional array of CCDs is fabricated on one face of a thin slice of semiconductor. This is sketched in Fig. 11.22. All the individual devices are biased with a small negative potential relative to the n-type semiconductor substrate. Now a light pattern is focused on the underside of the semiconductor slab, creating hole-electron pairs in the material in proportion to the local light intensity. The holes will be attracted to the semiconductor surface under the nearest negative charged metal electrode, with the charge there representing an analog of the incident light pattern. The various magnitudes of charge stored in a single line of CCD elements in the matrix can be "read" by rapidly shifting these charges to the end of the row in a serial manner, in the shift-register mode, by sequentially changing the potentials on the adjacent electrodes. Then the next row can be read and serialized after the first row in a manner which is a solid-state analog of a TV raster. The envelope of this pulse train can now be transmitted; it represents a video-signal translation of the original focused light pattern. Of course, it is necessary that the storage time of the trapped holes at the CCD elements be longer than the video frame time for successful vidicon operation.

11.5 CONCLUSION

An attempt has been made in these pages to describe the physics of electronic materials and devices so as to permit an appreciation of the wonderland of semiconductor devices available today and of those which will be invented tomorrow. This final chapter was intended to demonstrate the use that can be made of this new information to explain the operation of a few of the newer semiconductor devices whose fullest potential is yet to be realized. It has been perhaps, as one student expressed it, "a magical mystery tour!"

PROBLEMS

11.1 Find approximately how many 5.0 mm^2 integrated circuit chips can be obtained from a slice of a silicon crystal 6.0 cm in diameter.

11.2 Using the data in Figs. 11.3 and 11.4 and Table 11.1, determine by calculation whether or not successful n-pocket isolation is obtained by boron diffusion through a 10 μm thick n-type silicon epitaxial layer containing 1.0×10^{20} donor impurity atoms/m^3 grown onto a p-type silicon substrate, if the diffusion takes place

a) at 1200°C for 2 hours,
b) at 1100°C for 6 hours.

(*Hint*: For successful isolation diffusion the density of boron impurities which penetrate the 10 μm thick epitaxial layer must be greater than the density of n-type impurities everywhere in the penetrated region of that layer, where conversion to p-type is necessary.)

11.3 Determine approximately the diffusion coefficient of boron in silicon dioxide at 1200°C using the data in Table 11.2 and Fig. 11.3. Then calculate roughly the thickness of oxide required for masking against a 2-hour boron diffusion by estimating the diffusion length $\sqrt{Dt}$ of boron in SiO_2.

11.4 The average resistivity of the diffused p-layer used to fabricate a monolithic integrated resistor is 0.001 Ω-m, and the depth of this layer is 5.0 μm. Assume that the width of the diffused resistor is limited to no less than 10 μm by the ability to etch a narrow line in the oxide (caused by the basic resolution of the photoresist).

a) Calculate the length of resistor line required to produce a 10,000 Ω resistor.
b) If the change of this diffused resistor value with temperature is 3000 ppm/°C, by what percentage will the resistor value change for a temperature rise of 10°C?

11.5 Determine the percentage chip surface area of a 5.0 mm^2 chip occupied by a monolithic integrated capacitor whose value is 50 pF. Assume the active dielectric material is a silicon dioxide layer whose thickness is 1000 Å.

11.6 a) Show that for space-charge domain formation in the Gunn oscillator mode the nL product (electron density times active sample length) is of the order of 10^{16} m^{-2} for gallium arsenide, using the data given following Eq. (11.6).
b) For $nL \approx 10^{16}$ m^{-2}, is the inequality of Eq. (11.7) satisfied for the LSA mode? Take the oscillation period to be about one-third the transit time.

11.7 Calculate the ratio of the density of states at some energy E in the satellite valley as compared with the central valley of GaAs at 300°K, using Eq. (5.1) and the information given in Fig. 11.10.

11.8 a) Estimate the dc power dissipated in a GaAs Gunn-diode oscillator chip 1.0 mm^2 in area, whose active region has a resistivity of 0.10 Ω-m and is 70 μm thick, just at the threshold of oscillation.
b) Calculate the time between the pulses generated by this device operating in the Gunn mode.

11.9 Lasers have been produced yielding radiation from 0.32 μm to 16.5 μm in wavelength. Estimate the range of values of energy gap for the semiconductor materials used to fabricate these devices.

11.10 A GaAs laser chip is 100 μm × 250 μm in area and 50 μm thick.

a) Estimate the power dissipated in the device at 77°K if the current density threshold is 4×10^7 A/m^2 and the diode requires about a 1 V drop across it for lasing.

b) For pulsed operation using 1 μsec wide pulses at a repetition rate of 1 kHz, calculate the time average power dissipated in the device.

11.11   Derive Eq. (11.10), which is based on the constructive interference of internally reflected light rays.

11.12   a) Determine the energy $h\nu$ corresponding to the peak radiation indicated in Fig. 11.16b. What does this tell you about the value of the energy gap of the material from which this semiconductor laser is fabricated?

b) Calculate the spacing in angstrom units of the oscillation modes in a GaAs injection laser, according to Eq. (11.10), if the distance between the front and rear reflecting surfaces is 250 μm.

11.13   a) For a GaAs junction laser whose length $L = 250$ μm, determine the time for the radiation to traverse the sample length. (*Note*: The velocity of propagation of electromagnetic radiation in GaAs equals the velocity in vacuum divided by the index of refraction of GaAs.)

b) Calculate the reflectivity R of a GaAs-air interface using the Fresnel law, $R = [(n_0 - 1)/(n_0 + 1)]^2$, where n_0 is the index of refraction of GaAs.

c) If α, the loss factor, is 1200 m^{-1} at 10°K, find the gain G_{th} necessary for lasing using Eq. (11.9).

11.14   Using Eq. (11.11) and the results of Problem 11.13, estimate the threshold current density J_{th} at 10°K for the GaAs junction laser whose output is shown in Fig. 11.16b. Take $w = 4$ μm and $\eta = 0.8$.

11.15   a) Estimate the capacitance in the steady state of a silicon CCD of the type illustrated in Fig. 11.18, with electrode area 10 μm by 50 μm, if the silicon dioxide thickness is 1 μm and the depletion-layer width in the silicon under the electrode is also about 1 μm. (*Hint*: This capacitance may be calculated by determining separately the capacitances due to the oxide and the depletion layer and treating the structure as two capacitors in series.)

b) If these electrodes are spaced 2.0 μm apart, calculate the maximum number of such devices that can be constructed on a silicon chip 5.0 mm^2 in area.

11.16   The capacitance of a silicon CCD of the type shown in Fig. 11.18 is 0.015 pF in the steady state, when -6 V is applied to the metal electrode relative to the n-type silicon substrate. The electrode dimensions are 20 μm by 100 μm and the silicon dioxide thickness is 2.0 μm.

a) Calculate the charge on the metal electrode.

b) Estimate the depletion-layer width in the silicon. Assume it extends into the silicon only in a direction normal to the silicon surface. (*Hint*: See the hint in Problem 11.15a.)

c) Speculate on the capacitance value of this device immediately on application of this voltage (at $t = 0^+$) compared with the steady-state ($t = \infty$) capacitance value.

d) Estimate the density of hole charges which, if trapped in the depletion region at $t = 0^+$, would significantly affect the capacitance value at that time.

11.17   Redraw the energy band diagrams of Fig. 11.17 for the case of a p-type silicon substrate.

Appendixes

APPENDIX A RATIO OF DRIFT TO DIFFUSION CURRENT IN BIPOLAR TRANSISTORS

TO PROVE   At low injection levels the minority hole current in the n-region near the junction of a p^+-n diode is almost exclusively by diffusion, with the hole drift current being small in comparison.

PROOF   Calling R the numerical ratio of the hole drift to diffusion current, we can write, from Eqs. (5.28) and (5.31),

$$R = \frac{q\mu_p p \mathcal{E}}{q D_p \, dp/dx}.$$ (A.1)

Here the subscript on p_n is left off since the derivation refers only to the n-region. The electron current portion of Eq. (6.1b) may be rewritten as

$$\frac{dn}{dx} = \frac{J_n - q\mu_n n \mathcal{E}}{q D_n}.$$ (A.2)

Now the charge-neutrality assumption requires equality of injected holes and electrons. Hence

$$p - p_0 = n - n_0,$$ (A.3)

where again the subscript 0 refers to equilibrium conditions. Differentiation of Eq. (A.3) shows the equality required of the hole and electron gradients; that is

$$\frac{dp}{dx} = \frac{dn}{dx}.$$ (A.4)

Introducing Eq. (A.4) into Eq. (A.2), we can write

$$\frac{dp}{dx} = \frac{J_n}{q D_n} - \frac{\mu_n}{D_n} n \mathcal{E}.$$ (A.5)

Substituting this value for dp/dx into Eq. (A.1) yields

$$R = \frac{(\mu_p/D_p) p \mathcal{E}}{J_n/q D_n - (\mu_n/D_n) n E}.$$ (A.6)

Dividing numerator and denominator by $n \mathcal{E}$ then gives

$$R = \frac{p/n}{J_n/q\mu_n n \mathcal{E} - 1}.$$ (A.7)

Now the total electron current J_n consists in general of drift and diffusion components, $J_{\text{drift}} + J_{\text{diffusion}}$. Introducing this into Eq. (A.7) and using Eq. (A.2), we can write

$$R = \frac{(J_{\text{drift}})_n}{(J_{\text{diffusion}})_n} \frac{p}{n}.$$ (A.8)

Now for a p^+-n diode the current in the n-region near the junction is primarily hole current and so the electron current there is $J_n = (J_{drift})_n + (J_{diffusion})_n = 0$. Hence

$$\frac{|(J_{drift})_n|}{|(J_{diffusion})_n|} \approx 1, \qquad\qquad\qquad (A.9)$$

and since by definition low-level injection means $p/n \ll 1$, Eq. (A.8) states that near the junction $R \ll 1$, and the proposition is proved. Note that far from the junction nearly all injected holes have recombined, there is little diffusion current left, and so the drift current now dominates.

APPENDIX B   FUNDAMENTAL PHYSICAL CONSTANTS

Quantity	Symbol	Value	Units
Electronic charge (magnitude)	q	1.60×10^{-19}	coulombs
Electronic mass (magnitude)	m	9.11×10^{-31}	kilograms
Unit nucleonic mass	m_{nucl}	1.66×10^{-27}	kilograms
Avogadro's number	N_A	6.02×10^{26}	atoms per kilogram-atom
Boltzmann's constant	k	1.38×10^{-23}	joules/per degree Kelvin
Thermal energy (300°K)	kT	0.026	electron-volts
Thermal voltage (300°K)	kT/q	0.026	volts
Speed of light (vacuum)	c	3.0×10^8	meters per second
Planck's constant	h	6.63×10^{-34}	joule-seconds
Reduced Planck's constant $(h/2\pi)$	$\hbar$	1.06×10^{-34}	joule-seconds
Frequency (cycles/second)	Hz	1.0	hertz
Permittivity of free space	ϵ_0	8.85×10^{-12}	farads per meter
Relative dielectric constants:			
germanium	$(\epsilon_r)_{\text{Ge}}$	16	—
silicon	$(\epsilon_r)_{\text{Si}}$	11.8	—
gallium arsenide	$(\epsilon_r)_{\text{GaAs}}$	10.9	—
silicon dioxide	$(\epsilon_r)_{\text{SiO}_2}$	3.8	
Lattice constants (300°K):			
	a_{Ge}	5.66	angstroms
	a_{Si}	5.43	angstroms
	a_{GaAs}	5.65	angstroms
Energy gap (300°K):	$(E_g)_{\text{Ge}}$	0.66	electron-volts
	$(E_g)_{\text{Si}}$	1.15	electron-volts
	$(E_g)_{\text{GaAs}}$	1.43	electron-volts
Micrometer (micron)	μ, μm	1.0×10^{-6}	meters
Angstrom unit	Å	1.0×10^{-10}	meters
Electron-volt	eV	1.6×10^{-19}	joules

Prefixes:

milli (m) = 10^{-3} kilo (k) = 10^3

micro (μ) = 10^{-6} mega (M) = 10^6

nano (n) = 10^{-9} giga (G) = 10^9

pico (p) = 10^{-12}

TYPES 1N914, 1N914A, 1N914B, 1N915, 1N916, 1N916A, 1N916B, 1N917
SILICON SWITCHING DIODES

BULLETIN NO. DL-S 7311954, MARCH 1973

FAST SWITCHING DIODES

- **Rugged Double-Plug Construction**

Electrical Equivalents

1N914 . . . 1N4148 . . . 1N4531
1N914A . . . 1N4446
1N914B . . . 1N4448
1N916 . . . 1N4149
1N916A . . . 1N4447
1N916B . . . 1N4449

mechanical data

Double-plug construction affords integral positive contacts by means of a thermal compression bond. Moisture-free stability is ensured through hermetic sealing. The coefficients of thermal expansion of the glass case and the dumet plugs are closely matched to allow extreme temperature excursions. Hot-solder-dipped leads are standard.

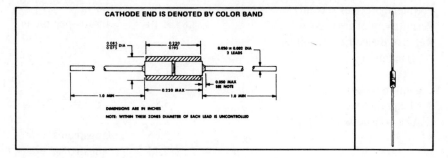

CATHODE END IS DENOTED BY COLOR BAND

DIMENSIONS ARE IN INCHES
NOTE: WITHIN THESE ZONES DIAMETER OF EACH LEAD IS UNCONTROLLED

absolute maximum ratings at specified free-air temperature

		1N914 1N914A 1N914B	1N915	1N916 1N916A 1N916B	1N917	UNIT
Working Peak Reverse Voltage from −65°C to 150°C		75*	50*	75*	30*	V
Average Rectified Forward Current (See Note 1)	at (or below) 25°C	75*	75*	75*	50*	mA
	at 150°C	10*	10*	10*	10*	
Peak Surge Current, 1 Second at 25°C (See Note 2)		500*	500	500*	300	mA
Continuous Power Dissipation at (or below) 25°C (See Note 3)		250*	250	250*	250	mW
Operating Free-Air Temperature Range		−65 to 175				°C
Storage Temperature Range		−65 to 200*				°C
Lead Temperature 1/16 Inch from Case for 10 Seconds		300				°C

NOTES:

1. These values may be applied continuously under a single-phase 60-Hz half-sine-wave operation with resistive load.
2. These values apply for a one-second square-wave pulse with the devices at nonoperating thermal equilibrium immediately prior to the surge.
3. Derate linearly to 175°C free-air temperature at the rate of 1.67 mW/°C.

TYPES 1N914, 1N914A, 1N914B, 1N915, 1N916, 1N916A, 1N916B, 1N917
SILICON SWITCHING DIODES

1N914 SERIES AND 1N915

*electrical characteristics at 25°C free-air temperature (unless otherwise noted)

PARAMETER		TEST CONDITIONS	1N914		1N914A		1N914B		1N915		UNIT
			MIN	MAX	MIN	MAX	MIN	MAX	MIN	MAX	
$V_{(BR)}$	Reverse Breakdown Voltage	I_R = 100 μA	100		100		100		65		V
I_R	Static Reverse Current	V_R = 10 V								25	nA
		V_R = 20 V		25		25		25			
		V_R = 20 V, T_A = 100°C						3		5	
		V_R = 20 V, T_A = 150°C		50		50		50			μA
		V_R = 50 V								5	
		V_R = 75 V		5		5		5			
V_F	Static Forward Voltage	I_F = 5 mA					0.62	0.72	0.6	0.73	
		I_F = 10 mA		1							
		I_F = 20 mA See Note 4				1					V
		I_F = 50 mA								1	
		I_F = 100 mA						1			
C_T	Total Capacitance	V_R = 0, f = 1 MHz	4		4		4		4		pF

1N916 SERIES AND 1N917

*electrical characteristics at 25°C free-air temperature (unless otherwise noted)

PARAMETER		TEST CONDITIONS	1N916		1N916A		1N916B		1N917		UNIT
			MIN	MAX	MIN	MAX	MIN	MAX	MIN	MAX	
$V_{(BR)}$	Reverse Breakdown Voltage	I_R = 100 μA	100		100		100		40		V
I_R	Static Reverse Current	V_R = 10 V								50	nA
		V_R = 20 V		25		25		25			
		V_R = 20 V, T_A = 100°C						3		25	
		V_R = 20 V, T_A = 150°C		50		50		50			μA
		V_R = 75 V		5		5		5			
V_F	Static Forward Voltage	I_F = 0.25 mA								0.64	
		I_F = 1.5 mA								0.74	
		I_F = 3.5 mA								0.83	
		I_F = 5 mA					0.63	0.73		1	V
		I_F = 10 mA		1							
		I_F = 20 mA See Note 4				1					
		I_F = 30 mA						1			
C_T	Total Capacitance	V_R = 0, f = 1 MHz	2		2		2		2.5		pF

NOTE 4: These parameters must be measured using pulse techniques. t_w = 300 μs, duty cycle ⩽ 2%.

TYPES 1N914, 1N914A, 1N914B, 1N915, 1N916, 1N916A, 1N916B, 1N917
SILICON SWITCHING DIODES

operating characteristics at 25°C free-air temperature

PARAMETER		TEST CONDITIONS	1N914 1N914A 1N914B 1N916 1N916A 1N916B		1N915		1N917		UNIT
			MIN	MAX	MIN	MAX	MIN	MAX	
t_{rr}	Reverse Recovery Time	$I_F = 10\,mA$, $I_{RM} = 10\,mA$, $i_{rr} = 1\,mA$, $R_L = 100\,\Omega$, See Figure 1 (Condition 1)		8		10*		3*	ns
		$I_F = 10\,mA$, $V_R = 6\,V$, $i_{rr} = 1\,mA$, $R_L = 100\,\Omega$, See Figure 1 (Condition 2)		4*					ns
$V_{FM(rec)}$	Forward Recovery Voltage	$I_F = 50\,mA$, $R_L = 50\,\Omega$, See Figure 2		2.5*					V
η_r	Rectification Efficiency	$V_r = 2\,V$, $R_L = 5\,k\Omega$, $C_L = 20\,pF$, $Z_{source} = 50\,\Omega$, $f = 100\,MHz$	45*						%

PARAMETER MEASUREMENT INFORMATION

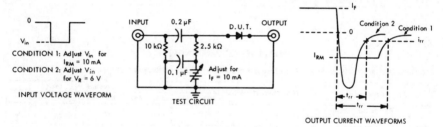

CONDITION 1: Adjust V_{in} for $I_{RM} = 10\,mA$
CONDITION 2: Adjust V_{in} for $V_R = 6\,V$

INPUT VOLTAGE WAVEFORM

TEST CIRCUIT

OUTPUT CURRENT WAVEFORMS

FIGURE 1 — REVERSE RECOVERY TIME

NOTES: a. The input pulse is supplied by a generator with the following characteristics: $Z_{out} = 50\,\Omega$, $t_r \leq 0.5\,ns$, $t_w = 100\,ns$.
     b. Output waveforms are monitored on an oscilloscope with the following characteristics: $t_r \leq 0.6\,ns$, $Z_{in} = 50\,\Omega$.

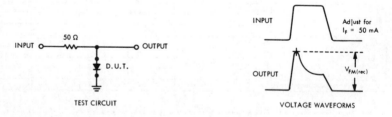

TEST CIRCUIT

VOLTAGE WAVEFORMS

FIGURE 2 — FORWARD RECOVERY VOLTAGE

NOTES: c. The input pulse is supplied by a generator with the following characteristics: $Z_{out} = 50\,\Omega$, $t_r \leq 30\,ns$, $t_w = 100\,ns$, PRR = 5 to 100 kHz.
     d. The output waveform is monitored on an oscilloscope with the following characteristics: $t_r \leq 15\,ns$, $R_{in} \geq 1\,M\Omega$, $C_{in} \leq 5\,pF$.

2N3946 (SILICON)
2N3947

NPN silicon annular transistors, designed for general purpose switching and amplifier applications. The 2N3946 and 2N3947 are complementary with PNP types 2N3250 and 2N3251, respectively.

CASE 22
(TO-18)

Collector connected to case

STYLE 1:
PIN 1. EMITTER
2. BASE
3. COLLECTOR

*MAXIMUM RATINGS (T_A = 25°C unless otherwise noted)

Rating	Symbol	Value	Unit
Collector-Base Voltage	V_{CB}	60	Vdc
Collector-Emitter Voltage	V_{CEO}	40	Vdc
Emitter-Base Voltage	V_{EB}	6.0	Vdc
Collector Current	I_C	200	mAdc
Total Device Dissipation @ T_C = 25°C	P_D	1.2	Watts
Derating Factor Above 25°C		6.9	mW/°C
Total Device Dissipation @ T_A = 25°C	P_D	0.36	Watt
Derating Factor Above 25°C		2.06	mW/°C
Thermal Resistance Junction to Ambient	θ_{JA}	0.49	°C/mW
Junction to Case	θ_{JC}	0.15	°C/mW
Junction Operating Temperature	T_J	200	°C
Storage Temperature Range	T_{stg}	-65 to +200	°C

*ELECTRICAL CHARACTERISTICS (T_A = 25°C unless otherwise noted)

Characteristic	Symbol	Min	Max	Unit
OFF CHARACTERISTICS				
Collector-Base Breakdown Voltage (I_C = 10 μAdc, I_E = 0)	BV_{CBO}	60	—	Vdc
Collector-Emitter Breakdown Voltage(1) (I_C = 10 mAdc)	BV_{CEO}*	40	—	Vdc
Emitter-Base Breakdown Voltage (I_E = 10 μAdc, I_C = 0)	BV_{EBO}	6.0	—	Vdc
Collector-Cutoff Current (V_{CE} = 40 Vdc, V_{OB} = 3 Vdc) (V_{CE} = 40 Vdc, V_{OB} = 3 Vdc, T_A = 150°C)	I_{CEX}	— —	.010 15	μAdc
Base Cutoff Current (V_{CE} = 40 Vdc, V_{OB} = 3 Vdc)	I_{BL}	—	.025	μAdc

(1)Pulse Test: PW ≦ 300 μs , Duty Cycle ≦ 2% V_{OB} = Base-Emitter Reverse Bias
* Indicates JEDEC Registered Data

2N3946, 2N3947 (continued)

*ELECTRICAL CHARACTERISTICS (continued)

(T_A = 25°C unless otherwise noted)

Characteristic	Symbol	Min	Max	Unit
ON CHARACTERISTICS				
DC Current Gain [1]	h_{FE}			—
(I_C = 0.1 mAdc, V_{CE} = 1 Vdc)　　　　　2N3946		30	—	
2N3947		60	—	
(I_C = 1.0 mAdc, V_{CE} = 1 Vdc)　　　　　2N3946		45	—	
2N3947		90	—	
(I_C = 10 mAdc, V_{CE}= 1 Vdc)　　　　　2N3946		50	150	
2N3947		100	300	
(I_C = 50 mAdc, V_{CE} = 1 Vdc)　　　　　2N3946		20	—	
2N3947		40	—	
Collector Saturation Voltage [1]	$V_{CE(sat)}$			Vdc
(I_C = 10 mAdc, I_B = 1 mAdc)		—	0.2	
(I_C = 50 mAdc, I_B = 5 mAdc)		—	0.3	
Base-Emitter Saturation Voltage [1]	$V_{BE(sat)}$			Vdc
(I_C = 10 mAdc, I_B = 1 mAdc)		0.6	0.9	
(I_C = 50 mAdc, I_B = 5 mAdc)		—	1.0	
TRANSIENT CHARACTERISTICS				
Output Capacitance	C_{ob}			pF
(V_{CB} = 10 Vdc, I_E = 0, f = 100 kHz)		—	4.0	
Input Capacitance	C_{ib}			pF
(V_{BE} = 1 Vdc, I_C = 0, f = 100 kHz)		—	8.0	
Current-Gain - Bandwidth Product	f_T			MHz
(I_C = 10 mAdc, V_{CE} = 20 Vdc, f = 100 MHz)　2N3946		250	—	
2N3947		300	—	
Delay Time　　V_{CC} = 3 Vdc, V_{OB} = 0.5 Vdc	t_d	—	35	ns
Rise Time　　　I_C = 10 mAdc, I_{B1} = 1 mA	t_r	—	35	ns
Storage Time　V_{CC} = 3 V, I_C = 10 mA,　2N3946	t_s	—	300	ns
2N3947		—	375	
Fall Time　　　I_{B1} = -I_{B2} = 1 mAdc	t_f	—	75	ns
SMALL SIGNAL CHARACTERISTICS				
Small-Signal Current Gain	h_{fe}			—
(I_C = 1.0 mA, V_{CE} = 10 V, f = 1 kHz)　　2N3946		50	250	
2N3947		100	700	
Voltage Feedback Ratio	h_{re}			X10⁻⁴
(I_C = 1.0 mA, V_{CE} = 10 V, f = 1 kHz)　　2N3946		—	10	
2N3947		—	20	
Input Impedance	h_{ie}			kohms
(I_C = 1.0 mA, V_{CE} = 10 V, f = 1 kHz)　　2N3946		0.5	6.0	
2N3947		2.0	12	
Output Admittance	h_{oe}			μmhos
(I_C = 1.0 mA, V_{CE} = 10 V, f = 1 kHz)　　2N3946		1.0	30	
2N3947		5.0	50	
Collector-Base Time Constant	$r_b'C_C$			ps
(I_C = 10 mA, V_{CE} = 20 V, f = 31.8 MHz)		—	200	
Wide Band Noise Figure	NF			dB
(I_C = 100 μA, V_{CE} = 5 V, R_g = 1 kΩ, f = 10 Hz to 15.7 kHz)		—	5.0	

[1]Pulse Test: PW ≤ 300 μs, Duty Cycle ≤ 2%　　　V_{OB} = Base-Emitter Reverse Bias
* Indicates JEDEC Registered Data

2N3250, A (SILICON)
2N3251, A

2N3250A JAN,JTX,JTXV AVAILABLE
2N3251A JAN,JTX AVAILABLE

PNP silicon annular transistors for high-speed switching and amplifier applications.

STYLE 1:
PIN 1. EMITTER
2. BASE
3. COLLECTOR

CASE 22
(TO-18)

Collector connected to case

*MAXIMUM RATINGS

Rating	Symbol	2N3250 2N3251	2N3250A 2N3251A	Unit
Collector-Base Voltage	V_{CB}	50	60	Vdc
Collector-Emitter Voltage	V_{CEO}	40	60	Vdc
Emitter-Base Voltage	V_{EB}	5.0		Vdc
Collector Current	I_C	200		mAdc
Total Device Dissipation @ 25°C Case Temperature Derating Factor Above 25°C	P_D	1.2 6.9		Watts mW/°C
Total Device Dissipation @ 25°C Ambient Temperature Derating Factor Above 25°C	P_D	0.36 2.06		Watts mW/°C
Junction Operating Temperature	T_J	200		°C
Storage Temperature Range	T_{stg}	−65 to +200		°C
Thermal Resistance, Junction to Ambient	θ_{JA}	0.49		°C/mW
Thermal Resistance, Junction to Case	θ_{JC}	0.15		°C/mW

*Indicates JEDEC Registered Data

Courtesy of Motorola Semiconductor, Inc.

2N3250, A, 2N3251, A (Continued)

*ELECTRICAL CHARACTERISTICS (T_A = 25°C unless otherwise noted)

Characteristic		Symbol	Min	Max	Unit
Collector Cutoff Current (V_{CE} = 40 Vdc, $V_{BE(off)}$ = 3 Vdc)		I_{CEX}	--	20	nAdc
Base Cutoff Current (V_{CE} = 40 Vdc, $V_{BE(off)}$ = 3 Vdc)		I_{BL}	--	50	nAdc
Collector-Base Breakdown Voltage (I_C = 10 μAdc)	2N3250, 2N3251 2N3250A, 2N3251A	BV_{CBO}	50 60	-- 	Vdc
Collector-Emitter Breakdown Voltage [1] (I_C = 10 mAdc)	2N3250, 2N3251 2N3250A, 2N3251A	BV_{CEO}	40 60	-- 	Vdc
Emitter-Base Breakdown Voltage (I_E = 10 μAdc)		BV_{EBO}	5.0	--	Vdc
Collector Saturation Voltage [1] (I_C = 10 mAdc, I_B = 1 mAdc)		$V_{CE(sat)}$	--	0.25	Vdc
(I_C = 50 mAdc, I_B = 5 mAdc)			--	0.5	
Base-Emitter Saturation Voltage [1] (I_C = 10 mAdc, I_B = 1 mAdc)		$V_{BE(sat)}$	0.6	0.9	Vdc
(I_C = 50 mAdc, I_B = 5 mAdc)			--	1.2	
DC Forward Current Transfer Ratio [1] (I_C = 0.1 mAdc, V_{CE} = 1 Vdc)	2N3250, 2N3250A 2N3251, 2N3251A	h_{FE}	40 80	-- --	--
(I_C = 1 mAdc, V_{CE} = 1 Vdc)	2N3250, 2N3250A 2N3251, 2N3251A		45 90	-- --	
(I_C = 10 mAdc, V_{CE} = 1 Vdc)	2N3250, 2N3250A 2N3251, 2N3251A		50 100	150 300	
(I_C = 50 mAdc, V_{CE} = 1 Vdc)	2N3250, 2N3250A 2N3251, 2N3251A		15 30	-- --	
Output Capacitance (V_{CB} = 10 Vdc, I_E = 0, f = 100 kHz)		C_{ob}	--	6.0	pF
Input Capacitance (V_{CB} = 1 Vdc, I_C = 0, f = 100 kHz)		C_{ib}	--	8.0	pF
Current-Gain - Bandwidth Product (I_C = 10 mAdc, V_{CE} = 20 Vdc, f = 100 MHz)	2N3250, 2N3250A 2N3251, 2N3251A	f_T	250 300	-- --	MHz

SMALL SIGNAL CHARACTERISTICS

Characteristic		Symbol	Min	Max	Unit
Small Signal Current Gain (I_C = 1.0 mA, V_{CE} = 10 V, f = 1 kHz)	2N3250, 2N3250A 2N3251, 2N3251A	h_{fe}	50 100	200 400	--
Voltage Feedback Ratio (I_C = 1.0 mA, V_{CE} = 10 V, f = 1 kHz)	2N3250, 2N3250A 2N3251, 2N3251A	h_{re}	-- --	10 20	X10⁻⁴
Input Impedance (I_C = 1.0 mA, V_{CE} = 10 V, f = 1 kHz)	2N3250, 2N3250A 2N3251, 2N3251A	h_{ie}	1.0 2.0	6.0 12	kohms
Output Admittance (I_C = 1.0 mA, V_{CE} = 10 V, f = 1 kHz)	2N3250, 2N3250A 2N3251, 2N3251A	h_{oe}	4.0 10	40 60	μ mhos
Collector-Base Time Constant (I_C = 10 mA, V_{CE} = 20 V)		$r'_b C_C$	--	250	ps
Noise Figure (I_C = 100 μA, V_{CE} = 5 V, R_S = 1 kΩ, f = 100 Hz)		NF	--	6.0	dB

* Indicates JEDEC Registered Data

[1] Pulse Test: PW = 300 μs, Duty Cycle = 2%

2N**4391** (SILICON)   2N**4392**   2N**4393**

SILICON N-CHANNEL
JUNCTION FIELD-EFFECT TRANSISTORS

Depletion Mode (Type A) Junction Field-Effect Transistors designed
for chopper and high-speed switching applications.

- Low Drain-Source "On" Resistance —
  $r_{ds(on)}$ = 30 Ohms (Max) @ f = 1.0 kHz (2N4391)

- Low Gate Reverse Current —
  I_{GSS} = 0.1 nAdc (Max) @ V_{GS} = 20 Vdc

- Guaranteed Switching Characteristics

MAXIMUM RATINGS

Rating	Symbol	Value	Unit
Drain-Source Voltage	V_{DS}	40	Vdc
Drain-Gate Voltage	V_{DG}	40	Vdc
Gate-Source Voltage	V_{GS}	40	Vdc
Forward Gate Current	$I_{G(f)}$	50	mAdc
Total Device Dissipation @T_C = 25°C Derate above 25°C	P_D	1.8 10	Watts mW/°C
Operating & Storage Junction Temperature Range	T_J, T_{stg}	-65 to +200	°C

FIGURE 1 — SWITCHING TIMES TEST CIRCUIT

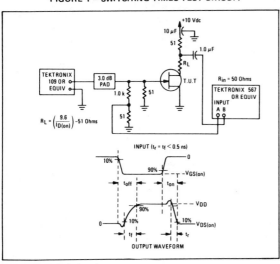

N-CHANNEL
JUNCTION FIELD-EFFECT
TRANSISTORS
(Type A)

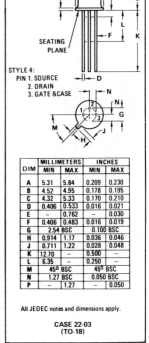

SEATING
PLANE

STYLE 4:
PIN 1. SOURCE
2. DRAIN
3. GATE &CASE

DIM	MILLIMETERS		INCHES	
	MIN	MAX	MIN	MAX
A	5.31	5.84	0.209	0.230
B	4.52	4.95	0.178	0.195
C	4.32	5.33	0.170	0.210
D	0.406	0.533	0.016	0.021
E	–	0.762	–	0.030
F	0.406	0.483	0.016	0.019
G	2.54 BSC		0.100 BSC	
H	0.914	1.17	0.036	0.046
J	0.711	1.22	0.028	0.048
K	12.70	–	0.500	–
L	6.35	–	0.250	–
M	45° BSC		45° BSC	
N	1.27 BSC		0.050 BSC	
P	–	1.27	–	0.050

All JEDEC notes and dimensions apply.

**CASE 22-03
(TO-18)**

2N4391, 2N4392, 2N4393 (continued)

ELECTRICAL CHARACTERISTICS (T_A = 25°C unless otherwise noted)

Characteristic		Symbol	Min	Max	Unit
OFF CHARACTERISTICS					
Gate-Source Breakdown Voltage (I_G = 1.0 μAdc, V_{DS} = 0)		$V_{(BR)GSS}$	40	-	Vdc
Gate-Source Forward Voltage (I_G = 1.0 mAdc, V_{DS} = 0)		$V_{GS(f)}$	-	1.0	Vdc
Gate-Source Voltage (V_{DS} = 20 Vdc, I_D = 1.0 nAdc)	2N4391 2N4392 2N4393	V_{GS}	4.0 2.0 0.5	10 5.0 3.0	Vdc
Gate Reverse Current (V_{GS} = 20 Vdc, V_{DS} = 0)		I_{GSS}	-	0.1	nAdc
(V_{GS} = 20 Vdc, V_{DS} = 0, T_A = 150°C)			-	0.2	μAdc
Drain-Cutoff Current (V_{DS} = 20 Vdc, V_{GS} = 12 Vdc)	2N4391	$I_{D(off)}$	-	0.1	nAdc
(V_{DS} = 20 Vdc, V_{GS} = 7.0 Vdc)	2N4392		-	0.1	
(V_{DS} = 20 Vdc, V_{GS} = 5.0 Vdc)	2N4393		-	0.1	
(V_{DS} = 20 Vdc, V_{GS} = 12 Vdc, T_A = 150°C)	2N4391		-	0.2	μAdc
(V_{DS} = 20 Vdc, V_{GS} = 7.0 Vdc, T_A = 150°C)	2N4392		-	0.2	
(V_{DS} = 20 Vdc, V_{GS} = 5.0 Vdc, T_A = 150°C)	2N4393		-	0.2	
ON CHARACTERISTICS					
Zero-Gate Voltage Drain Current (1) (V_{DS} = 20 Vdc, V_{GS} = 0)	2N4391 2N4392 2N4393	I_{DSS}	50 25 5.0	150 75 30	mAdc
Drain-Source "ON" Voltage (I_D = 12 mAdc, V_{GS} = 0)	2N4391	$V_{DS(on)}$	-	0.4	Vdc
(I_D = 6.0 mAdc, V_{GS} = 0)	2N4392		-	0.4	
(I_D = 3.0 mAdc, V_{GS} = 0)	2N4393		-	0.4	
Static Drain-Source "ON" Resistance (I_D = 1.0 mAdc, V_{GS} = 0)	2N4391 2N4392 2N4393	$r_{DS(on)}$	- - -	30 60 100	Ohms
SMALL-SIGNAL CHARACTERISTICS					
Drain-Source "ON" Resistance (V_{GS} = 0, I_D = 0, f = 1.0 kHz)	2N4391 2N4392 2N4393	$r_{ds(on)}$	- - -	30 60 100	Ohms
Input Capacitance (V_{DS} = 20 Vdc, V_{GS} = 0, f = 1.0 MHz)		C_{iss}	-	14	pF
Reverse Transfer Capacitance (V_{DS} = 0, V_{GS} = 12 Vdc, f = 1.0 MHz)	2N4391	C_{rss}	-	3.5	pF
(V_{DS} = 0, V_{GS} = 7.0 Vdc, f = 1.0 MHz)	2N4392		-	3.5	
(V_{DS} = 0, V_{GS} = 5.0 Vdc, f = 1.0 MHz)	2N4393		-	3.5	
SWITCHING CHARACTERISTICS					
Turn-On Time (See Figure 1) ($I_{D(on)}$ = 12 mAdc)	2N4391	t_{on}	-	15	ns
($I_{D(on)}$ = 6.0 mAdc)	2N4392		-	15	
($I_{D(on)}$ = 3.0 mAdc)	2N4393		-	15	
Rise Time (See Figure 1) ($I_{D(on)}$ = 12 mAdc)	2N4391	t_r	-	5.0	ns
($I_{D(on)}$ = 6.0 mAdc)	2N4392		-	5.0	
($I_{D(on)}$ = 3.0 mAdc)	2N4393		-	5.0	
Turn-Off Time (See Figure 1) ($V_{GS(off)}$ = 12 Vdc)	2N4391	t_{off}	-	20	ns
($V_{GS(off)}$ = 7.0 Vdc)	2N4392		-	35	
($V_{GS(off)}$ = 5.0 Vdc)	2N4393		-	50	
Fall Time (See Figure 1) ($V_{GS(off)}$ = 12 Vdc)	2N4391	t_f	-	15	ns
($V_{GS(off)}$ = 7.0 Vdc)	2N4392		-	20	
($V_{GS(off)}$ = 5.0 Vdc)	2N4393		-	30	

(1) Pulse Test: Pulse Width ≤ 100 μs, Duty Cycle ≤ 1.0%.

2N**4342** (SILICON)

SILICON P-CHANNEL
JUNCTION FIELD-EFFECT TRANSISTOR

Depletion Mode (Type A) Junction Field-Effect Transistor designed
primarily for high-gain audio frequency applications.

- High Forward Transadmittance —
  $|y_{fs}|$ = 2.0 mmhos (Min) @ V_{DS} = -10 Vdc (2N4342)

- Low Noise Figure —
  NF = 1.5 dB (Max) @ f = 100 Hz

- Low Drain-Source "ON" Resistance —
  $r_{ds(on)}$ = 700 Ohms (Max) @ f = 1.0 kHz (2N4342)

P-CHANNEL
JUNCTION FIELD-EFFECT
TRANSISTORS

(Type A)

*MAXIMUM RATINGS

Rating	Symbol	Value	Unit
Drain-Source Voltage	V_{DS}	-25	Vdc
Drain-Gate Voltage	V_{DG}	-25	Vdc
Reverse Gate-Source Voltage	$V_{GS(r)}$	25	Vdc
Forward Gate Current	I_{GF}	50	mAdc
Total Device Dissipation @ T_A = 25°C Derate above 25°C	P_D	200 2.0	mW mW/°C
Operating and Storage Junction Temperature Range	T_J, T_{stg}	-55 to +125	°C

*Indicates JEDEC Registered Data.

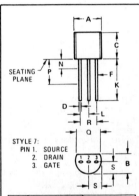

STYLE 7:
PIN 1.   SOURCE
     2.   DRAIN
     3.   GATE

DIM	MILLIMETERS		INCHES	
	MIN	MAX	MIN	MAX
A	4.450	5.200	0.175	0.205
B	3.180	4.190	0.125	0.165
C	4.320	5.330	0.170	0.210
D	0.407	0.533	0.016	0.021
F	0.407	0.482	0.016	0.019
K	12.700	–	0.500	–
L	1.150	1.390	0.045	0.055
N	–	1.270	–	0.050
P	6.350	–	0.250	–
Q	3.430	–	0.135	–
R	2.410	2.670	0.095	0.105
S	2.030	2.670	0.080	0.105

CASE 29-02
TO-92

2N4342 (continued)

*ELECTRICAL CHARACTERISTICS (T_A = 25°C unless otherwise noted)

Characteristic	Symbol	Min	Max	Unit
OFF CHARACTERISTICS				
Gate-Source Breakdown Voltage (I_G = 10 μAdc, V_{DS} = 0)	$V_{(BR)GSS}$	25	–	Vdc
Gate-Source Cutoff Voltage (V_{DS} = –10 Vdc, I_D = 1.0 μAdc)	$V_{GS(off)}$	1.0	5.5	Vdc
Gate Reverse Current (V_{GS} = 15 Vdc, V_{DS} = 0) (V_{GS} = 15 Vdc, V_{DS} = 0, T_A = 65°C)	I_{GSS}	– –	10 0.5	nAdc μAdc
ON CHARACTERISTICS				
Zero-Gate Voltage Drain Current (V_{DS} = –10 Vdc, V_{GS} = 0)	I_{DSS}	4.0	12	mAdc
Gate-Source Voltage (V_{DS} = –10 Vdc, I_D = 0.4 mAdc) (V_{DS} = –10 Vdc, I_D = 1.0 mAdc)	V_{GS}	0.7	5.0	Vdc
SMALL-SIGNAL CHARACTERISTICS				
Drain-Source "ON" Resistance (V_{GS} = 0, I_D = 0, f = 1.0 kHz)	$r_{ds(on)}$	–	700	Ohms
Forward Transadmittance (V_{DS} = –10 Vdc, V_{GS} = 0, f = 1.0 kHz)	Y_{fs}	2000	6000	μmhos
Forward Transconductance (V_{DS} = –10 Vdc, V_{GS} = 0, f = 1.0 MHz)	$Re(y_{fs})$	1500	–	μmhos
Output Admittance (V_{DS} = –10 Vdc, V_{GS} = 0, f = 1.0 kHz)	Y_{os}	–	75	μmhos
Input Capacitance (V_{DS} = –10 Vdc, V_{GS} = 0, f = 1.0 MHz)	C_{iss}	–	20	pF
Reverse Transfer Capacitance (V_{DS} = –10 Vdc, V_{GS} = 0, f = 1.0 MHz)	C_{rss}	–	5.0	pF
Common-Source Noise Figure (V_{DS} = –10 Vdc, V_{GS} = 0, R_G = 1.0 Megohm, f = 100 Hz, BW = 15 Hz)	NF	– –	1.5	dB
Equivalent Short-Circuit Input Noise Voltage (V_{DS} = –10 Vdc, V_{GS} = 0, f = 100 Hz, BW = 15 Hz)	E_n	–	0.08	$\mu V/\sqrt{Hz}$

*Indicates JEDEC Registered Data.

2N4351 (SILICON)

CASE 20-02
(TO-72)

STYLE 2
PIN 1. SOURCE
     2. GATE
     3. DRAIN
     4. SUBSTRATE AND
          CASE LEAD

Silicon N-channel MOS field effect transistors, designed for enhancement-mode operation in low power switching applications. The 2N4351 is complementary with type 2N4352.

MAXIMUM RATINGS (T_A = 25°C unless otherwise noted)

Rating	Symbol	Value	Unit
Drain-Source Voltage	V_{DS}	25	Vdc
Drain-Gate Voltage	V_{DG}	30	Vdc
Gate-Source Voltage	V_{GS}	± 30	Vdc
Drain Current	I_D	30	mAdc
Power Dissipation at T_A = 25°C Derate above 25°C	P_D	300 1.7	mW mW/°C
Power Dissipation at T_C = 25°C Derate about 25°C	P_D	800 4.56	mW mW/°C
Operating Junction Temperature	T_J	200	°C
Storage Temperature Range	T_{stg}	-65 to +200	°C

HANDLING PRECAUTIONS:

MOS field-effect transistors have extremely high input resistance. They can be damaged by the accumulation of excess static charge. Avoid possible damage to the devices while handling, testing, or in actual operation, by following the procedures outlined below:

1. To avoid the build-up of static charge, the leads of the devices should remain shorted together with a metal ring except when being tested or used.
2. Avoid unnecessary handling. Pick up devices by the case instead of the leads.
3. Do not insert or remove devices from circuits with the power on because transient voltages may cause permanent damage to the devices.

Courtesy of Motorola Semiconductor, Inc.

2N4351 (continued)

ELECTRICAL CHARACTERISTICS (T_A = 25°C unless otherwise noted)
Substrate connected to source.

Characteristic	Figure No.	Symbol	Min	Max	Unit
OFF CHARACTERISTICS					
Drain-Source Breakdown Voltage (I_D = 10 μA, V_{GS} = 0)	—	$V_{(BR)DSS}$	25	—	Vdc
Gate Leakage Current (V_{GS} = ±30 Vdc, V_{DS} = 0)	—	I_{GSS}	—	10	pAdc
Zero-Gate-Voltage Drain Current (V_{DS} = 10 V, V_{GS} = 0)	—	I_{DSS}	—	10	nAdc
ON CHARACTERISTICS					
Gate-Source Threshold Voltage (V_{DS} = 10 V, I_D = 10 μA)	—	$V_{GS(TH)}$	1.0	5.0	Vdc
"ON" Drain Current (V_{GS} = 10 V, V_{DS} = 10 V)	3	$I_{D(on)}$	3.0	—	mAdc
Drain-Source "ON" Voltage (I_D = 2 mA, V_{GS} = 10 V)	—	$V_{DS(on)}$	—	1.0	Vdc
SMALL SIGNAL CHARACTERISTICS					
Drain-Source Resistance (V_{GS} = 10 V, I_D = 0, f = 1 kHz)	4	$r_{ds(on)}$	—	300	ohms
Forward Transfer Admittance (V_{DS} = 10 V, I_D = 2 mA, f = 1 kHz)	1	$\lvert y_{fs} \rvert$	1000	—	μmho
Reverse Transfer Capacitance (V_{DS} = 0, V_{GS} = 0, f = 140 kHz)	2	C_{rss}	—	1.3	pF
Input Capacitance (V_{DS} = 10 V, V_{GS} = 0, f = 140 kHz)	2	C_{iss}	—	5.0	pF
Drain-Substrate Capacitance ($V_{D(SUB)}$ = 10 V, f = 140 kHz)	—	$C_{d(sub)}$	—	5.0	pF
SWITCHING CHARACTERISTICS					
Turn-On Delay I_D = 2.0 mAdc, V_{DS} = 10 Vdc,	6, 10	t_{d1}	—	45	ns
Rise Time V_{GS} = 10 Vdc	7, 10	t_r	—	65	ns
Turn-Off Delay (See Figure 10; Times	8, 10	t_{d2}	—	60	ns
Fall Time Circuit Determined)	9, 10	t_f	—	100	ns

FIGURE 1 — FORWARD TRANSFER ADMITTANCE   FIGURE 2 — CAPACITANCE

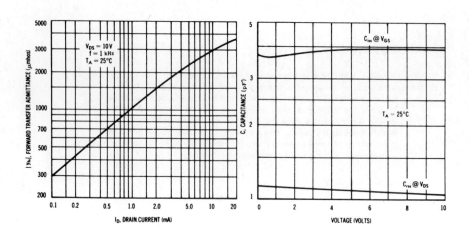

2N4352 (SILICON)

CASE 20-02
(TO-72)

STYLE 2
PIN 1. SOURCE
2. GATE
3. DRAIN
4. SUBSTRATE AND
CASE LEAD

Silicon P-channel MOS field-effect transistor designed for enhancement-mode operation in low-power switching applications. The 2N4352 is complementary with type 2N4351.

MAXIMUM RATINGS (T_A = 25°C unless otherwise noted)

Rating	Symbol	Value	Unit
Drain-Source Voltage	V_{DS}	25	Vdc
Drain-Gate Voltage	V_{DG}	30	Vdc
Gate-Source Voltage	V_{GS}	± 30	Vdc
Drain Current	I_D	30	mAdc
Power Dissipation at T_A = 25°C Derate above 25°C	P_D	300 1.7	mW mW/°C
Power Dissipation @ T_C = 25°C Derate above 25°C	P_D	800 4.56	mW mW/°C
Operating Junction Temperature	T_J	200	°C
Storage Temperature Range	T_{stg}	–65 to + 175	°C

HANDLING PRECAUTIONS:

MOS field-effect transistors have extremely high input resistance. They can be damaged by the accumulation of excess static charge. Avoid possible damage to the devices while handling, testing, or in actual operation, by following the procedures outlined below:
1. To avoid the build-up of static charge, the leads of the devices should remain shorted together with a metal ring except when being tested or used.
2. Avoid unnecessary handling. Pick up devices by the case instead of the leads.
3. Do not insert or remove devices from circuits with the power on because transient voltages may cause permanent damage to the devices.

Courtesy of Motorola Semiconductor, Inc.

2N4352 (continued)

ELECTRICAL CHARACTERISTICS (T_A = 25°C unless otherwise noted)
Substrate connected to source.

Characteristic	Figure No.	Symbol	Min	Max	Unit
OFF CHARACTERISTICS					
Drain-Source Breakdown Voltage (I_D = -10 μA, V_{GS} = 0)	—	$V_{(BR)DSS}$	25	—	Vdc
Gate Leakage Current (V_{GS} = ±30 V, V_{DS} = 0)	—	I_{GSS}	—	10	pAdc
Zero-Gate Voltage Drain Current (V_{DS} = -10 V, V_{GS} = 0)	—	I_{DSS}	—	10	nAdc
ON CHARACTERISTICS					
Gate-Source Threshold Voltage (V_{DS} = -10 V, I_D = -10 μA)	—	$V_{GS(TH)}$	1.0	5.0	Vdc
"ON" Drain Current (V_{GS} = -10 V V_{DS} = -10 V)	3	$I_{D(on)}$	3.0	—	mA
Drain-Source "ON" Voltage (I_D = -2.0 mA, V_{GS} = -10 V)	5	$V_{DS(on)}$	—	1.0	V
SMALL SIGNAL CHARACTERISTICS					
Drain-Source Resistance (V_{GS} = -10 V, I_D = 0, f = 1 kHz)	4	$r_{ds(on)}$	—	600	ohms
Forward Transfer Admittance (V_{DS} = -10 V, I_D = 2 mA, f = 1 kHz)	1	$\|y_{fs}\|$	1000	—	μ mho
Reverse Transfer Capacitance (V_{DS} = 0, V_{GS} = 0, f = 140 kHz)	2	C_{rss}	—	1.3	pF
Input Capacitance (V_{DS} = -10 V, V_{GS} = 0, f=140 kHz)	2	C_{iss}	—	5.0	pF
Drain-Substrate Capacitance ($V_{D(SUB)}$= -10 V, f=140 kHz)	—	$C_{d(sub)}$	—	4.0	pF
SWITCHING CHARACTERISTICS					
Turn-On Delay	6, 10	t_{d1}	—	45	ns
Rise Time	7, 10	t_r	—	65	ns
Turn-Off Delay	8, 10	t_{d2}	—	60	ns
Fall Time	9, 10	t_f	—	100	ns

Switching conditions: I_D = -2.0 mAdc, V_{DS} = -10 Vdc, V_{GS} = -10 V (See Figure 10, Times Circuit Determined)

FIGURE 1 — FORWARD TRANSFER ADMITTANCE   FIGURE 2 — CAPACITANCE

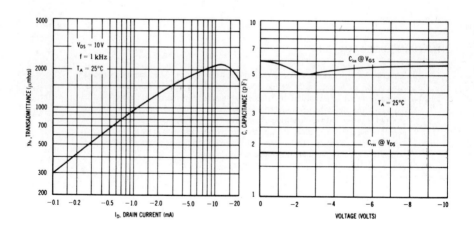

Index

Boldface numbers indicate tables.

Acceptor-type impurities in silicon, 128–132, 332

Atomic states s, p, d, f, . . . , 83n

Avalanche breakdown, 196, 252, 296, 300, 371
    junction voltage, 177, 181–183, 231
    multiplication factor (M), 177, 181–182, 248, 251, 253
    resistivity dependence, 183

Band bending, 148, 188

Bandwidth, bipolar transistor (f_β), 279–280

Base width stretching, 251

Binding energy of solids, 43, **44**, 45

Blackbody radiation, 51–52

Bloch function, 76, 107

Bohr atom, 53

Boltzmann relation, *see* Junction law

Bonding forces in solids, 41–43, 145
    attractive, 43
    covalent, 43–44, 118, 126, 128, 140
    ionic, 43, 45
    metallic, 43, 44
    repulsive, 42, 45
    tight-binding, 79
    van der Waals, 43, 44

Bonds, broken ("dangling"), 148, 150

"Bottoming" (saturation), *see* Transistor, bipolar junction, operating modes

Bragg condition for electrons, 92, 107

Breakdown, junction voltage. *See also*

Avalanche breakdown BV_{CBO}, BV_{CEO}, BV_{CER}, $V_{CE\,(\text{sus})}$, 253

Brillouin zone, 80, 92, 97, 101–103

Built-in field, 166, 186, 193
    in transistor base region, 254

Built-in voltage, *see* Contact potential

Capacitance
    collector overlap, 278
    depletion-layer (space-charge)
        in diodes, 159, 166, 177, 190, 209–212
        in transistors, 247, 263, 280–282
    diffusion, 210–212, 277, 280–281
    emitter depletion-layer, 276, 282
    measurement of, 320
    MOS gate, 298, 302, 305, 318

CCD, *see* Charge-coupled devices

Central-field problem, *see* Hydrogen atom

Charge carriers
    continuity of flow, 170, 234
    drift of, 136, 139
    excess, 139, 142
    intrinsic density of (n_i), 173, 175, 178, 231
    majority carriers, 140, 156, 160–162, 169, 173, 190
    minority carriers, 156, 160–162, 173, 185–186, 215
    mobility of, *see* Mobility, carrier

Charge-control model, 230, 234, 286
    for bipolar transistor, 244, 247
    for p-n junction diodes, 214

Charge-coupled devices (CCD), 146, 166, 364–373
    capacitance of depletion layer, 367, 372